T0139260

MESO- TO MICRO-
ACTUATORS

A Theoretical and Practical Approach

MESO- TO MICRO-
ACTUATORS

A Theoretical and Practical Approach

ALBERTO BORBONI

CRC Press
Taylor & Francis Group
Boca Raton   London   New York

CRC Press is an imprint of the
Taylor & Francis Group, an **informa** business

CRC Press
Taylor & Francis Group
6000 Broken Sound Parkway NW, Suite 300
Boca Raton, FL 33487-2742

© 2008 by Taylor & Francis Group, LLC
CRC Press is an imprint of Taylor & Francis Group, an Informa business

---

Library of Congress Cataloging-in-Publication Data

Borboni, Alberto.
    Meso- and micro-actuators : a theoretical and practical approach / Alberto Borboni.
        p. cm.
    Includes bibliographical references and index.
    ISBN-13: 978-0-8493-9089-0 (alk. paper)
    ISBN-10: 0-8493-9089-3 (alk. paper)
    1. Actuators. I. Title.

TJ223.A25B87 2008
621--dc22   2007047565

Visit the Taylor & Francis Web site at
http://www.taylorandfrancis.com

and the CRC Press Web site at
http://www.crcpress.com

This work is dedicated particularly to those who are subject to abuses, tortures, and homicide as a result of their opinions, and to those who research the truth. Hence, I want to mention here my principal fellows in this latter, fundamental activity: my colleagues, my friends, my family, my wife Scilla, and my son Allesandro.

Contents

Preface

The concepts of the micro- and meso-actuator are introduced in this monograph through a multidisciplinary approach that combines aspects of mechanics and of functional kinematical production. General concepts are described in the first chapter, are practically expressed in the second chapter, are deepened with modern software instruments in the third chapter and, finally, are seasoned with a practical example in the fourth chapter. The approach adopted is based on continuity between theoretical and practical viewpoints on micro- and meso-actuators to provide an understanding of physical phenomena and suggestions for their application to the real world.

This approach emerges only in a rather implicit form within the monograph that, therefore, can represent only a first version or a first hint of a wider activity of research, organization, and collocation of knowledge. Different errors and information gaps are certainly present and the author apologizes for them, relying on the goodwill of readers to highlight errors and suggest possible extensions of the work in e-mails to alberto.borboni@ing.unibs.it.

Acknowledgments

I am grateful to MIUR (Italian Ministry of the University and Research) for the PRIN 2003 grant related to the MINIPAR project. I would like to thank Michael Slaughter, Mohamed Gad-el-Hak, and Rodolfo Faglia, and the anonymous referees for the important contributions to the editorial project, proposing corrections, improvements, and suggestions; to Jessica Vakili and Marsha Hecht for coordination and production of this book; Kathy Johnson for the layout; Joe Mendez for his meticulous linguistic review; Maurizio Mor for his contribution on pneumatics; Diego DeSantis for his contribution on microcontinua; and Andrea Dassa for his contribution on flexure hinges. Last but not least, I am really grateful to my colleagues, to my family, to my friends, to my wife Scilla, and to my son Allesandro, who tolerated my absence and supported me when I was working on this monograph.

Author

Alberto Borboni received a master's degree in mechanical engineering in 1997 and a Ph.D. in applied mechanics in 2002 from the University of Brescia, Italy, where he conducts research in the Laboratory of Applied Mechanics. Current research interests are applied micromechanics and actuation systems.

1

General Knowledge

The purpose of this chapter is to briefly expound on some basic notions that are useful in dealing with the subject of meso- and micro-actuators in a unitary manner. In addition, an additional and hidden purpose is pursued: to propose a coherent notation that accompanies the reader through the entire length of this monograph. Although these tasks seem apparently easy, various obstacles arise between us and our objectives. Regarding the possibility of realizing a unitary, clear, and concise synthesis, we observe that human limits and the knowledge of the author do not allow writing the last word on such a wide and complex subject. Furthermore, the author is interested in illustrating the theme emphasizing the functional characteristics of an actuator, that is, first, its ability to produce a movement and, second, its ability to generate a mechanical work. In this sense, the notions that are introduced in this chapter, except for kinematic and dynamic ones such as structural, electromagnetic, and thermal knowledge and others, are considered only to properly represent the kinematic and dynamic behavior of meso- and micro-actuators and to allow their description, design, or correct selection. In fact, in these types of machines, we can seldom discuss matters that are not strictly related to kinematics and dynamics because they are energy transformers, and the high degree of integration in their design involves different physical phenomena. These adjoined nonmechanical notions can only be briefly summarized because, if deeply treated, they would require a series of specialized treatises. Because of the interaction between different disciplines and with the cornerstone fields of kinematics and dynamics, the latter problem has to be faced. A simple solution can be expounding distinct notations in distinct chapters, but that does not allow acquisition of an overall view of the technical issues regarding meso- and micro-actuators; therefore, we also undertake this difficult task, avoiding an excessive discontinuity with standard nomenclatures that are widely analyzed and proposed by scientific and technical communities. Various authors have directed their research activities to the correct selection of a proper notation, and we cannot set aside the large number of works composed by many scientific societies and institutions; among different results, an interesting reference is the standard ISO 31, which represents a significant proposal on technical and scientific notations. The research teams that have been involved in interdisciplinary problems have usually achieved useful formulations; therefore, the notations proposed by developers of multiphysics analysis software can provide additional interesting references for selecting a proper notation. However, the indicated references should be considered only as sources of suggestions for our choices on symbolism because we are driven by the necessity of adopting an approach that is suitable for properly describing the functional and design characteristics of a device from a kinematic and dynamic viewpoint and, additionally, an approach that considers interdisciplinary aspects of the proposed subject of this monograph. For this reason, the order of illustration of fundamental knowledge introduced in this chapter should be considered as a hierarchy to be respected also in the selection of a proper notation, with a penalization of the possible use of non-standard notation in non-mechanical disciplines.

This chapter deals first, with some general problems on geometry and on systems of observation; these topics should be not considered as mere academic digressions but rather some observations allowing a coherent vision of every topic. Attention is then shifted to the reference disciplines, kinematics and dynamics, which allow a proper description of behavior and performances of a mechanical machine, focusing on their functional characteristics in reference to their ability to generate movements and exert mechanical actions. In investigating in depth the mechanical behavior of a body, deformations, internal actions, and material properties have to be taken into account. During the study of the modality of energetic conversion of an actuator, some additional knowledge fields are useful. These can involve electromagnetic phenomena and thermal effects or some special characteristics of materials, such as piezoelectricity, shape memory effect, Joule effect, or Peltier effect. When the actuator dimensions are very reduced and comparable with the dimension of the fundamental material structures that compose the system, some modern general theories, such as the microcontinua approach, can be introduced to improve modeling ability. A further scale reduction, or the presence of uncertainty on some parameters of the system, can open the way to statistic mechanics or analogous theories.

1.1   Context

1.1.1   Geometrical Domain

Within this monograph, the real space is represented through a three-dimensional Euclidean space composed of a positive defined scalar product and a homogeneous space on a real field. This unobvious type of choice, on the one hand, allows reusing knowledge of real Euclidean geometry, and, on the other, a homogeneous space affords an opportunity for a nimbleness in kinematic and dynamic representations. Generally, orthogonal Cartesian reference systems are used but, sometimes, alternative reference systems can also be used. Also, in the case of problems that can be represented in a bidimensional space, we will try to use three-dimensional models to guarantee notational coherence. A geometrical point in the Euclidean

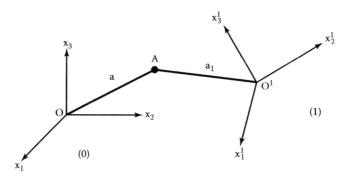

FIGURE 1.1
Geometrical representation of the point A referred to in two reference systems.

space is represented with a column vector composed of its coordinates in the associated homogeneous space referred to in a conventional reference system; the example in Figure 1.1 shows that the point A is seen from two different orthogonal Cartesian reference systems.

The point A is associated with the vector OA of coordinates more simply called a, or with the vector O^1A of coordinates. If the vector OA is referred to in the reference system (0), it is indicated as $a_{0(0)}$, where the first subscript designates the origin of the vector, and the bracketed subscript denotes the reference system used to observe the point. Specifically, the origin O and the reference system (0) hold a simplified notation that allows their omission, namely a vector, which joins the origin O of the triad (0) with the point A, and which can be denoted with a in place of $a_{0(0)}$. This possibility allows some expressive simplifications in this monograph. If the same vector OA is observed in the reference system (1), it can be designated with the symbol $a_{(1)}$, where only the origin of the vector, and not the reference system, is omitted. Analogously, the vector O^1A can be indicated by the symbol a_1 when it is related to the reference system (0), but it is indicated by the symbol $a_{1(1)}$ when it is seen in the reference system (1). Generally, the position of the point A can be identified by a set $a_{i(j)}$ of coordinates that represent the vector O^iA with respect to the reference system (j). This general notation can also be used for the vector OA seen in the reference system (0), but the subscript omission is preferred to simplify formulations proposed in this monograph.

Example 1.1: Set of Coordinates Described in Different Reference Systems

Two reference systems are considered: (0) is located by the origin O and by the triad (x_1, x_2, x_3) of Cartesian axes, whereas (1) is located by the origin O^1 and by the triad (x^1_1, x^1_2, x^1_3) of Cartesian axes. For the sake of simplicity, the axis x_i is considered as equidirectional to the axis x^1_i, and the origin O^1 of (1) does not coincide with the origin O of (0), and particularly, O^1 can be represented by a vector that is applied in O and by a set of homogeneous coordinates, referred to (0), as:

$$OO^1 = o^1_{0(0)} = o^1_{(0)} = o^1 = \begin{bmatrix} 2 \\ 3 \\ 4 \\ 1 \end{bmatrix}$$

As better described in the rest of this paragraph, the matrix M_{01} transforms the vision of the space from a viewpoint (1) into the vision of the same space from a viewpoint (0).

$$M_{01} = \begin{bmatrix} 1 & 0 & 0 & 2 \\ 0 & 1 & 0 & 3 \\ 0 & 0 & 1 & 4 \\ 0 & 0 & 0 & 1 \end{bmatrix}$$

Analogously, the matrix M_{10} carries out the opposite function, i.e., it transforms the vision of the space from a viewpoint (0) into the vision of the same space from a viewpoint (1).

$$M_{10} = M_{01}^{-1} = \begin{bmatrix} 1 & 0 & 0 & -2 \\ 0 & 1 & 0 & -3 \\ 0 & 0 & 1 & -4 \\ 0 & 0 & 0 & 1 \end{bmatrix}$$

Once the relation between (0) and (1) is defined, the point A is considered, and its position in (0) is described by a vector of homogeneous coordinates, which is applied and is equal to

$$OA = a_{0(0)} = a_{(0)} = a = \begin{bmatrix} 1 \\ 2 \\ 1 \\ 1 \end{bmatrix}$$

The position of A in the reference system (1) can be described with the vector O^1A, which is applied in O^1.

$$O^1A = a_{1(1)} = M_{10} \cdot a = \begin{bmatrix} 1 & 0 & 0 & -2 \\ 0 & 1 & 0 & -3 \\ 0 & 0 & 1 & -4 \\ 0 & 0 & 0 & 1 \end{bmatrix} \begin{bmatrix} 1 \\ 2 \\ 1 \\ 1 \end{bmatrix} = \begin{bmatrix} -1 \\ -1 \\ -3 \\ 1 \end{bmatrix}$$

Some intermediate representations of the position of A, referred to (0) and (1), can be considered. That is, the vector O^1A can be described in the reference system (0), obtaining the vector $a_{1(0)} = a_1$ with the same coordinates of $a_{1(1)}$ but with the application point in O^1, which assumes non-null coordinates:

$$OO^1 = o^1_{0(0)} = o^1_{(0)} = o^1 = M_{01} \begin{bmatrix} 0 \\ 0 \\ 0 \\ 1 \end{bmatrix} = \begin{bmatrix} 2 \\ 3 \\ 4 \\ 1 \end{bmatrix}$$

Another intermediate representation consists in the description of OA in the reference system (1), achieving a vector $a_{0(1)} = a_{(1)}$ with the same coordinates of a but with non-null coordinates of the application point in O:

$$O^1O = o_{0(1)} = o_{(1)} = M_{10} \begin{bmatrix} 0 \\ 0 \\ 0 \\ 1 \end{bmatrix} = \begin{bmatrix} -2 \\ -3 \\ -4 \\ 1 \end{bmatrix}$$

The sets of coordinates describing the position of the point A are connected by the position matrix M_{ij}, as defined in (1.1) in a compact form or, more extensively, in (1.2), where the wide use of homogeneous coordinates can be observed, allowing a simple construction of the same matrix.

$$a_{i(i)} = M_{ij} a_{j(j)} \tag{1.1}$$

$$\begin{pmatrix} a_{1i(i)} \\ a_{2i(i)} \\ a_{3i(i)} \\ 1 \end{pmatrix} = \begin{pmatrix} u^j_{11i(i)} & u^j_{21i(i)} & u^j_{31i(i)} & o^j_{1i(i)} \\ u^j_{12i(i)} & u^j_{22i(i)} & u^j_{32i(i)} & o^j_{2i(i)} \\ u^j_{13i(i)} & u^j_{23i(i)} & u^j_{33i(i)} & o^j_{3i(i)} \\ 0 & 0 & 0 & 1 \end{pmatrix} \begin{pmatrix} a_{1j(j)} \\ a_{2j(j)} \\ a_{3j(j)} \\ 1 \end{pmatrix} \tag{1.2}$$

The formulation expressed in (1.2) emphasizes the structure of the matrix M_{ij} and its geometrical meaning, allowing the possibility of a direct construction. In fact, the first column represents the direction of the first Cartesian axis of the reference system (j) referred to the reference system (i), and an analogous meaning is associated to the second and third columns, whereas the fourth represents the position of the origin of the reference system (j) referred to the reference system (i). More explicitly, the term $u^j_{hki(i)}$ represents the k-th component of a unit vector on the h-th axis of the reference system (j) expressed in the reference system (i), whereas the term $o^j_{hi(i)}$ represents the h-th coordinate of the vector, which joints the origins O^i and O^j, seen in the reference system (i).

Example 1.2: Construction of the Position Matrix

Two reference systems are considered: (0) is located by the origin O and by the triad (x_1, x_2, x_3) of Cartesian axes, whereas (1) is located by the origin O^1 and by the triad (x^1_1, x^1_2, x^1_3) of Cartesian axes; particularly, the origin O^1 can be represented by a vector that is applied in O and with a set of homogeneous coordinates, referred to (0), equal to

$$OO^1 = o^1_{0(0)} = o^1_{(0)} = o^1 = \begin{bmatrix} 2 \\ 5 \\ 3 \\ 1 \end{bmatrix}$$

Furthermore, three points, A_1, A_2, and A_3, are considered, respectively, on the axes x^1_1, x^1_2, and x^1_3 and with unit distance from O^1; these three points are described by three vectors of homogeneous coordinates in the reference system (0) equal to

$$a_1 = \begin{bmatrix} 2 \\ 6 \\ 3 \\ 1 \end{bmatrix}; \quad a_2 = \begin{bmatrix} 2+\sin(\alpha) \\ 5 \\ 3+\cos(\alpha) \\ 1 \end{bmatrix}; \quad a_3 = \begin{bmatrix} 2+\cos(\alpha) \\ 5 \\ 3-\sin(\alpha) \\ 1 \end{bmatrix}$$

The i-th column of the matrix M_{01}, which transforms the vision of the space from a viewpoint (1) into the vision of the same space from a viewpoint (0), is associated to the vector a_i, and, to obtain it, we can subtract the vector o_1 from the vector a_i.

$$u^1_1 = a_1 - o^1 = \begin{bmatrix} 0 \\ 1 \\ 0 \\ 0 \end{bmatrix}; \quad u^1_2 = a_2 - o^1 = \begin{bmatrix} \sin(\alpha) \\ 0 \\ \cos(\alpha) \\ 0 \end{bmatrix}; \quad u^1_3 = a_3 - o^1 = \begin{bmatrix} \cos(\alpha) \\ 0 \\ \sin(\alpha) \\ 0 \end{bmatrix}$$

If the three vectors u^1_i are not unit vectors, a normalization is necessary; however, this is not our situation because O^1A_i was chosen as unit vector. Finally, the fourth column of M_{01} is represented by o^1.

$$M_{01} = \begin{bmatrix} u^1_1 & u^1_2 & u^1_3 & o^1 \end{bmatrix} = \begin{bmatrix} 0 & \sin(\alpha) & \cos(\alpha) & 2 \\ 1 & 0 & 0 & 5 \\ 0 & \cos(\alpha) & -\sin(\alpha) & 3 \\ 0 & 0 & 0 & 1 \end{bmatrix}$$

The position matrix has, furthermore, two interesting properties identified by (1.3), which allows an easy inversion of the reference transformation, and by (1.4), which allows a sequence of changes of reference systems.

$$M_{ij}^{-1} = M_{ji} \qquad (1.3)$$

$$M_{ik} = M_{ij}M_{jk} \qquad (1.4)$$

Example 1.3: Change of Reference System

Three reference systems are considered: (0) is located by the origin O and by the triad (x_1, x_2, x_3) of Cartesian axes, (1) is located by the origin O^1 and by the triad (x^1_1, x^1_2, x^1_3) of Cartesian axes, and (2) is located by the origin O^2 and by the triad (x^2_1, x^2_2, x^2_3) of Cartesian axes, and the relative positions and orientations of these reference systems are described by the matrixes M_{01} and M_{12}.

$$M_{01} = \begin{bmatrix} 0 & 0 & -1 & 0 \\ 1 & 0 & 0 & 0 \\ 0 & -1 & 0 & 0 \\ 0 & 0 & 0 & 1 \end{bmatrix}; \quad M_{12} = \begin{bmatrix} 1 & 0 & 0 & 0 \\ 0 & 0 & -1 & 0 \\ 0 & 1 & 0 & 1 \\ 0 & 0 & 0 & 1 \end{bmatrix}$$

Furthermore, the point A is considered in the space, and its position referred to (0) is represented by the vector a.

$$a = \begin{bmatrix} 1 \\ 2 \\ 3 \\ 1 \end{bmatrix}$$

With (1.3), the position of A can be referred to (1), and it is expressed by the vector $a_{1(1)}$.

$$a_{1(1)} = M_{10}a = M_{01}^{-1}a = \begin{bmatrix} 0 & 1 & 0 & 0 \\ 0 & 0 & -1 & 0 \\ -1 & 0 & 0 & 0 \\ 0 & 0 & 0 & 1 \end{bmatrix} a = \begin{bmatrix} 2 \\ -3 \\ -1 \\ 1 \end{bmatrix}$$

In a similar way, with (1.3) and (1.4), the position of A can be referred to (2) and the vector $a_{2(2)}$ is obtained.

$$a_{2(2)} = M_{20}a = \left(M_{01}M_{12}\right)^{-1} a = \begin{bmatrix} 0 & 1 & 0 & 0 \\ -1 & 0 & 0 & -1 \\ 0 & 0 & 1 & 0 \\ 0 & 0 & 0 & 1 \end{bmatrix} a = \begin{bmatrix} 2 \\ -2 \\ 3 \\ 1 \end{bmatrix}$$

The vector $a_{2(2)}$ can also be obtained from the vector $a_{1(1)}$.

$$a_{2(2)} = M_{21}a_{1(1)} = M_{12}^{-1}a_{1(1)} = \begin{bmatrix} 1 & 0 & 0 & 0 \\ 0 & 0 & 1 & -1 \\ 0 & -1 & 0 & 0 \\ 0 & 0 & 0 & 1 \end{bmatrix} a_{1(1)} = \begin{bmatrix} 2 \\ -2 \\ 3 \\ 1 \end{bmatrix}$$

Numerous other observations can be performed on the transformation matrixes and in literature are present in many treatises, which appertain to the disciplines of algebra (Wickless, 2004), of analytic geometry (Talpaert, 2000), of theoretic kinematics (Malek et al., 1998), of kinematics of mechanisms (Vinogradov, 2000), and of kinematics of manipulators (Murray et al., 1994; Legnani, 2003).

1.1.2   Method of Observation

After a discussion of the representation of the physical space from a geometric viewpoint, suggesting specific treatises on the mechanics of the continuous (Jog, 2002) and of mathematical physics (Kaushal and Parashar 2000), we consider the presence of a material particle in such space. We mention two ways to describe the evolution of the space–matter structure in time, with the objective of delineating a specific notation and putting forward some topics involving the fields of physical properties, which are considered in depth in the following paragraphs. A definition of a material particle is not outlined here; it is simply considered an elementary concept and is denoted by the symbol B, and we remark that it is possible to locate the position of this particle with a couple of spatial–temporal coordinates. A temporal instant is taken as reference and conventionally denoted by the temporal zero, and therefore the particle B is identified with the couple $(b, 0)$ in the initial configuration. In the same instant, the particle B occupies a geometrical point P_0, which can be identified with the same coordinates of the body B. The body B is supposed to be subjected to a displacement and to go to another position after a time lapse t, so the value of the coordinate b changes and B can be indicated with the couple (b, t) of coordinates occupying the geometric point P_t (p, t). More generally, it is possible to associate with the body B not only a position but a wide class of physical properties, which are all recapitulated in the concept of configuration. Therefore, the displacement χ of a body B can be generalized to an operator that is able to turn a configuration c_0 of the

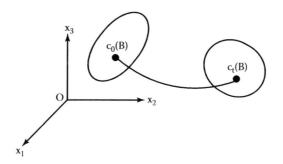

FIGURE 1.2
Displacement of a body.

body B into another configuration c_t of the same body, as indicated in (1.5) and as represented in Figure 1.2.

$$c_0(B) \xrightarrow{\chi} c_t(B) \tag{1.5}$$

Equation 1.5 can be written referring only to the space–time configuration as expressed in (1.6), where the body B is identified with the space coordinates but omitting the time coordinate because the latter is already present in the subscript of the notation of the configuration c_t.

$$c_0(b) \xrightarrow{\chi} c_t(b) \tag{1.6}$$

Considering the relationship between the coordinates of the body B and the coordinates of the geometric points that are occupied by B during its displacement, (1.7) can be obtained.

$$p = \chi(b, t) \tag{1.7}$$

Let us suppose that owing to the impenetrability of bodies, (1.7) is invertible. As a matter of fact, a geometric point occupied in a time instant always corresponds to a unique occupying body. If a physical property F is described with a system (b, t) of coordinates, they are referred to a Lagrangian observer or material observer; whereas, when the same physical property is described with a system (p, t) of coordinates, they are referred to an Eulerian observer or local observer. However, in both cases, the value of such physical property is equivalent (1.8).

$$F_{Lag}(b, t) = F_{Eul}(p, t) \tag{1.8}$$

A point to be noted is that the derivatives of a physical property, seen by the two observers, i.e., the partial derivative of equation 1.8 with respect to time can be expressed in three different notations (1.9), where the definition of the material derivative appears, which is equivalent to the total derivative as underlined by (1.10).

$$\frac{\partial F_{Lag}}{\partial t} = \frac{\partial F_{Eul}}{\partial t} + \sum_i \frac{\partial F_{Eul}}{\partial p_i} \cdot \frac{\partial p_i}{\partial t}$$

$$= \frac{\partial F_{Eul}}{\partial t} + F_{Eul,i} \frac{\partial p_i}{\partial t} \tag{1.9}$$

$$= \frac{\partial F_{Eul}}{\partial t} + \nabla_p F_{Eul} \circ \frac{\partial p}{\partial t}$$

$$\frac{\partial F_{Lag}}{\partial t} = \frac{dF_{Eul}}{dt} \tag{1.10}$$

Practically, Lagrangian observers are useful to *know*, i.e., where a body is positioned, and they are used traditionally to describe the kinematics and dynamics of solids, whereas Eulerian observers are useful to *determine*, i.e., what quantity of a fluid passes through the section of a tube, and they are used traditionally to describe the kinematics and dynamics of fluids. In fact, they are particularly convenient for the definition of scalar or vectorial field of physical properties, i.e., electromagnetic fields. In this monograph, the traditional habits of the reader are maintained in order to allow the reutilization of knowledge, whereas the utilization of a single typology allows a uniform description common among different authors.

1.2 Kinematics

This paragraph traces an outline of the free motion of particles, rigid bodies, and deformable bodies. In addition, some fundamental kinematic constraints are presented to allow a delineation of the constrained motion of such material entities. Accurate and systematic examinations can be executed with regard to the kinematics of continua (Jog, 2002), the kinematic chains of rigid or deformable bodies (Murray et al., 1994), and the kinematics of constraints (Lobontiu, 2003) or to vibrating bodies (Meirovitch, 1986); recent studies can be found in different scientific journals.

1.2.1 Particle Kinematics

The motion of a P particle (Figure 1.3), that moves in the space through a set of geometrical points X, in a time lapse T, may be described with the coordinates x of the geometric points X.

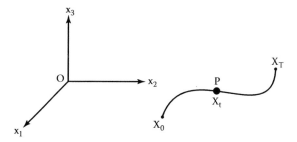

FIGURE 1.3
Motion of a particle P through geometric points X.

A first description of the motion of P along X may be represented directly by (1.11), which expresses a temporal relationship between homogeneous coordinates. Otherwise, the trajectory may be described intrinsically, leaving aside the temporal coordinate: if a reference point X_0 is fixed and the distance between X and X_0 is indicated by s, a curvilinear abscissa measured along the trajectory, the motion of P is described by (1.12).

$$p(t) = x(t) \tag{1.11}$$

$$p(s) = x(s)$$
$$s = s(t) \tag{1.12}$$

Once the motion of the particle P has been sufficiently defined in spatial–temporal terms, the concepts of speed and acceleration can be introduced to improve this kinematic description. The scalar speed and acceleration are defined by the first-time derivative and by the second-time derivative of the curvilinear abscissa (1.13), whereas the respective vectorial quantity is obtained by computing the time derivative of equation 1.12 (1.14).

$$v(t) \equiv \dot{s}(t) \quad a(t) \equiv \ddot{s}(t) \tag{1.13}$$

$$v(t) \equiv \dot{x}(t) \quad a(t) \equiv \ddot{x}(t) \tag{1.14}$$

Equation 1.15 is particularly interesting for the design activity because it expresses the relationship between scalar speed and vectorial speed through the geometrical speed x', which represents the derivative of the vectorial position $x(s)$ with respect to the curvilinear abscissa s.

$$v(t) \equiv x'(s) \cdot v(t) \tag{1.15}$$

Some simple principal concepts about the kinematics of particles have been presented, and they have various practical applications. A more complex problem is shown in the next section—the kinematics of the rigid body—to allow the tractability of a wider class of practical problems.

1.2.2 Kinematics of a Rigid Body

The position and orientation of a rigid body in space can be represented by a set of six unrelated coordinates. These may be selected in several ways; in this section, a reference system integral with the body is considered, and its position and orientation are described with respect to another reference system by a position matrix M_{ij}. Every particle of the body is unambiguously located; in fact, a particle of a rigid body cannot move relatively to a reference system on the body. Furthermore, the displacement of a rigid body in space can be geometrically led back to an equivalent helicoidal rototranslation, which can be represented with homogeneous coordinates. Therefore, the study of a generic movement of a rigid body can be related, from a geometric viewpoint, with the definition of the helicoidal axis of the motion, the pitch of the helix, and the rotation amplitude (Figure 1.4).

Generically, a rigid body can be localized by six independent parameters and, less restrictively, it can be localized in space by three out-of-line points: in the initial position, these three points are denoted as P^1_1, P^1_2, and P^1_3, whereas in the final position they are denoted as P^2_1, P^2_2, and P^2_3. The points P^2_i correspond to the points P^1_i after the movement, and the distances among the points of the body and their relative positions remain constant during the considered time period because the body moves rigidly. Then, the homogeneous coordinates p^1_i that describe the points P^1_i in the initial position can be related with the homogeneous coordinates p^2_i that describe the points P^2_i in the final position through the matrix $Q_{(0)}$ of helicoidal motions, which is written with respect to the reference system (0) as shown in equation (1.16).

$$p^1_{i(0)} = Q_{(0)} \cdot p^1_{i(0)} \tag{1.16}$$

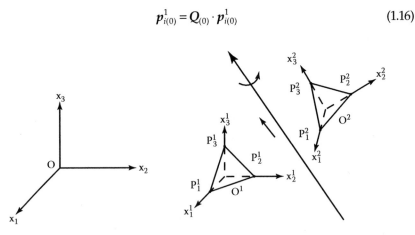

FIGURE 1.4
Helicoidal motion of a rigid body.

The general definition of $Q_{(k)}$, written with reference to system (k), can be expressed in terms of position matrixes of the initial and final configurations (1.17) with simple geometrical and algebraic observations.

$$Q_{(k)} = M_{k2} \cdot M_{1k} \qquad (1.17)$$

A matrix $Q_{(k)}$ may be expressed in another reference system with a proper elaboration of (1.17), and the expression (1.18) for the change of reference can be obtained.

$$Q_{(h)} = M_{hk} \cdot Q_{(k)} \cdot M_{kh} \qquad (1.18)$$

Three interesting matrixes of rototranslations, Q^1, Q^2, and Q^3, are explicitly expressed in (1.19): they describe a rototranslation around the axes x_1, x_2, and x_3 of the reference system (0). In the matrixes Q^i, there is a first block that represents a rotation ϑ_i around the axis x_i, whereas a second block describes a translation along the same axis for a quantity equal to h_i. If there are n sequential rototranslations with respect to the reference system (0), the total equivalent rototranslation can be calculated by a product of rototranslation matrixes, in which, generally, the order of calculations is not modifiable because the proposed matrix product is not commutative (1.20).

$$
Q^1 = Rtras(x_1, \vartheta_1, h_1) =
\left[
\begin{array}{ccc|c}
1 & 0 & 0 & h_1 \\
0 & \cos(\vartheta_1) & -\sin(\vartheta_1) & 0 \\
0 & \sin(\vartheta_1) & \cos(\vartheta_1) & 0 \\
0 & 0 & 0 & 1
\end{array}
\right]
$$

$$
Q^2 = Rtras(x_2, \vartheta_2, h_2) =
\left[
\begin{array}{ccc|c}
\cos(\vartheta_2) & 0 & \sin(\vartheta_2) & 0 \\
0 & 1 & 0 & h_2 \\
-\sin(\vartheta_2) & 0 & \cos(\vartheta_2) & 0 \\
0 & 0 & 0 & 1
\end{array}
\right]
\qquad (1.19)
$$

$$
Q^3 = Rtras(x_3, \vartheta_3, h_3) =
\left[
\begin{array}{ccc|c}
\cos(\vartheta_3) & -\sin(\vartheta_3) & 0 & 0 \\
\sin(\vartheta_3) & \cos(\vartheta_3) & 0 & 0 \\
0 & 0 & 1 & h_3 \\
0 & 0 & 0 & 1
\end{array}
\right]
$$

$$
\begin{aligned}
Q_{(0)} &= Q^n_{(0)} Q^{n-1}_{(0)} \cdots Q^i_{(0)} Q^{i-1}_{(0)} \cdots Q^2_{(0)} Q^1_{(0)} \\
&= Q^1_{(0)} Q^2_{(1)} \cdots Q^{i-1}_{(i-2)} Q^i_{(i-1)} \cdots Q^{n-1}_{(n-2)} Q^n_{(n-1)}
\end{aligned}
\qquad (1.20)
$$

It is interesting, from an applicative viewpoint, to consider the case of rototranslations around axes of the local reference system; in these conditions the product of rototranslation matrixes can be obtained from (1.18), as shown in (1.20).

Now it is possible to treat the infinitesimal rototranslation problem, which allows us to define a new class of matrixes; these are useful in dealing with the kinematics problems of speed and acceleration. Observing the rototranslation matrixes around the axes of the reference system (1.19) and considering infinitesimal displacements, either in rotation terms or in translation terms, it is easy to obtain the matrix of infinitesimal rototranslations (1.21).

$$dQ = \left[\begin{array}{ccc|c} 1 & -d\vartheta_3 & d\vartheta_2 & dh_1 \\ d\vartheta_3 & 1 & -d\vartheta_1 & dh_2 \\ -d\vartheta_2 & d\vartheta_1 & 1 & dh_3 \\ \hline 0 & 0 & 0 & 1 \end{array}\right] \tag{1.21}$$

Equation 1.21 may be expressed in the form of equation 1.22 with the identity matrix I and the matrix L, which describes the helicoidal axis of the infinitesimal displacement. The matrix L might underline the properties of the axis of helicoidal motion, but in this monograph, equation 1.22 is conveniently considered as a method to define the matrix L.

$$dQ = I + L \cdot d\vartheta \tag{1.22}$$

In the case of infinitesimal rototranslations, the commutative property is valid, and each rototranslation is independent of the execution order. In fact, the formula for the composition of infinitesimal rototranslations is represented by a summation notation without considering the infinitesimals of an order higher than first and referring to the absolute reference system (1.23).

$$dQ = I + \sum_{i=1}^{n} L_i \cdot d\vartheta_i \tag{1.23}$$

Equation 1.24 is obtained by replacing (1.22) in (1.18) and by elaborating the expressions. This allows us to describe a matrix L in different reference systems, introducing possible simplifications in the study of infinitesimal movements.

$$L_{(h)} = M_{hk} \cdot L_{(k)} \cdot M_{kh} \tag{1.24}$$

As we just mentioned, we do not consider the general method to calculate a matrix L, but we consider some elementary matrixes L that describe rototranslations around the axes of the absolute reference system (1.25), and from them, by the formula of reference change (1.24) and by (1.23) for the composition of infinitesimal rototranslations, many matrixes L of applicative interest can be obtained.

$$
L^{t1} = \begin{bmatrix} 0 & 0 & 0 & 1 \\ 0 & 0 & 0 & 0 \\ 0 & 0 & 0 & 0 \\ 0 & 0 & 0 & 0 \end{bmatrix} \qquad
L^{r1} = \begin{bmatrix} 0 & 0 & 0 & 0 \\ 0 & 0 & -1 & 0 \\ 0 & 1 & 0 & 0 \\ 0 & 0 & 0 & 0 \end{bmatrix}
$$

$$
L^{t2} = \begin{bmatrix} 0 & 0 & 0 & 0 \\ 0 & 0 & 0 & 1 \\ 0 & 0 & 0 & 0 \\ 0 & 0 & 0 & 0 \end{bmatrix} \qquad
L^{r2} = \begin{bmatrix} 0 & 0 & 1 & 0 \\ 0 & 0 & 0 & 0 \\ -1 & 0 & 0 & 0 \\ 0 & 0 & 0 & 0 \end{bmatrix} \tag{1.25}
$$

$$
L^{t3} = \begin{bmatrix} 0 & 0 & 0 & 0 \\ 0 & 0 & 0 & 0 \\ 0 & 0 & 0 & 1 \\ 0 & 0 & 0 & 0 \end{bmatrix} \qquad
L^{r3} = \begin{bmatrix} 0 & -1 & 0 & 0 \\ 1 & 0 & 0 & 0 \\ 0 & 0 & 0 & 0 \\ 0 & 0 & 0 & 0 \end{bmatrix}
$$

The matrixes L^{ti} and L^{ri} indicated in (1.25) correspond, respectively, to translations and rotation around the x_i axis. To understand the importance of matrixes L, we consider two bodies B_i and B_j. B_j can only be moved with a helicoidal motion with respect to B_i, and B_i is fixed (this body is called frame in machine theory) to the reference system (k). Furthermore, the position of a generic point P of the body B_j can be represented either in the reference system (k) or (j) by defining a curvilinear abscissa q along the coupling helix (1.26 and 1.27).

$$
p_{k(k)} = M_{kj}p_{j(j)} \tag{1.26}
$$

A method for the evaluation of the geometrical speed (1.27) is proposed by computing the partial derivative of equation 1.26 with respect to the q curvilinear abscissa, as shown better in the following.

$$
\frac{\partial p_{k(k)}}{\partial q} = \frac{\partial M_{kj}}{\partial q} p_{j(j)} \tag{1.27}
$$

Besides, the validity of (1.28) can be demonstrated, thus obtaining an algorithm to calculate automatically the derivative of the position matrix with the help of the L matrixes. Then it is easy to understand their great importance both within the computational kinematic and dynamic, and, in general, in multiphysics problems.

$$\frac{\partial M_{kj}}{\partial q} = L_{ij(k)} M_{kj} \tag{1.28}$$

In order to introduce the speed kinematics for a rigid body B with respect to a reference frame (1), a point P of the same body is considered. Then, the temporal derivative of the homogeneous coordinates of P is associated with the same coordinates by the speed matrix W; this relation is expressed by (1.29) or more explicitly by (1.30).

$$\dot{p} = W \cdot p \tag{1.29}$$

$$\begin{pmatrix} \dot{p}_1 \\ \dot{p}_2 \\ \dot{p}_3 \\ 0 \end{pmatrix} = \left[\begin{array}{ccc|c} 0 & -\omega_3 & \omega_2 & v_1 \\ \omega_3 & 0 & -\omega_1 & v_2 \\ -\omega_2 & \omega_1 & 0 & v_3 \\ \hline 0 & 0 & 0 & 0 \end{array} \right] \begin{pmatrix} p_1 \\ p_2 \\ p_3 \\ 1 \end{pmatrix} = \left[\begin{array}{c|c} \boldsymbol{\omega} & \mathbf{v} \\ \hline 0 & 0 \end{array} \right] \cdot \begin{pmatrix} p_1 \\ p_2 \\ p_3 \\ 1 \end{pmatrix} \tag{1.30}$$

The speed matrix can also be expressed with respect to different reference systems. The formula for the change of reference is denoted by (1.31), in which the relation between two "visions" of W is represented with respect to two reference frames (i) and (j), which are not in relative movement.

$$W_{(i)} = M_{ij} \cdot W_{(j)} \cdot M_{ji} \tag{1.31}$$

Considering a similar situation, the acceleration matrix H is defined by simply examining the ratio between the second-order temporal derivative of the homogeneous coordinates of the same point P and the coordinates of P, and then we assert (1.32) or more explicitly (1.33).

$$\ddot{p} = H \cdot p \tag{1.32}$$

$$\begin{pmatrix} \ddot{p}_1 \\ \ddot{p}_2 \\ \ddot{p}_3 \\ 0 \end{pmatrix} = \left[\begin{array}{ccc|c} & & & a_1 \\ & \dot{\boldsymbol{\omega}} + \boldsymbol{\omega}^2 & & a_2 \\ & & & a_3 \\ \hline 0 & 0 & 0 & 0 \end{array} \right] \tag{1.33}$$

Even the acceleration matrix H can be expressed, as well as the speed matrix W (1.31), with respect to different reference frames that are not in relative movement (1.34). The acceleration matrix H can also be obtained directly from the speed matrix W by (1.35).

$$H_{(i)} = M_{ij} \cdot H_{(j)} \cdot M_{ji} \tag{1.34}$$

$$H = \dot{W} + W^2 \tag{1.35}$$

Moreover, we note that the definitions (1.30) and (1.33) of the matrixes W and H can be obtained directly by computing the time derivative of equation 1.26, thus obtaining a relation that involves the M position matrixes (1.36). Such a relation is shown in the following text, and it allows the use of L matrixes for a simplified calculation of the speed and acceleration matrixes in different cases of applicative interest.

$$\dot{M} = W \cdot M \qquad \ddot{M} = H \cdot M \tag{1.36}$$

Considering all the hypotheses that have led to (1.26), calculating the time derivative of the homogeneous coordinates $p_{k(k)}$, and comparing the result with (1.29), a relationship is achieved among the speed matrix W, the time derivative of the curvilinear abscissa q, and the matrix L of the kinematic pair, as expressed in (1.37).

$$W_{(k)} = \dot{q} \cdot L_{ij(k)} \tag{1.37}$$

Calculating the second-order time derivative of $p_{k(k)}$ and following a similar procedure, it is possible to obtain the relationship among the acceleration matrix H, the curvilinear abscissa q, and the matrix L of the kinematic pair, as shown more clearly in the treatment of kinematic constraints.

In this section we have considered the displacement of a rigid body with regard to a fixed reference frame. We briefly introduce, in the following text, some comments on the kinematic of bodies in relative movement. We consider an absolute reference system (r), the position and the speed of three bodies B_i, B_j, and B_k are referred to (r), and the three reference systems, (i), (j), and (k), respectively, are situated on these bodies. The body B_i is immobile, and the body B_j is moving with respect to B_i, whereas the body B_k is moving with respect to B_j. As far as positional kinematic is concerned, there is nothing new with respect to the previous situations; in particular, the just-shown conditions can be represented by (1.38). The speed kinematic, which defines the absolute speed of (k) with regard to (i) in the reference system (r), needs the Rivals theorem, which allows the composition of the relative motion of (k) with respect to (j) and the composition of the drag motion of (j) with respect to (i), as shown in (1.39). As far as the acceleration kinematic is concerned, it is necessary to consider the Coriolis theorem, which shows that absolute acceleration comes from the composition of relative acceleration,

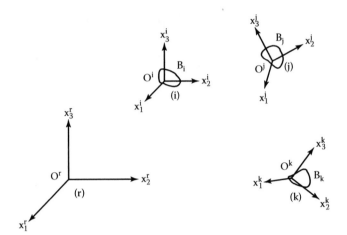

FIGURE 1.5
Relative motion of the body B_k with respect to the body B_j, which moves with respect to the body B_i.

drag acceleration, and a new term called Coriolis acceleration or centrifugal acceleration or centripetal acceleration (1.40) (Figure 1.5).

$$M_{ik} = M_{ij} \cdot M_{jk} \tag{1.38}$$

$$W_{ik(r)} = W_{ij(r)} + W_{jk(r)} \tag{1.39}$$

$$H_{ik(r)} = H_{ij(r)} + H_{jk(r)} + 2W_{ij(r)}W_{jk(r)} \tag{1.40}$$

1.2.3   Kinematics of a Flexible Beam

We consider a deformable body with a prevalent dimension. For the sake of simplicity, it can be considered to be a beam without giving other details about this definition. To describe the movement of this beam, the global motion is decomposed in a rigid component and in another one due to the body deformation. For the purpose of describing the movement of the rigid beam B, a point O^i of the body B is considered, and a reference frame (i) is fixed on this point. Therefore, the rigid component of the motion is perfectly described by the movement of (i). To describe the deformation displacement, a point O^j of the body B is considered, and a reference system (j) is assumed on this point. If there is no deformation, the reference frame (j) always coincides with the reference frame (i), but, because there is a deformation, (j) moves. The description of the motion of (j) is not enough to represent the absolute motion of the beam because, during the deformation, the beam changes its shape. If we are interested in the motion of a specific point of the body B, for example, the motion of a free end of the beam with respect to the

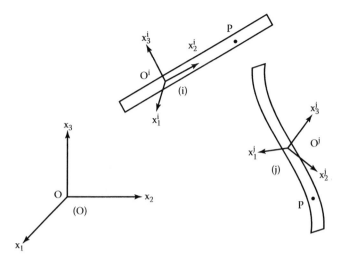

FIGURE 1.6
Position of a deformable beam.

other one, we can be satisfied by a simplified description; otherwise, it is necessary to remind ourselves that the matrixes that represent the deformation have only a single local value and must be recalculated or parameterized depending on the body local coordinates (Figure 1.6). Assuming an absolute reference frame (0), the spatial configuration of the reference frame (i) with respect to the reference frame (0) can be described by the position matrix M_{0i}, and the spatial configuration of the reference system (j) with respect to the reference (i) system can be described by the position matrix M_{ij}. Thus, with (1.4), the absolute spatial configuration of the reference system (j) can be computed with respect to the absolute reference system (0). If we consider a point P on the body B, and we assume a reference system (p) on it, the local positional configuration of B can be described in P by a proper position matrix. Because we have decomposed the motion of B into a rigid component and a deformation component, we can consider a configuration that has been subjected only to a rigid motion, and it can be called B'. This configuration has not undergone a deformation, whereas the configuration B has been subjected to a rigid and deforming motion. Therefore, the same considerations are also valid for the point P on B, i.e., an undeformed configuration P' and a deformed configuration P are assumed. Then it is possible to describe either the deformed positional configuration or the undeformed positional configuration of the point P with respect to the reference systems (0), (i), and (j). The matrix $M_{0p'}$ describes the absolute undeformed positional configuration, and the matrix $M_{ip'}$ describes the positional configuration P' with respect to the (i) reference system that moved rigidly. Furthermore, the matrix M_{jp} describes the deformed configuration P with respect to the reference system (j) because it has also been subjected to a deformation. Even at this point of the description, the system kinematics can be represented, also considering some observations in the previous sections; particularly, in the common case of small deformations, it is favorable to consider a simplified form of the matrix M_{jp}. Systematic treatises on deformable mechanisms can be found in the literature (Leitmann et al., 1999), we will show, in the following text, only some specific cases of practical interest.

1.2.4   Kinematics of a Continuum Body

A general case in kinematics is the class of deformable continua in which there are problems of rigid motion and deformations that can also be small in simplified situations, but there is prevalent dimension. This situation can be considered for its generality in order to outline a basic skeleton for a formal kinematics or to deal with complex subjects that cannot be represented in a simplified manner without losing the desired precision. In section 1.1.2, we introduced the continuous body concept and showed the possibility of representing its kinematics by Lagrangian or Eulerian terms. Without detailing a general theory of continua (Jog, 2002), only some subjects are outlined in this section; detailed applicative examples are proposed in the following chapters. The position and orientation of a continuous body can be defined as a local property of every infinitesimal element of the body and, if an infinitesimal element moves, this local property depends also on time. The most-used derived properties, speed and acceleration, can be gained from the position if appropriate hypotheses of the position functional shape are considered and if appropriate differentiation modalities of such derived properties are defined. Also, in the case of deformable continua, simplified hypotheses, such as the presence of small deformations, the absence of deformation vortexes in the case of solid bodies, and the absence of lacerations, can be considered. Furthermore, if a precise application is identified, other hypotheses can often be added to simplify the mathematical formalism and to permit a convenient tractability of physical models. A set of common simplifying conditions is represented by a deformable continuum indicated in Figure 1.7 and described in the following text.

Referring to Figure 1.7, the body B occupies the volume V in the space where it is present, and it is separated from it by a limit surface S, which is, in its turn, subdivided in two portions with different functional and

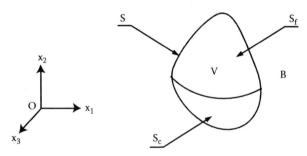

FIGURE 1.7
Deformable continuums.

kinematic characteristics. The free surface S_f can freely move in the space, whereas the constrained surface S_c is subjected to kinematic constrains; furthermore, both surfaces are provided with an exiting unitary vector $\boldsymbol{n}$, which is defined locally and expresses its local spatial configuration in an unambiguous manner. The term *deformable continuum*, which indicates the body in Figure 1.7, is adopted to pointing out two aspects: deformability, that is, the ability to change its shape, and continuity, which recalls the possibility of describing every displacement of every point of the body B with mathematical functions that are characterized by their continuity and by the continuity of their derivative to the appropriate order. In addition, from a purely kinematic viewpoint, we assume reversible displacements; consequently, there is a biunique correspondence between the initial and final positions of every point of the body B, and lacerative and compenetrating phenomena are excluded. For the sake of simplicity, the hypothesis of monoconnectivity of B is also included, that is, the ability of being separated into two distinct parts with a unique cut executed along a generic surface. Together with the hypothesis of deformable continuum, the Cauchy hypotheses are included on its kinetic–dynamic nature, so that they will be properly described in the context of the dynamics of continuous bodies. To limit the field of analysis, we decompose the movement of a body B into the sum of a rigid movement of every point of B and a deforming movement, which express the relative movement between different points of B. Because we only examined rigid motions, which deal with the motion of rigid bodies, we focus the attention, in this section, only on deforming motions to describe the change of lengths or the variations of angular positions of components of the same body. As anticipated in subsection 1.1.2, we opt for the use of a Lagrangian observer, which is widely adopted within studies of solid mechanics.

A point P belonging to the body B is considered in the initial, or reference, configuration Γ_0, when the same point P lies on the geometrical point P_0, identified in the reference frame $Ox_1x_2x_3$ by the vector OP_0 indicated by the coordinates of x_0 (Figure 1.8). After a lapse of time t, the same point P of the body B is in another configuration Γ_i, called the final configuration, and it is situated in the geometrical point P_i, identified in the reference frame $Ox_1x_2x_3$ by the vector OP_i indicated by the coordinates of x_i. To characterize the configuration Γ_i with respect to the configuration Γ_0, we refer to the

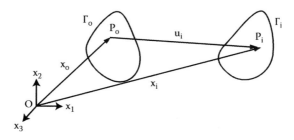

FIGURE 1.8
Motion of a point P of the body B.

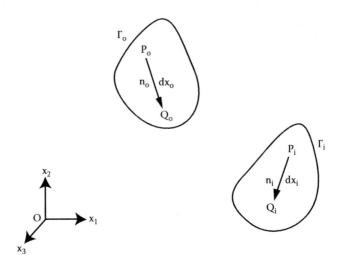

FIGURE 1.9
Movement of a fiber PQ of the body B.

vector of displacement u_i that describes the difference between these two configurations, as expressed in (1.41), or (1.42), omitting the index i because it is assumed to be generic.

$$u_i(x_0, t) = x_i(x_0, t) - x_0; \quad \forall i \in N \tag{1.41}$$

$$u(x_0, t) = x(x_0, t) - x_0 \tag{1.42}$$

The displacement generally includes rigid movements, which are not interesting for the study of the deformations in the continuum body; therefore, we examine only the relative movements of P and of its neighborhood. To allow this analysis, a generic fiber of the body B is considered with an infinitesimal length joining the point P with a point Q in its neighborhood (Figure 1.9).

Analogous to the point P, the point Q lies on the geometrical point Q_0 in the initial configuration Γ_0 and, after a time lapse t, the same point Q is in the configuration Γ_i occupying the point Q_i; hence the distance between the points P and Q is described, in the initial configuration Γ_0, by the vector dx_0, and in the final configuration Γ_i by the vector dx_i. Generically, two useful pieces of information can be extracted from the vector dx_i in the configuration Γ_i; these are its length dL_i and its direction, which is expressed by the unitary vector dn_i (1.43), or, omitting the index i because it assumes a generic value, by (1.44).

$$dL_i = \sqrt{dx_i^t \cdot dx_i}; \quad n_i = \frac{dx_i}{dL_i}; \quad \forall i \in N \tag{1.43}$$

$$dL = \sqrt{dx^t \cdot dx}\,; \quad n = \frac{dx}{dL} \tag{1.44}$$

The definition of the total differential can be used to describe synthetically the movement of the fiber PQ, obtaining the expression (1.45), where F is the gradient tensor of deformations and it can be represented with (1.46). In (1.46), I is the identity matrix and Ψ the derivative tensor of the displacement vector. The gradient tensor of deformations allows us to express the length of the fiber PQ with (1.47).

$$dx_i(x_0,t) = \left[\frac{\partial x_i(x_0,t)}{\partial x_0}\right]^T dx_0 = F \cdot dx_0 \tag{1.45}$$

$$F = \left[\frac{\partial x_0}{\partial x_0}\right]^T + \left[\frac{\partial u}{\partial x_0}\right]^T = I + \Psi \tag{1.46}$$

$$dL_i = \sqrt{dx_i^T dx_i} = \sqrt{dx_0^T F^T F dx_0} = \sqrt{dx_0^T C dx_0} \tag{1.47}$$

where C is the right Cauchy–Green tensor and it represents deforming phenomena on the body B. An interesting characteristic of the tensor C is its symmetry due to its definition as a product of a tensor with its transpose. Therefore, to measure the stretch of the fiber PQ, we can directly refer to the definition expressed in (1.48) or we can use (1.49), where the absence of square roots allows a simplified computation.

$$\lambda = dL_0/dL_i \tag{1.48}$$

$$\lambda^2 = n_0^T \cdot C \cdot n_0 \tag{1.49}$$

In the presence of only rigid motions, the tensor C is reduced to the identity matrix I, obtaining a unitary stretch λ. To know the deformation state in the neighborhood of P, that is, to know the stretch of every fiber PQ, with Q close to P, we consider the principal stretches and principal stretch directions. These are obtained maximizing or minimizing expression (1.49) under the condition that n_0 is a unitary vector; consequently, with the stationary condition and with the help of Lagrangian multipliers, they result from the solution of equation (1.50).

$$\det\left(C - \lambda^2 I\right)\lambda^2 = 0 \tag{1.50}$$

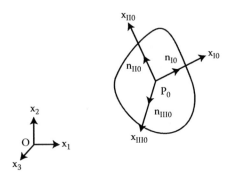

FIGURE 1.10
Principal stretches reference frame.

The annulment of (1.50) produces three eigenvalues of the matrix C, indicating the squares of the three principal stretches in the neighborhood of P, then three unitary and orthogonal eigenvectors, n_{I0}, n_{II0} e n_{III0}, can be associated to these eigenvalues to represent the three principal stretch directions. Therefore, a new reference frame can be defined with its origin coinciding with P and its Cartesian axes exiting from P and directed as the eigenvectors of C, obtaining the principal stretch reference frame. For every different point P of the continuum B, generally, there is a different value of the tensor C, and thus the principal stretch reference frame and the relative principal stretches also depend on the point P. On the other hand, with a fixed point P, the principal stretches do not depend on the reference system $Ox_1x_2x_3$ because they are invariant characteristics of the tensor C and depend only on the boundary conditions of the problem (Figure 1.10).

Returning to the example of Figure 1.8 in which the displacement of the point P of the body B is described as passing from configuration Γ_0 to configuration Γ_i, or simply, to configuration Γ, it is interesting to observe how the principal stretch reference frame is modified after the displacement. The expression (1.51) can be obtained using (1.43) and applying the transformation (1.45) for every principal stretch fiber, as shown in Figure 1.11.

$$n_I = F \cdot n_{I0}/\lambda_I; \quad n_{II} = F \cdot n_{II0}/\lambda_{II}; \quad n_{III} = F \cdot n_{III0}/\lambda_{III} \qquad (1.51)$$

Therefore, the stretch can be defined in terms of the final configuration Γ, starting from the definition in (1.48) and applying the transformation in (1.45), to obtain, after proper calculations, the relation (1.52).

$$\lambda^2 = \left[n^T \cdot \left(F \cdot F^T \right)^{-1} \cdot n \right]^{-1} = \left[n^T \cdot B^{-1} \cdot n \right]^{-1} \qquad (1.52)$$

where B is the left Cauchy–Green tensor. After the occurred deformation, the axes of the principal stretch reference frame remain orthogonal;

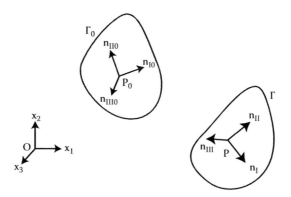

FIGURE 1.11
Principal stretch reference frame after the displacement.

consequently, with respect to the initial principal stretch reference frame, we can represent the change of reference system with a rototranslation matrix M, which is composed by a rotational term R and a translational term T as expressed in (1.53), analogously to the results of section 1.2.2.

$$M = \left[\begin{array}{c|c} R & T \\ \hline 0 & 1 \end{array}\right]$$

(1.53)

Particularly, introducing the right stretch tensor V_R described in (1.54), the matrix R can be obtained as in (1.55), and a relation with the tensor C can be recognized in (1.56).

$$V_R = \lambda_I n_{I0} n_{I0}^T + \lambda_{II} n_{II0} n_{II0}^T + \lambda_{III} n_{III0} n_{III0}^T$$

(1.54)

$$R = F \cdot V_R^{-1}$$

(1.55)

$$V_R^2 = C$$

(1.56)

Likewise, in the final configuration, the matrix R can be obtained in (1.58) with the help of the left stretch tensor V_L described in (1.57), recognizing a relation with the tensor B (1.59).

$$V_L = \lambda_I n_I n_I^T + \lambda_{II} n_{II} n_{II}^T + \lambda_{III} n_{III} n_{III}^T$$

(1.57)

$$R = V_L^{-1} \cdot F$$

(1.58)

$$V_L^2 = B$$

(1.59)

For practical purposes, the stretch does not express in a satisfying manner the level of deformation; in fact, the strain ε is preferred as described by (1.60), which indicates a family of common mathematical functions used to represent it.

$$\varepsilon = f_{(m)}(\lambda) = \begin{cases} \dfrac{1}{m}\left(\lambda^m - 1\right), & m \in Z - \{0\} \\ \ln(\lambda), & m = 0 \end{cases} \tag{1.60}$$

When the parameter m assumes a null value, the logarithmic strain is obtained from (1.60), and it is characterized by a high level of precision in describing the real deformation but needs a certain computational effort. If m assumes a value equal to one, it results in the nominal, or Biot, strain, which is widely used for the measure of small deformations but often also for large deformations. If, finally, m assumes a value equal to two, it produces the Green–Lagrange strain, which is widely used in problems of nonlinear mechanics. More generally, we can define a strain tensor in a generic reference frame, deriving it from the left or right (1.61) stretch tensor, generalizing (1.60).

$$\varepsilon = f(V_R) \tag{1.61}$$

Particularly, in the principal stretch reference frame of the initial configuration, applying (1.60 and 1.61) to (1.54), the strain tensor can be explicitly expressed with (1.62).

$$\varepsilon = f\left(\lambda_I\right) n_{I0} n_{I0}^T + f\left(\lambda_{II}\right) n_{II0} n_{II0}^T + f\left(\lambda_{III}\right) n_{III0} n_{III0}^T \tag{1.62}$$

Furthermore, (1.62) calculates the eigenvalues and eigenvectors of the tensor C in every point P of the continuum, but, selecting the relation between the stretch tensor and strain tensor with the Green–Lagrange option, the strain tensor can be expressed by (1.63) without calculation of the eigenvalues or eigenvectors.

$$E = \frac{1}{2}(C - I) \tag{1.63}$$

More generally, every function of the family (1.60) in which m is an even integer (also with negative value) is able to simplify (1.62), avoiding the computation of eigenvalues or eigenvectors, as expressed in (1.64).

$$\varepsilon_{(m)} = \frac{1}{2}\left(C^{(m/2)} - I\right) \tag{1.64}$$

Besides, the measures of deformation are all derived from the displacement, i.e., with the Green–Lagrange strain tensor, (1.63) can be expressed explicitly in the form indicated by (1.65).

$$E = \frac{1}{2}\left(\Psi + \Psi^T + \Psi^T\Psi\right) = \frac{1}{2}\left[\left(\frac{\partial u}{\partial x_0}\right)^T + \frac{\partial u}{\partial x_0} + \frac{\partial u}{\partial x_0}\left(\frac{\partial u}{\partial x_0}\right)^T\right] \qquad (1.65)$$

After this mention of the strain tensor, some observations can be made on the speed of deformation; for this reason, the act of motion is considered in the configuration Γ, described with respect to the configuration Γ_0, to achieve the definition of the speed as in (1.66).

$$v(x_0, t) = \frac{\partial x(x_0, t)}{\partial t} \qquad (1.66)$$

From the definition of speed, the speed gradient can be easily calculated in the configuration Γ, allowing the definition of the speed gradient tensor L (1.67).

$$\left(\frac{\partial v}{\partial x}\right)^T = \frac{\partial F}{\partial t}F^{-1} = L \qquad (1.67)$$

The tensor L is just purged from the components of rigid translational speed of the neighborhood of P, but it still exhibits components of spin speed, which can be removed with (1.68).

$$D = \frac{1}{2}\left(L + L^T\right); \quad W = \frac{1}{2}\left(L - L^T\right) \qquad (1.68)$$

where D is the (symmetric) strain rate tensor, and W is the (antisymmetric) spin tensor; these tensors can also be written in an explicit form (1.69), using the definition proposed in (1.67). Furthermore, with the hypothesis of small deformations, which has wide practical interest, the displacement is related to speed with the relation (1.70); consequently, a tensor φ of small strain and a tensor θ of small spin can be defined, starting from (1.69), to obtain the simplified form expressed in (1.71).

$$D = \frac{1}{2}\left[\left(\frac{\partial v}{\partial x}\right)^T + \frac{\partial v}{\partial x}\right]; \quad W = \frac{1}{2}\left[\left(\frac{\partial v}{\partial x}\right)^T - \frac{\partial v}{\partial x}\right] \qquad (1.69)$$

$$u = v\,dt \qquad (1.70)$$

$$\varphi = D\,dt = \frac{1}{2}\left[\left(\frac{\partial u}{\partial x}\right)^T + \frac{\partial u}{\partial x}\right]; \quad \theta = W\,dt = \frac{1}{2}\left[\left(\frac{\partial u}{\partial x}\right)^T - \frac{\partial u}{\partial x}\right] \qquad (1.71)$$

The tensor φ of small strain can also be obtained, in an alternative way, as a particularization of the Green–Lagrange tensor (1.65), observing that square terms are negligible and that the derivatives with respect to x_0 can be substituted by derivatives with respect to x because the configurations Γ_0 and Γ can be approximately superimposed in the condition of small displacements, obtaining (1.72), which represents the congruence equation of the continuum.

$$\varphi = \frac{1}{2}\left[\left(\frac{\partial u}{\partial x}\right)^T + \frac{\partial u}{\partial x}\right] \tag{1.72}$$

To grasp the physical meaning of the tensor φ, the stretch λ_i of the fiber PQ can be considered, in the first instance, with its initial direction at undeformed state, directed as the axis x_i. Then, with the hypothesis of small deformations, λ_i can be related with the term φ_{ii} of the tensor of small strain as in (1.73). Furthermore, (1.74) can be obtained considering the angles α_i and α_{j}, which represent, respectively, the projections on the plane Ox_ix_j of the angles formed by the deformed fiber PQ with its initial direction (along the axis x_i) and by a deformed fiber PR with its initial direction (along the axis x_j). Consequently, the tensor of small strain can be rewritten with (1.75).

$$\lambda_i = \sqrt{1 + 2\frac{\partial u_i}{\partial x_i} + \sum_{j=1}^{3}\left(\frac{\partial u_j}{\partial x_j}\right)^2} \simeq 1 + \frac{\partial u_i}{\partial x_i} = 1 + \varphi_{ii}; \quad \forall i \in \{1,2,3\} \tag{1.73}$$

$$\begin{cases} \tan(\alpha_i) = \dfrac{\dfrac{\partial u_j}{\partial x_i}}{1 + \dfrac{\partial u_i}{\partial x_j}} \simeq \dfrac{\partial u_j}{\partial x_i} \simeq \alpha_i \\[4mm] \varphi_{ij} = \dfrac{1}{2}\left(\dfrac{\partial u_i}{\partial x_j} + \dfrac{\partial u_j}{\partial x_i}\right) \simeq \dfrac{1}{2}(\alpha_i + \alpha_j) = \dfrac{1}{2}\gamma_{ij} \end{cases} ; \quad i,j \in \{1,2,3\} \tag{1.74}$$

$$\varphi = \begin{bmatrix} \varepsilon_1 & \dfrac{1}{2}\gamma_{12} & \dfrac{1}{2}\gamma_{13} \\[3mm] \dfrac{1}{2}\gamma_{21} & \varepsilon_2 & \dfrac{1}{2}\gamma_{23} \\[3mm] \dfrac{1}{2}\gamma_{31} & \dfrac{1}{2}\gamma_{32} & \varepsilon_3 \end{bmatrix} \tag{1.75}$$

The principal strains can be searched also for the tensor φ with a direct computation or with a particularization of problems previously solved;

then, observing from a principal strain reference frame, we are interested in decomposing the deformation process into a component of volumetric deformation and a component of shape deformation in the neighborhood of point P. For this purpose, a tensor ε_V is defined to represent only the volumetric deformation (1.76), and then a deviatoric tensor ε_D is obtained with a subtraction from the tensor φ (1.76).

$$\varepsilon_V = \frac{1}{3} tr(\varphi) I; \quad \varepsilon_D = \varphi - \varepsilon_V \qquad (1.76)$$

where I is the identity matrix.

Finally, the kinematic analysis of continua lead to a mathematical problem with a differential nature, i.e., the external congruence equations (1.72) or the internal congruence equations, which can be obtained with derivatives of (1.72), eliminating the displacement components with the Schwarz theorem. The knowledge of these equations is often insufficient to answer problems of continuum mechanics; to remedy these deficiencies, external loads can be usefully introduced in the boundary conditions, in the form of surface or volume forces. These extensions will be properly considered in continuum dynamics.

1.2.5 Kinematics with Theoretical Constraints

The motion of a body in space can be free or restricted to one or more paths by some kinematic constraints, which in this monograph are expressed by impediments of some movements. From a mathematical viewpoint, such constraints can be shown as inequalities that involve the positions and speed of a body (or of a part) and the temporal variable. In particular, when constraints are expressed by equalities, they are called *bilateral constraints*; when the time does not appear explicitly in the relation, the constraint is called *scleronomous*, and when time appears, the constraint is called *rheonomous*. Furthermore, when a constraint is expressed by a mathematical relation without the position derivatives of a body or of a body part, it is called *geometrical or holonomic or positional*. In the following text, we refer to bilateral holonomic and scleronomous constraints, and we point out some fundamental properties through the representation of kinematic impediments with two connected rigid bodies in space. A first body T is always motionless and is called a *frame*, whereas a latter body B moves relatively to the first one; movement constraints of the body B are represented by kinematic pairs between it and the frame T, and so the B–T group is also called a kinematic pair. The result of a kinematic coupling is the reduction of the mobility of the body B. The coupling is represented by a pair of coupled surfaces, or lines or points; the first one belongs to the body B, whereas the second belongs to the body T. These kinematic elements are also called *conjugate*. Conjugate kinematic elements are not necessarily in contact, but their connection assumes a kinematic nature.

For a first analysis of kinematic constraint between B and T, we refer to the simplified case of plane relative movement, then a finite movement of B is observed to find a point of B whose location does not change as a result of this displacement. Such a point is called the *pole of the finite movement*. In the case of an infinitesimal movement, a corresponding concept can be defined: the instantaneous center (of velocity) C, where the speed relative to a frame of reference is zero at a given instant, and, moreover, it is possible to show that the point C always exists under reasonable hypotheses. Furthermore, a finite movement of B can be decomposed into the following infinitesimal displacements that have respective instantaneous centers of velocity; the locus of these points seen from a reference system on the fixed body T is called *fixed centrode*, whereas the same locus seen from a reference frame on the mobile body B is called *moving centrode*. To better expose the proposed definitions, we consider the example of a revolute pair, which admits only circular movements of B in the plane Ox_1x_2 around a hinge O, described in a fixed reference frame (0) rigidly connected with the frame T, as shown in Figure 1.12; in these conditions, the unique instant center of B can be easily identified with O. Another interesting example is the prismatic pair, which allows movements of B only along an axis, i.e., axis x_1, as shown in Figure 1.13, where the unique center of B can be identified with the improper point of the axis x_2.

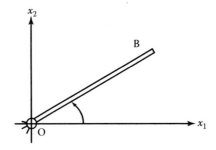

FIGURE 1.12
Revolute pair.

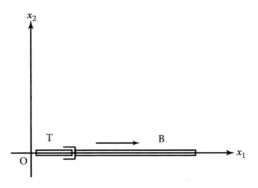

FIGURE 1.13
Prismatic pair.

To perform a deeper description of kinematic pairs, we can observe that the position of a rigid body in the plane can be represented by three independent parameters associated with the corresponding three degrees of freedom or, in other words, to three independent ways of movement. Therefore, we can affirm that prismatic and revolute pairs introduce two kinematic degrees of constraint and one degree of freedom of the body B; prismatic and revolute kinematic pairs are also called lower pairs, and they are formed by contact lines in the plane, or by contact surfaces in space (Ionescu, 2003). Using only lower pairs, which connect rigid bodies, a wide class of mechanical systems can be deduced to deal with different real-world applications; in order to widen such classes and consider further systems with a practical interest, a higher kinematic pair can be introduced. This pair is characterized by contact points in the plane and contact points or contact lines in space (Ionescu, 2003) and, usually, it does not allow an equivalent representation with a finite number of inferior pairs. An example of a higher pair is the cam mechanism, which is characterized by two bodies, T and B, shaped in such a way as to impart the desired relative motions, and which are constrained to have a point of their profiles in relative contact considering only bilateral constraints (Figure 1.14).

After the previous description of planar pairs, in the following text, kinematic pairs are described in three-dimensional space, where unconstrained rigid bodies have six degrees of freedom, three more than in the planar situation. We focus our attention on a specific class of kinematic pairs to easily observe some specific and general properties, i.e., a revolute pair between the bodies T and B in space. The body B exhibits only a rotary degree of freedom relative to T, and therefore it has five degrees of constraint relative to T; three degrees of freedom are added to the planar situation because the plane of relative B-T motion is constrained to be rigidly connected to T in the space. Furthermore, the revolute pair acts in the same way on all the planes that are parallel to the considered relative motion plane; more precisely, the revolute motion takes place in a sheaf of parallel planes, and therefore the concepts of center of velocity and centrode should be extended. Particularly, connecting all the instantaneous centers of the sheaf, we obtain an instantaneous screw axis, which is the locus of the points of B, whose linear velocity is parallel to the angular velocity vector of B at a given instant (Ionescu, 2003). This

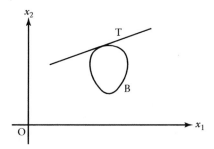

FIGURE 1.14
Example of a higher pair.

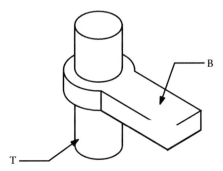

FIGURE 1.15
Revolute pair.

definition allows contact with the mathematical treatment proposed in section 1.2.2 and allows us to use powerful matricial tools to study kinematic pairs. To improve our example, we can also consider the prismatic pair acting, together with the revolute pair, between T and B with the same instantaneous screw axis; so, a screw motion, consisting in a combination of a rotation with a translation, which is parallel to the rotation, can be described with the same matricial approach. Then, for a generic relative displacement of B with respect to T, the locus of the instantaneous screw axes described in a reference system connected with T is called *fixed axode*, whereas the same locus described in a reference system connected with B is called *mobile axode*.

We stressed the presence of conceptual analogies between three-dimensional and bidimensional situations, intuiting some properties of spatial kinematic constraints, in spite of difficulties in graphical representation and in abstraction activity. Furthermore, we suggested the possibility of a mathematical representation that allows the use of formulations and models just presented in previous sections. Now we deal with a brief description of the same kinematic pairs with applicative involvements. A revolute pair between the frame T and a moving body B is examined; the reference frame (0) is rigidly connected with T, maintaining its origin on the axis of the revolute pair, whereas, the reference frame (1) is positioned on the body B with its origin O^1 coinciding with O, which is the origin of the fixed reference frame (Figure 1.15). To simplify the calculations without loss of generality, the axis of the revolute pair is assumed coincident with the x_3-axis of the reference system (0); therefore, the matrix M_{01} is represented by 1.41 and describes the position of the reference frame (1) with respect to the reference frame (0) or, in the same manner, describes the position of B with respect to T.

$$M_{01} = \left[\begin{array}{ccc|c} \cos(\alpha) & -\sin(\alpha) & 0 & 0 \\ \sin(\alpha) & \cos(\alpha) & 0 & 0 \\ 0 & 0 & 1 & 0 \\ \hline 0 & 0 & 0 & 1 \end{array}\right] \qquad (1.77)$$

$$L_{01} = \begin{bmatrix} 0 & -1 & 0 & | & 0 \\ 1 & 0 & 0 & | & 0 \\ 0 & 0 & 0 & | & 0 \\ \hline 0 & 0 & 0 & | & 0 \end{bmatrix}$$

(1.78)

$$W_{01} = L_{01} \cdot \dot{\alpha} = \begin{bmatrix} 0 & -\dot{\alpha} & 0 & | & 0 \\ \dot{\alpha} & 0 & 0 & | & 0 \\ 0 & 0 & 0 & | & 0 \\ \hline 0 & 0 & 0 & | & 0 \end{bmatrix};$$

(1.79)

$$H_{01} = L_{01}^{2} \cdot \dot{\alpha}^{2} + L_{01} \cdot \ddot{\alpha} = \begin{bmatrix} -\dot{\alpha}^{2} & -\ddot{\alpha} & 0 & | & 0 \\ \ddot{\alpha} & -\dot{\alpha}^{2} & 0 & | & 0 \\ 0 & 0 & 0 & | & 0 \\ \hline 0 & 0 & 0 & | & 0 \end{bmatrix}$$

The corresponding matrix L, which represents the kinematic pair, is composed only of the constant terms indicated in (1.78), and the speed and acceleration matrixes can be deduced as in 1.43, where the presence of a single scalar parameter α is enough to describe the motion of B with respect to T. Thus, the revolute pair allows a degree of freedom, and it suppresses five of them with five respective degrees of constraint. We consider an initial instant of time conventionally indicated with t_0 and a further instant t_1, where the scalar parameter α assumes the respective values α_0 and α_1. Using a notation relative to the initial instant and, consequently, to the initial position, t_1 can be expressed in terms of t_0 by means of the differential Δt, and likewise, α_1 can be expressed in terms of α_0 by means of the differential $\Delta \alpha$. Without details on the value of these differentials but avoiding any trivial or infinity conditions, (1.80) expresses the two positions of B with respect to T. Therefore, the rototranslation, which is able to pass from the first spatial configuration of B to its second spatial configuration, can be represented with (1.81), with respect to the reference system (0).

$$M_{01}\big|_{t=t_0} = \begin{bmatrix} \cos(\alpha_0) & -\sin(\alpha_0) & 0 & | & 0 \\ \sin(\alpha_0) & \cos(\alpha_0) & 0 & | & 0 \\ 0 & 0 & 1 & | & 0 \\ \hline 0 & 0 & 0 & | & 1 \end{bmatrix};$$

(1.80)

$$M_{01}\big|_{t=t_0+\Delta t} = \begin{bmatrix} \cos(\alpha_0+\Delta\alpha) & -\sin(\alpha_0+\Delta\alpha) & 0 & | & 0 \\ \sin(\alpha_0+\Delta\alpha) & \cos(\alpha_0+\Delta\alpha) & 0 & | & 0 \\ 0 & 0 & 1 & | & 0 \\ \hline 0 & 0 & 0 & | & 1 \end{bmatrix}$$

$$Q_{(0)} = \left[\begin{array}{c|c} R_{(0)} & \Delta T_0 \\ \hline 0 & 1 \end{array} \right] = \left(M_{01} \big|_{t=t_0+\Delta t} \right) \cdot \left(M_{01} \big|_{t=t_0} \right)^T$$

$$= \left[\begin{array}{ccc|c} \cos(\Delta\alpha) & -\sin(\Delta\alpha) & 0 & 0 \\ \sin(\Delta\alpha) & \cos(\Delta\alpha) & 0 & 0 \\ 0 & 0 & 1 & 0 \\ \hline 0 & 0 & 0 & 1 \end{array} \right] \qquad (1.81)$$

Consequently, the parameters of the helicoidal movement can be extracted directly from the rototranslation matrix and, in connection with this, although in this particular case powerful calculation tools are not necessary, a computational approach is introduced here to deal properly with more difficult cases shown in this section or further sections. As a matter of fact, the rotation ϑ around the helicoidal axis is calculated simply by the general formula indicated in (1.82), which allows the value $\Delta\alpha$ to be obtained from the components of the rototranslation matrix $Q_{(0)}$. Furthermore, the inhomogeneous components of the vector u, which represents the ideal point of the motion axis and therefore also its direction, are deduced from (1.83), where there are two distinct formulations that can be used alternatively. After having verified that the respective denominator is not null, then the ideal point of the x_3-axis can be obtained with the substitution of the elements of $Q_{(0)}$, as indicated in (1.81).

$$c = \left(Q_{(0)11} + Q_{(0)22} + Q_{(0)33} - 1 \right) / 2 \; ;$$

$$s = \pm \sqrt{\left(Q_{(0)32} - Q_{(0)23} \right)^2 + \left(Q_{(0)31} - Q_{(0)13} \right)^2 + \left(Q_{(0)21} - Q_{(0)12} \right)^2} \Big/ 2 \qquad (1.82)$$

$$\vartheta = \arctan(s, c)$$

$$u_1 = \frac{Q_{(0)32} - Q_{(0)23}}{2\sin(\vartheta)} = \pm\sqrt{\frac{Q_{(0)11} - 1}{1 - \cos(\vartheta)} + 1}$$

$$u_2 = \frac{Q_{(0)13} - Q_{(0)31}}{2\sin(\vartheta)} = \pm\sqrt{\frac{Q_{(0)22} - 1}{1 - \cos(\vartheta)} + 1} \qquad (1.83)$$

$$u_3 = \frac{Q_{(0)21} - Q_{(0)12}}{2\sin(\vartheta)} = \pm\sqrt{\frac{Q_{(0)33} - 1}{1 - \cos(\vartheta)} + 1}$$

In a revolute pair there are no translational movements along the axis of motion, but to present a comprehensive approach, (1.84) can be used to calculate the helicoidal motion pitch for obtaining a null value. Furthermore, if we consider only the three inhomogeneous coordinates of the vector p, a point P of the motion axis can be represented with (1.85), that is a three-dimensional vector of the inhomogeneous coordinates of p; then, setting the arbitrary parameter λ to zero, we obtain the origin of the fixed reference system (0); thus, we verify that the x_3-axis is the instantaneous screw axis for every $\Delta\alpha$.

$$h = u^T \left[\frac{\Delta T_0}{1} \right] \tag{1.84}$$

$$\frac{\left(I - R_{(0)}^{-1}\right)\Delta T_0}{2\cos(\vartheta)} + \lambda u \tag{1.85}$$

The vectors p and u represent the motion axis with respect to the fixed reference frame, and because for every $\Delta\alpha$ or every Δt, p and u remain constant, the fixed axode is always represented by a single axis. With respect to the moving reference frame (1), we can describe the axode from the moving object viewpoint (1.86), realizing that, in our case, in the moving reference system also there is only one axis with the same coordinate values.

$$u_{(1)} = M_{10} \cdot u_{(0)}; \qquad p_{(1)} = M_{10} \cdot p_{(0)} \tag{1.86}$$

After a schematic description of a revolute pair, a prismatic pair is examined between the fixed frame T and a moving body B. A reference system (0) is rigidly connected to the fixed frame with the x_3-axis, which is coincident with the prismatic motion axis, and a reference system (1) is rigidly connected to the body B with the x_3^1-axis, which is coincident with x_3 (Figure 1.16), and then the position matrix M_{01} is represented by (1.87).

$$M_{01} = \begin{bmatrix} 1 & 0 & 0 & 0 \\ 0 & 1 & 0 & 0 \\ 0 & 0 & 1 & z \\ 0 & 0 & 0 & 1 \end{bmatrix} \tag{1.87}$$

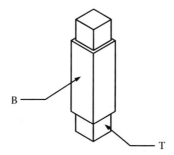

FIGURE 1.16
Prismatic pair.

The constant matrix L shown in (1.88) represents the kinematic pair, and the speed and acceleration matrixes can be deduced from it (1.89). The presence of a single scalar parameter z in (1.89) is enough to describe the motion of B with respect to T, but then also the prismatic pair allows only one degree of freedom and suppresses five of them with five respective degrees of constraint. Also in this case we consider two positions associated with two temporal instants, t_0 and t_1, where t_1 results from the sum of t_0 and a temporal interval Δt (1.90); therefore, the rototranslation that transforms a spatial configuration into the other is indicated in (1.91).

$$L_{01} = \left[\begin{array}{ccc|c} 0 & 0 & 0 & 0 \\ 0 & 0 & 0 & 0 \\ 0 & 0 & 0 & 1 \\ \hline 0 & 0 & 0 & 0 \end{array}\right] \tag{1.88}$$

$$W_{01} = L_{01}\dot{z} = \left[\begin{array}{ccc|c} 0 & 0 & 0 & 0 \\ 0 & 0 & 0 & 0 \\ 0 & 0 & 0 & \dot{z} \\ \hline 0 & 0 & 0 & 0 \end{array}\right]; \quad H_{01} = L_{01}^2\dot{z}^2 + L_{01}\ddot{z} = \left[\begin{array}{ccc|c} 0 & 0 & 0 & 0 \\ 0 & 0 & 0 & 0 \\ 0 & 0 & 0 & \ddot{z} \\ \hline 0 & 0 & 0 & 0 \end{array}\right] \tag{1.89}$$

$$M_{01}\big|_{t=t_0} = \left[\begin{array}{cccc} 1 & 0 & 0 & 0 \\ 0 & 1 & 0 & 0 \\ 0 & 0 & 1 & z_0 \\ 0 & 0 & 0 & 1 \end{array}\right]; \quad M_{01}\big|_{t=t_0+\Delta t} = \left[\begin{array}{cccc} 1 & 0 & 0 & 0 \\ 0 & 1 & 0 & 0 \\ 0 & 0 & 1 & z_0+\Delta z \\ 0 & 0 & 0 & 1 \end{array}\right] \tag{1.90}$$

$$Q_{(0)} = \left[\begin{array}{cccc} 1 & 0 & 0 & 0 \\ 0 & 1 & 0 & 0 \\ 0 & 0 & 1 & z_0+\Delta z \\ 0 & 0 & 0 & 1 \end{array}\right] \cdot \left[\begin{array}{cccc} 1 & 0 & 0 & 0 \\ 0 & 1 & 0 & 0 \\ 0 & 0 & 1 & z_0 \\ 0 & 0 & 0 & 1 \end{array}\right]^{-1} = \left[\begin{array}{cccc} 1 & 0 & 0 & 0 \\ 0 & 1 & 0 & 0 \\ 0 & 0 & 1 & \Delta z \\ 0 & 0 & 0 & 1 \end{array}\right] \tag{1.91}$$

Expression (1.91) can be used to extract the helicoidal motion parameters. The rotation ϑ around the helicoidal axis is calculated by (1.82), obtaining a null value, and moreover, (1.83) expressions cannot be used for the calculation of the components of the inhomogeneous vector u, which represents the ideal point of the screw axis because the denominator is always null. This situation indicates that this is a case of pure translation, and therefore the components of the inhomogeneous vector u are obtained with a normalization of the vector ΔT_0, and the result is exactly the vector x_3. Furthermore, (1.84) shows that the helicoidal displacement pitch is Δz. With pure translation, (1.85) is not

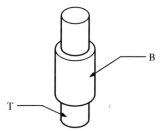

FIGURE 1.17
Cylindrical pair.

properly effective for the research of the homogeneous points of the heli-coidal rototranslation axis; in fact, the fixed axode is a sheaf of straight lines through the ideal point of the x_3-axis, and therefore, any homogeneous point can be chosen. Conventionally, we select the origin O of the reference frame (0), and so the screw axis can be deduced and identified with x_3. With respect to the reference system (1), the mobile screw axis and the mobile axode can be described by (1.86), realizing that the resulting axis is the same, even if it is translated on itself.

Another kinematic pair can be considered to slightly improve the complex-ity of the physical system: the cylindrical pair, which allows rotations and translations along an axis of the mobile body B with respect to the frame T (Figure 1.17). A reference system (0) is rigidly connected to the fixed frame with the x_3-axis, which is coincident with the prismatic motion axis, and a ref-erence system (1) is rigidly connected to the body B with the x_3^1-axis, which is coincident with x_3 (Figure 1.16). Then the position matrix M_{01} is represented by (1.87). A reference system (0) is rigidly connected to the fixed frame T with the x_3-axis on the axis of the cylindrical pair, whereas a reference system (1) is rigidly connected to the body B with the x_3^1-axis on the same axis of the cylindrical pair. Then the position matrix M_{01} is represented by (1.92), which shows that the scalar parameters α and z are enough to describe the motion of B with respect to T; therefore, the cylindrical pair allows two degrees of freedom and it abolishes four of them through four degrees of constraint.

$$
M_{01} = \begin{bmatrix} \cos(\alpha) & -\sin(\alpha) & 0 & 0 \\ \sin(\alpha) & \cos(\alpha) & 0 & 0 \\ 0 & 0 & 1 & 0 \\ 0 & 0 & 0 & 1 \end{bmatrix} \cdot \begin{bmatrix} 1 & 0 & 0 & 0 \\ 0 & 1 & 0 & 0 \\ 0 & 0 & 1 & z \\ 0 & 0 & 0 & 1 \end{bmatrix}
$$

$$
= \begin{bmatrix} \cos(\alpha) & -\sin(\alpha) & 0 & 0 \\ \sin(\alpha) & \cos(\alpha) & 0 & 0 \\ 0 & 0 & 1 & z \\ 0 & 0 & 0 & 1 \end{bmatrix} \tag{1.92}
$$

The constant matrixes L, which are associated with all degrees of freedom, are indicated in (1.93), and they allow a convenient calculation of the speed and acceleration matrixes (1.94), in accordance with (1.39 and 1.40). As for the previous kinematic pairs, two different spatial configurations are examined, and these are associated with two different temporal instants (1.95). The purpose of this approach is to calculate the rototranslation matrix, which is able to turn one configuration into the other (1.96). Furthermore, the same result expressed in (1.96) can be deduced by the composition of singular simple rototranslations that are associated with any degree of freedom, i.e., by the composition of (1.81) and (1.91) in accordance with the properties of (1.20), as will be indicated in (1.97).

$$L_\alpha = \begin{bmatrix} 0 & -1 & 0 & 0 \\ 1 & 0 & 0 & 0 \\ 0 & 0 & 0 & 0 \\ 0 & 0 & 0 & 0 \end{bmatrix}; \quad L_z = \begin{bmatrix} 0 & 0 & 0 & 0 \\ 0 & 0 & 0 & 0 \\ 0 & 0 & 0 & 1 \\ 0 & 0 & 0 & 0 \end{bmatrix} \tag{1.93}$$

$$W_{01} = L_\alpha \cdot \dot\alpha + L_z \cdot \dot z = \begin{bmatrix} 0 & -\dot\alpha & 0 & 0 \\ \dot\alpha & 0 & 0 & 0 \\ 0 & 0 & 0 & \dot z \\ 0 & 0 & 0 & 0 \end{bmatrix};$$

$$H_{01} = L_\alpha^2 \cdot \dot\alpha^2 + L_z^2 \cdot \dot z^2 + L_\alpha \cdot \ddot\alpha + L_z \cdot \ddot z + 2 \cdot L_\alpha \cdot L_z \cdot \dot\alpha\dot z \tag{1.94}$$

$$= \begin{bmatrix} -\dot\alpha^2 & -\ddot\alpha & 0 & 0 \\ \ddot\alpha & -\dot\alpha^2 & 0 & 0 \\ 0 & 0 & 0 & \ddot z \\ 0 & 0 & 0 & 0 \end{bmatrix}$$

$$M_{01}\big|_{t=t_0} = \begin{bmatrix} \cos(\alpha_0) & -\sin(\alpha_0) & 0 & 0 \\ \sin(\alpha_0) & \cos(\alpha_0) & 0 & 0 \\ 0 & 0 & 1 & z_0 \\ 0 & 0 & 0 & 1 \end{bmatrix};$$

$$M_{01}\big|_{t=t_0+\Delta t} = \begin{bmatrix} \cos(\alpha_0+\Delta\alpha) & -\sin(\alpha_0+\Delta\alpha) & 0 & 0 \\ \sin(\alpha_0+\Delta\alpha) & \cos(\alpha_0+\Delta\alpha) & 0 & 0 \\ 0 & 0 & 1 & z_0+\Delta z \\ 0 & 0 & 0 & 1 \end{bmatrix} \tag{1.95}$$

$$Q_{(0)} = \left(M_{01} \big|_{t=t_0+\Delta t} \right) \cdot \left(M_{01} \big|_{t=t_0} \right)^{-1} = \begin{bmatrix} \cos(\Delta\alpha) & -\sin(\Delta\alpha) & 0 & 0 \\ \sin(\Delta\alpha) & \cos(\Delta\alpha) & 0 & 0 \\ 0 & 0 & 1 & \Delta z \\ 0 & 0 & 0 & 1 \end{bmatrix} \tag{1.96}$$

$$Q_{(0)} = Q^\alpha_{(0)} Q^z_{(0)} = Q^z_{(0)} Q^\alpha_{(0)}$$

$$= \begin{bmatrix} \cos(\Delta\alpha) & -\sin(\Delta\alpha) & 0 & 0 \\ \sin(\Delta\alpha) & \cos(\Delta\alpha) & 0 & 0 \\ 0 & 0 & 1 & 0 \\ 0 & 0 & 0 & 1 \end{bmatrix} \begin{bmatrix} 1 & 0 & 0 & 0 \\ 0 & 1 & 0 & 0 \\ 0 & 0 & 1 & \Delta z \\ 0 & 0 & 0 & 1 \end{bmatrix} \tag{1.97}$$

The rototranslation matrix can be used to extract the helicoidal motion parameters. Expression (1.82) allows us to obtain a rotation equal to $\Delta\alpha$; according to (1.83), the calculation of the vector u permits us to achieve the ideal point of the x_3-axis; then the helicoidal pitch, which is determined by (1.84), is Δz. As in the prismatic pair, any homogeneous point can be chosen: conventionally, we select the origin O of the reference frame (0), and so the screw axis can be deduced and identified with the x_3-axis. With respect to the reference system (1), the mobile screw axis and the mobile axode can be describe by (1.86), realizing that the resulting axis is the same even if it is translated on itself.

After schematic description of a cylindrical pair, a spherical pair is examined between the fixed frame T and a moving body B, which is constrained to have a point O^1, the spherical hinge, always coincident with the corresponding point O on the frame T. To describe the position of B with respect to T, a reference system (0) is rigidly connected to the fixed frame with the origin O in the spherical hinge, and a reference system (1) is rigidly connected to the body B with the origin O^1 in the spherical hinge. Therefore, the position of B can be represented with respect to T by (1.98), the terms c_i and s_i representing, respectively, the cosine and sine of the Cardan angle α_i (Figure 1.18). Expression (1.98) shows that the three parameters α_1, α_2, and

FIGURE 1.18
Spherical pair.

α_3 are enough to describe the motion of B with respect to T; therefore, the spherical pair allows three degrees of freedom, and it abolishes three of them through three degrees of constraint. The constant L matrixes that are associated to every degree of freedom are represented by (1.99) and allow the calculation of speed and acceleration matrixes (1.100).

$$
M_{01} = \begin{bmatrix} 1 & 0 & 0 & 0 \\ 0 & c_1 & -s_1 & 0 \\ 0 & s_1 & c_1 & 0 \\ 0 & 0 & 0 & 1 \end{bmatrix} \begin{bmatrix} c_2 & 0 & s_2 & 0 \\ 0 & 1 & 0 & 0 \\ -s_2 & 0 & c_2 & 0 \\ 0 & 0 & 0 & 1 \end{bmatrix} \begin{bmatrix} c_3 & -s_3 & 0 & 0 \\ s_3 & c_3 & 0 & 0 \\ 0 & 0 & 1 & 0 \\ 0 & 0 & 0 & 1 \end{bmatrix}
$$

$$
= \begin{bmatrix} c_2 c_3 & -c_2 s_3 & s_2 & 0 \\ s_1 s_2 c_3 + c_1 s_3 & -s_1 s_2 s_3 + c_1 s_3 & -s_1 c_2 & 0 \\ -c_1 s_2 c_3 + s_1 s_3 & c_1 s_2 s_3 + s_1 c_3 & c_1 c_2 & 0 \\ 0 & 0 & 0 & 1 \end{bmatrix}
$$

(1.98)

$$
L_1 = \begin{bmatrix} 0 & 0 & 0 & 0 \\ 0 & 0 & -1 & 0 \\ 0 & 1 & 0 & 0 \\ 0 & 0 & 0 & 0 \end{bmatrix} ; \quad L_2 = \begin{bmatrix} 0 & 0 & 1 & 0 \\ 0 & 0 & 0 & 0 \\ -1 & 0 & 0 & 0 \\ 0 & 0 & 0 & 0 \end{bmatrix} ; \quad L_3 = \begin{bmatrix} 0 & -1 & 0 & 0 \\ 1 & 0 & 0 & 0 \\ 0 & 0 & 0 & 0 \\ 0 & 0 & 0 & 0 \end{bmatrix}
$$

(1.99)

$$
W_{01} = \sum_{i=1}^{3} L_i \cdot \dot{\alpha}_i = \begin{bmatrix} 0 & -\dot{\alpha}_3 & \dot{\alpha}_2 & 0 \\ \dot{\alpha}_3 & 0 & -\dot{\alpha}_1 & 0 \\ -\dot{\alpha}_2 & \dot{\alpha}_1 & 0 & 0 \\ 0 & 0 & 0 & 0 \end{bmatrix}
$$

(1.100)

$$
H_{01} = \sum_{i=1}^{3} L_i \ddot{\alpha}_i + \sum_{j=1}^{3} \sum_{k=1}^{3} \left(L_k L_j \dot{\alpha}_k \dot{\alpha}_j \right)
$$

$$
= \begin{bmatrix} 0 & -\ddot{\alpha}_3 & \ddot{\alpha}_2 & 0 \\ \ddot{\alpha}_3 & 0 & -\ddot{\alpha}_1 & 0 \\ -\ddot{\alpha}_2 & \ddot{\alpha}_1 & 0 & 0 \\ 0 & 0 & 0 & 0 \end{bmatrix} + \begin{bmatrix} -\dot{\alpha}_2 - \dot{\alpha}_3 & 0 & 0 & 0 \\ 0 & -\dot{\alpha}_1^2 - \dot{\alpha}_3^2 & 0 & 0 \\ 0 & 0 & -\dot{\alpha}_1^2 - \dot{\alpha}_2^2 & 0 \\ 0 & 0 & 0 & 0 \end{bmatrix}
$$

$$
+ \begin{bmatrix} 0 & \dot{\alpha}_1 \dot{\alpha}_2 & \dot{\alpha}_1 \dot{\alpha}_3 & 0 \\ \dot{\alpha}_1 \dot{\alpha}_2 & 0 & 0 & 0 \\ \dot{\alpha}_1 \dot{\alpha}_3 & \dot{\alpha}_2 \dot{\alpha}_3 & 0 & 0 \\ 0 & 0 & 0 & 0 \end{bmatrix} = \begin{bmatrix} -\dot{\alpha}_2^2 - \dot{\alpha}_3^2 & \dot{\alpha}_2 \dot{\alpha}_1 - \ddot{\alpha}_3 & \dot{\alpha}_3 \dot{\alpha}_1 + \ddot{\alpha}_2 & 0 \\ \dot{\alpha}_1 \dot{\alpha}_2 + \ddot{\alpha}_3 & -\dot{\alpha}_1^2 - \dot{\alpha}_3^2 & \dot{\alpha}_3 \dot{\alpha}_2 - \ddot{\alpha}_1 & 0 \\ \dot{\alpha}_1 \dot{\alpha}_3 - \ddot{\alpha}_2 & \dot{\alpha}_2 \dot{\alpha}_3 + \ddot{\alpha}_1 & -\dot{\alpha}_1^2 - \dot{\alpha}_2^2 & 0 \\ 0 & 0 & 0 & 0 \end{bmatrix}
$$

As for the previous kinematic pairs, two different spatial configurations are examined, and these are associated to two different temporal instants (1.101). The purpose of this approach is to calculate the rototranslation matrix that is able to turn a configuration into another.

$$
M_{01}\Big|_{t=t_0} = \begin{bmatrix} c_{20}c_{30} & -c_{20}s_{30} & s_{20} & 0 \\ s_{10}s_{20}c_{30}+c_{10}s_{30} & -s_{10}s_{20}s_{30}+c_{10}s_{30} & -s_{10}c_{20} & 0 \\ -c_{10}s_{20}c_{30}+s_{10}s_{30} & c_{10}s_{20}s_{30}+s_{10}c_{30} & c_{10}c_{20} & 0 \\ 0 & 0 & 0 & 1 \end{bmatrix}
$$

$$
M_{01}\Big|_{t=t_0+\Delta t} = \begin{vmatrix} c_{20+2\Delta}c_{30+3\Delta} & -c_{20+2\Delta}s_{30+3\Delta} \\ s_{10+1\Delta}s_{20+2\Delta}c_{30+3\Delta}+c_{10+1\Delta}s_{30+3\Delta} & -s_{10+1\Delta}s_{20+2\Delta}s_{30+3\Delta}+c_{10+1\Delta}s_{30+3\Delta} \\ -c_{10+1\Delta}s_{20+2\Delta}c_{30+3\Delta}+s_{10+1\Delta}s_{30+3\Delta} & c_{10+1\Delta}s_{20+2\Delta}s_{30+3\Delta}+s_{10+1\Delta}c_{30+3\Delta} \\ 0 & 0 \end{vmatrix}
$$

$$
\begin{vmatrix} s_{20+2\Delta} & 0 \\ -s_{10+1\Delta}c_{20+2\Delta} & 0 \\ c_{10+1\Delta}c_{20+2\Delta} & 0 \\ 0 & 1 \end{vmatrix}
$$

(1.101)

where c_{10} and s_{10} are, respectively, the cosine and the sine of the angle α_i, and $c_{i+i\Delta}$ and $s_{i+i\Delta}$ are, respectively, the cosine and the sine of the angle α_i increased by a value equal to $\Delta\alpha_i$. The computation of the rototranslation matrix $Q_{(0)}$ (1.81) is simple, but this procedure is relatively long; therefore, to improve the clarity of the paragraph, the complete deduction is presented in appendix C.1. In the case of infinitesimal rototranslation, the matrix $Q_{(0)}$ assumes a simplified form, which is still indicated in appendix C.1 and which depends on the angular variation and on the initial position.

The rotation ϑ around the helicoidal axis is calculated by substituting proper values in (1.82), and for small angular variations, using values indicated in (C.11–C.19), we could obtain (1.102). Furthermore, the direction of the screw axis is identified by the vector u, and its inhomogeneous components are calculated in (1.103) for small angular variations.

$$
c = 1 - \Delta_1\Delta_3 s_{20} - \Delta_1\Delta_2\Delta_3 c_{20}/2
$$

$$
s = \pm\left(\Delta_1^2\Delta_2^2 - 2\Delta_1\Delta_2^2\Delta_3 s_{20} + \Delta_1^2\Delta_3^2 c_{20}^2 + 8\Delta_1\Delta_3 s_{20} + 4\Delta_2^2 + 4\Delta_3^2 + 4\Delta_1^2 +\right.
$$
$$
\left. + \Delta_2^2\Delta_3^2 + 4\Delta_1\Delta_2\Delta_3 c_{20}\right)^{1/2}
$$

(1.102)

$$
\vartheta = \arctan(s,c)
$$

$$u_1 = \left(2\Delta_3 s_{20} + 2\Delta_1 + \Delta_2 \Delta_3 c_{20}\right) \Big/ \left(\Delta_1^2 \Delta_2^2 - 2\Delta_1 \Delta_2^2 \Delta_3 s_{20} + \Delta_1^2 \Delta_3^2 c_{20}^2 + \right.$$

$$\left. + 8\Delta_1 \Delta_3 s_{20} + 4\Delta_1^2 + 4\Delta_2^2 + 4\Delta_3^2 + \Delta_2^2 \Delta_3^2 + 4\Delta_1 \Delta_2 \Delta_3 c_{20}\right)^{1/2}$$

$$u_2 = -\left(\Delta_1 \Delta_3 c_{20} c_{10} - 2\Delta_2 c_{10} + \Delta_1 \Delta_2 s_{10} + 2\Delta_3 s_{10} c_{20} - \Delta_2 \Delta_3 s_{10} s_{20}\right) \Big/$$

$$\left(\Delta_1^2 \Delta_2^2 - 2\Delta_1 \Delta_2^2 \Delta_3 s_{20} + \Delta_1^2 \Delta_3^2 c_{20}^2 + 8\Delta_1 \Delta_3 s_{20} + 4\Delta_1^2 + 4\Delta_2^2 + 4\Delta_3^2 + \right. \tag{1.103}$$

$$\left. + \Delta_2^2 \Delta_3^2 + 4\Delta_1 \Delta_2 \Delta_3 c_{20}\right)^{1/2}$$

$$u_3 = -\left(-2\Delta_2 s_{10} - 2\Delta_3 c_{10} c_{20} - \Delta_1 \Delta_2 c_{10} + \Delta_1 \Delta_3 c_{20} s_{10} + \Delta_2 \Delta_3 c_{10} s_{20}\right) \Big/$$

$$\left(\Delta_1^2 \Delta_2^2 - 2\Delta_1 \Delta_2^2 \Delta_3 s_{20} + \Delta_1^2 \Delta_3^2 c_{20}^2 + 8\Delta_1 \Delta_3 s_{20} + 4\Delta_1^2 + 4\Delta_2^2 + 4\Delta_3^2 + \right.$$

$$\left. + \Delta_2^2 \Delta_3^2 + 4\Delta_1 \Delta_2 \Delta_3 c_{20}\right)^{1/2}$$

Moreover, with (1.84), the absence of translating motions can be verified and, with (1.85), the spherical hinge O can be identified as a point of the screw axis. Finally, it is interesting to observe that, with regard to the previous kinematic pairs (the revolute one, the prismatic one, and the cylindrical one), the spherical pair has a moving screw axis. Thus, for different starting positions, even with the same angular motion, we find different screw axes.

After schematic description of a spherical pair, a planar pair (Figure 1.19) is examined between the fixed frame T and a moving body B; a reference system (0) is rigidly connected to the fixed frame T, and a (1) reference system is rigidly connected to the body B. The planar constraint allows only planar relative movement of B with respect to T. Therefore, the position of B can be represented by (1.104), where the same notation used in (1.98) identifies the sine and cosine of α_3. In (1.104), the three parameters x, y, and α_3 are enough to describe the motion of B with respect to T, and therefore the planar pair allows three degrees of freedom and the same number of degrees of

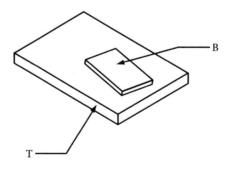

FIGURE 1.19
Planar pair.

constraint. The constant L matrixes, which are associated with every degree of freedom, are represented in (1.105) and can be used to perform the calculation of the speed and acceleration matrixes (1.106 and 1.107). As for the previous pairs, we consider two different spatial configurations that are associated to two different temporal instants (1.108) and we calculate the rototranslation matrix that is able to turn one configuration into the other.

$$
M_{01} = \begin{bmatrix} c_3 & -s_3 & 0 & x \\ s_3 & c_3 & 0 & y \\ 0 & 0 & 1 & 0 \\ 0 & 0 & 0 & 1 \end{bmatrix}
\tag{1.104}
$$

$$
L_1 = \begin{bmatrix} 0 & 0 & 0 & 1 \\ 0 & 0 & 0 & 0 \\ 0 & 0 & 0 & 0 \\ 0 & 0 & 0 & 0 \end{bmatrix}; \;
L_2 = \begin{bmatrix} 0 & 0 & 0 & 0 \\ 0 & 0 & 0 & 1 \\ 0 & 0 & 0 & 0 \\ 0 & 0 & 0 & 0 \end{bmatrix}; \;
L_3 = \begin{bmatrix} 0 & -1 & 0 & 0 \\ 1 & 0 & 0 & 0 \\ 0 & 0 & 0 & 0 \\ 0 & 0 & 0 & 0 \end{bmatrix}
\tag{1.105}
$$

$$
W_{01} = L_1 \dot{x} + L_2 \dot{y} + L_3 \dot{\alpha}_3 = \begin{bmatrix} 0 & -\dot{\alpha}_3 & 0 & \dot{x} \\ \dot{\alpha}_3 & 0 & 0 & \dot{y} \\ 0 & 0 & 0 & 0 \\ 0 & 0 & 0 & 0 \end{bmatrix}
\tag{1.106}
$$

$$
H_{01} = L_1 \ddot{x} + L_2 \ddot{y} + L_3 \ddot{\alpha}_3 + L_1^2 \dot{x}^2 + L_2^2 \dot{y}^2 + L_3^2 \dot{\alpha}_3^2
$$

$$
= \begin{bmatrix} -\dot{\alpha}_3^2 & -\ddot{\alpha}_3 & 0 & \ddot{x} \\ \ddot{\alpha}_3 & -\dot{\alpha}_3^2 & 0 & \ddot{y} \\ 0 & 0 & 0 & 0 \\ 0 & 0 & 0 & 0 \end{bmatrix}
\tag{1.107}
$$

$$
M_{01}\big|_{t=t_0} = \begin{bmatrix} c_{30} & -s_{30} & 0 & x_0 \\ s_{30} & c_{30} & 0 & y_0 \\ 0 & 0 & 1 & 0 \\ 0 & 0 & 0 & 1 \end{bmatrix};
$$

$$
M_{01}\big|_{t=t_0+\Delta t} = \begin{bmatrix} c_{30+3\Delta} & -s_{30+3\Delta} & 0 & x_{0+\Delta} \\ s_{30+3\Delta} & c_{30+3\Delta} & 0 & y_{0+\Delta} \\ 0 & 0 & 1 & 0 \\ 0 & 0 & 0 & 1 \end{bmatrix}
\tag{1.108}
$$

where c_{30} and s_{30} are, respectively, the cosine and sine of α_3; $c_{30+3\Delta}$ and $s_{30+3\Delta}$ are, respectively, the cosine and sine of α_3 increased to $\Delta\alpha_3$; x_0 and y_0 are, respectively, x and y at the initial temporal instant t_0; and $x_{0+\Delta}$ and $y_{0+\Delta}$ are, respectively, x and y increased to values Δx and Δy. As proposed in (1.81), the rototranslation matrix $Q_{(0)}$ (1.109) can be deduced from (1.108).

$$Q_{(0)} = \left(M_{01}\big|_{t=t_0+\Delta t}\right) \cdot \left(M_{01}\big|_{t=t_0}\right)^{-1} = \begin{bmatrix} c_{3\Delta} & -s_{3\Delta} & 0 & s_{3\Delta}y - c_{3\Delta}x + x + \Delta x \\ s_{3\Delta} & c_{3\Delta} & 0 & -c_{3\Delta}y - s_{3\Delta}x + y + \Delta y \\ 0 & 0 & 1 & 0 \\ 0 & 0 & 0 & 1 \end{bmatrix} \quad (1.109)$$

where $c_{3\Delta}$ and $s_{3\Delta}$ are, respectively, the cosine and sine of $\Delta\alpha_3$. The rotation around the helicoidal axis is calculated with (1.82) to obtain $\Delta\alpha_3$ by the substitution of proper values. With (1.83), the inhomogeneous components of the vector u can be achieved to identify the screw axis (1.110), which is always perpendicular to the motion plane, and the helicoidal displacement pitch is null (1.84) because every displacement is perpendicular to the screw axis.

$$u_1 = 0 \quad u_2 = 0 \quad u_3 = 1 \quad (1.110)$$

If we suppose the absence of rotary motions, we can easily identify two prismatic motions in the plane; therefore, considering the observations on the degeneration of (1.82 and 1.83) for prismatic pairs, the identification of h and the screw axis comes through a vectorial composition of the principal displacements, and the rotation angle is recognized to be null.

Finally, we consider the helicoidal pair that is represented by a cylindrical pair, where there is the (1.111) relation between the parameters α and z that is indicated in (1.92), so that the position matrix M_{01} of this pair takes the form indicated in (1.112).

$$z = p\frac{\alpha}{2\pi} \quad (1.111)$$

where p is a constant parameter that takes the name of the pitch of the helix and represents the pitch of the helicoidal displacement for a 2π rotation.

$$M_{01} = \begin{bmatrix} \cos(\alpha) & -\sin(\alpha) & 0 & 0 \\ \sin(\alpha) & \cos(\alpha) & 0 & 0 \\ 0 & 0 & 1 & p\frac{\alpha}{2\pi} \\ 0 & 0 & 0 & 1 \end{bmatrix} \quad (1.112)$$

The matrix M_{01}, which is identified by (1.112), exhibits only one degree of freedom associated to the parameter α and five degrees of constraint. Therefore, with (1.27), the matrix L of the pair can be computed (1.113) to allow a fast calculation of the speed and acceleration matrixes (1.114).

$$L_{01} = \begin{bmatrix} 0 & -1 & 0 & 0 \\ 1 & 0 & 0 & 0 \\ 0 & 0 & 0 & \dfrac{p}{2\pi} \\ 0 & 0 & 0 & 0 \end{bmatrix} \tag{1.113}$$

$$W_{01} = \begin{bmatrix} 0 & -\dot\alpha & 0 & 0 \\ \dot\alpha & 0 & 0 & 0 \\ 0 & 0 & 0 & p\dfrac{\dot\alpha}{2\pi} \\ 0 & 0 & 0 & 0 \end{bmatrix} ; \quad H_{01} = \begin{bmatrix} -\dot\alpha^2 & -\ddot\alpha & 0 & 0 \\ \ddot\alpha & -\dot\alpha^2 & 0 & 0 \\ 0 & 0 & 0 & p\dfrac{\ddot\alpha}{2\pi} \\ 0 & 0 & 0 & 0 \end{bmatrix} \tag{1.114}$$

Also in this case, even though we know how to obtain results similar to the cylindrical pair, two different positions are considered (1.115) to deduce the rototranslation matrix $Q_{(0)}$ (1.116).

$$M_{01}\big|_{t=t_0} = \begin{bmatrix} \cos(\alpha) & -\sin(\alpha) & 0 & 0 \\ \sin(\alpha) & \cos(\alpha) & 0 & 0 \\ 0 & 0 & 1 & p\dfrac{\alpha}{2\pi} \\ 0 & 0 & 0 & 1 \end{bmatrix} ;$$

$$M_{01}\big|_{t=t_0+\Delta t} = \begin{bmatrix} \cos(\alpha+\Delta\alpha) & -\sin(\alpha+\Delta\alpha) & 0 & 0 \\ \sin(\alpha+\Delta\alpha) & \cos(\alpha+\Delta\alpha) & 0 & 0 \\ 0 & 0 & 1 & p\dfrac{\alpha+\Delta\alpha}{2\pi} \\ 0 & 0 & 0 & 1 \end{bmatrix} \tag{1.115}$$

$$Q_{(0)} = \begin{bmatrix} \cos(\Delta\alpha) & -\sin(\Delta\alpha) & 0 & 0 \\ \sin(\Delta\alpha) & \cos(\Delta\alpha) & 0 & 0 \\ 0 & 0 & 1 & p\dfrac{\Delta\alpha}{2\pi} \\ 0 & 0 & 0 & 1 \end{bmatrix} \tag{1.116}$$

With (1.82), the rotation angle ϑ is obtained, and its value is $\Delta\alpha$; whereas, with (1.83), the inhomogeneous components of the vector u are gained, resulting in the same axis of the cylindrical pair. Finally, with (1.84), we deduce the helicoidal displacement pitch, as shown in (1.117).

$$h = p\,\frac{\Delta\alpha}{2\pi} \tag{1.117}$$

Besides the kinematic pairs already considered, there are some other pairs that could be examined, and we could expand on interesting problems about the composition and decomposition of the kinematic pair or about the local or global kinematic equivalence of pairs that are structurally different, but we prefer to refer to specific literature, i.e., (Shigley, Uicker, 1980). For this reason, after having examined the kinematics of six pairs (revolute, prismatic, cylindrical, spherical, planar, and helicoidal) we propose a summarizing table (Table 1.1), which allows a comparison of several kinematic properties for each considered pair. In this table, the matrix L of every kinematic pair is associated to the matrixes of the elementary pairs, which are indicated in (1.25). Furthermore, we observe that, in the case of the spherical joint, considering the length of the expressions of u and ϑ, we referred to the mathematical expressions already indicated in (1.103 and 1.104).

Table 1.1 represents a schematic comparison of different kinematic pairs and also a reference in order to verify if compliant kinematic pairs, which

TABLE 1.1

Comparison of Kinematic Properties for the Considered Pairs

Joint	L	Degree of Freedom	Degree of Constraint	u	h	ϑ
Revolute	L^{ri}	1	5	$u_i = 1$ $u_j = 0$ $i \neq j$	0	$\Delta\theta_i$
Prismatic	L^{ti}	1	5	$u_i = 1$ $u_j = 0$ $i \neq j$	Δx_i	0
Cylindrical	L^{ri}, L^{ti}	2	4	$u_i = 1$ $u_j = 0$ $i \neq j$	Δx_i	$\Delta\theta_i$
Spherical	L^{r1}, L^{r2}, L^{r3}	3	3	(1.103)	0	(1.102)
Planar	L^{ti}, L^{tj}, L^{rk}	3	3	$u_k = 1$, composed	$\Delta x_i \,\Delta x_j$	$\Delta\theta_k$
Helicoidal	$L^{ri} + (p\alpha/2\pi)$ L^{ti}	1	5	$u_i = 1$ $u_j = 0$ $i \neq j$	$p\,\Delta\theta_i/2\pi$	$\Delta\theta_i$

will be discussed in section 1.2.6, have kinematic similarities with the theoretical pairs that had been considered in this section.

1.2.6 Kinematics with Compliant Constraints

A theoretical kinematic pair, among those listed in Table 1.1, is selected and its number of degrees of freedom is increased, including a finite number of lumped compliances, to obtain a compliant kinematic pair that is the subject of this paragraph. These supplementary degrees of freedom can increase the mobility of the theoretical kinematic pair, removing one or more degree of constraint (i.e., in Figure 1.20 a revolute pair is furnished with a compliance along its screw axis, and therefore it is transformed into a cylindrical pair). A different situation is when the compliance acts on the motion trajectory of the mobile body B with respect to the fixed body T of the theoretical kinematic pair, introducing a kinematic redundancy (i.e., in Figure 1.21, a prismatic pair is furnished with a compliance along its screw axis, introducing a kinematic redundancy). To distinguish the two classes of compliances, we can introduce a functional attribute, speaking of structural compliance for

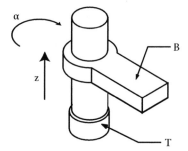

FIGURE 1.20
A revolute pair between the bodies B and T, with the degree of freedom α, is furnished with a compliance that introduces the degree of freedom z, producing a cylindrical pair.

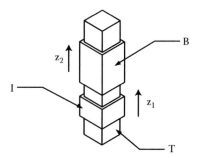

FIGURE 1.21
A prismatic pair between the bodies B and T, with the degree of freedom z_2, is furnished with an intermediate body I and a compliance z_1, producing a redundancy along the screw axis.

the first type, which removes a structural constraint of the theoretical kinematic pair, and speaking of kinematic compliance for the second type, which introduces a kinematic redundancy on the trajectory of the kinematic pair.

As in Figure 1.21, the introduction of a kinematic compliance is associated with the introduction of an intermediate fictitious body and, as will be shown in the dynamics of the compliant pair, different properties of the pair will be associated to it. The degrees of freedom produced by compliances, with respect to ordinary degrees of freedom, are characterized by an associated binding relation that expresses the level of compliance or the level of friction; but this aspect is postponed to a further examination during the exposition of the dynamics of kinematic pairs. Furthermore, usually but not in every situation, the degree of freedom associated with the compliances allows displacements smaller than common degree of freedom. Therefore, in this paragraph, because we did not define the concept of mechanical stiffness, the ratio between the amplitude of the maximal displacement of a degree of freedom due to a compliance and the amplitude of the maximal displacement of an ordinary degree of freedom is taken as an indicator of the compliance of a kinematic pair. The degree of compliance is defined in a relative manner as a ratio between maximal displacements; therefore, from a kinematic viewpoint, it is a negative or positive contribution to the functionality of the kinematic pair. An opinion was not expressed on the advantage or disadvantage of a higher or lower degree of compliance because, in the same applicative situations, an increase in stiffness is appropriate to reduce vibrating phenomena, whereas in other applications, a compliance can result in an alternative way to transform motions with a simple connection. This second opportunity begins to appear with a scale reduction and, for really small dimensions, it is a very interesting option.

After this discussion, which can be widely deepened in Lobontiu (2003), Howell (2001), and Readman (1994), a fixed joint and the kinematic pairs listed in Table 1.1 are taken as a starting point to introduce one or more compliances and verify the produced effects on the kinematic properties just considered in Table 1.1. The first example is based on a fixed joint between B and T; it consists of a statically determinate (isostatic) junction between the two bodies, which does not allow any relative degree of freedom. This pair is characterized, according to properties listed in Table 1.1, by a null matrix L, no degree of freedom, 6 degrees of constraint, h and ϑ are equal to zero, and u loses its meaning because no relative movements are allowed between B and T. A structural compliance can be introduced for every degree of constraint to increase the mobility of B associated with the acquisition of the respective degree of freedom; in fact kinematic pairs that are similar to the ones listed in Table 1.1 can be obtained by concentrating proper compliances in the fixed joint. We do not show all the analysis just exposed in the previous paragraph for these new six kinematic pairs produced with structural compliances; using a suitable notation, the symbol δ is used in place of the symbol Δ to distinguish between common degrees of freedom and degrees of freedom due to compliances, obtaining analogous results. Then, a slightly complex situation can be examined: the presence of a series of two or more compliances to remove a degree of constraint, introducing also kinematic redundancy. Therefore, even if kinematic compliances are not present, we need to use fictitious intermediate bodies to represent the kinematics of the pair. This result represents a natural extension of properties stressed in Figures 1.20 and 1.21; as a matter of fact, the finite number of fictitious intermediate bodies introduced for a removed constraint is equal to the number of the introduced structural compliances minus one.

To give some examples, we produce the kinematic pairs listed in Table 1.1 starting from a fixed joint and using n structural compliances in series to remove the desired constraint, where n is a natural number and is greater than or equal to two. As first example, a revolute pair is generated from a fixed joint using n structural compliances in series to obtain a surrogate of the kinematic pair described in Figure 1.15; therefore the position matrix M_{0n} indicated in (1.118) is analogous to (1.82).

$$
\left\{
\begin{aligned}
M_{i-1\ i} &= \begin{bmatrix} \cos(\alpha_i) & -\sin(\alpha_i) & 0 & 0 \\ \sin(\alpha_i) & \cos(\alpha_i) & 0 & 0 \\ 0 & 0 & 1 & 0 \\ 0 & 0 & 0 & 1 \end{bmatrix}; \qquad \alpha \equiv \sum_{i=1}^{n} \alpha_i \\
M_{0n} &= \prod_{i=1}^{n} M_{i-1\ i} = \begin{bmatrix} \cos(\alpha) & -\sin(\alpha) & 0 & 0 \\ \sin(\alpha) & \cos(\alpha) & 0 & 0 \\ 0 & 0 & 1 & 0 \\ 0 & 0 & 0 & 1 \end{bmatrix}
\end{aligned}
\right.
$$

$$(1.118)$$

In (1.118), n parameters are present to describe n degrees of freedom (n-1 degrees are redundant), which represent the displacement of B with respect to T; even if only an α summary parameter can be used, we can anticipate that the use of a high number of degrees of freedom often allows us to simplify the elastic dissipative relations associated with every compliance in order to describe completely the dynamic behavior of the system. In fact, either the mathematical modelization or the constructive design are benefited by it; therefore, we use all the n system parameters.

The matrix L_i, which is associated to the generic i-th degree of freedom, is the same as the one indicated in (1.78), which represents also the matrix L of the summary parameter α; this allows a simple computation of the speed matrix (1.119) and the acceleration matrix (1.120).

$$W_{0n} = \sum_{i=1}^{n} L_i \cdot \dot{\alpha}_i = L \cdot \sum_{i=1}^{n} \dot{\alpha}_i = L \cdot \dot{\alpha} \qquad (1.119)$$

$$H_{0n} = \sum_{i=1}^{n} \left(L_i \ddot{\alpha}_i + L_i^2 \dot{\alpha}_i^2 \right) + 2 \sum_{j=2}^{n} \sum_{k=1}^{j-1} \left(L_k L_j \dot{\alpha}_k \dot{\alpha}_j \right)$$

$$= L \sum_{i=1}^{n} \left(\ddot{\alpha}_i \right) + L^2 \left(\sum_{i=1}^{n} \dot{\alpha}_i^2 + 2 \sum_{j=2}^{n} \sum_{k=1}^{j-1} \dot{\alpha}_k \dot{\alpha}_j \right) \qquad (1.120)$$

$$= L \sum_{i=1}^{n} \left(\ddot{\alpha}_i \right) + L^2 \left(\sum_{i=1}^{n} \dot{\alpha}_i \right)^2$$

$$= L \cdot \ddot{\alpha} + L^2 \cdot \dot{\alpha}^2$$

To deduce the properties of the helicoidal transformation, which represents a motion along the examined axis of motion, we compute the derivatives of the matrixes (1.121 and 1.122), which are analogous to the ones indicated in (1.80).

$$\left\{ \begin{array}{l} \left. M_{i-1 \ i} \right|_{t=t_0} = \begin{bmatrix} \cos(\alpha_{0i}) & -\sin(\alpha_{0i}) & 0 & 0 \\ \sin(\alpha_{0i}) & \cos(\alpha_{0i}) & 0 & 0 \\ 0 & 0 & 1 & 0 \\ 0 & 0 & 0 & 1 \end{bmatrix}; \quad \alpha_0 \equiv \sum_{i=1}^{n} \alpha_{0i} \\[6mm] \left. M_{0n} \right|_{t=t_0} = \prod_{i=1}^{n} \left. M_{i-1 \ i} \right|_{t=t_0} = \begin{bmatrix} \cos(\alpha_0) & -\sin(\alpha_0) & 0 & 0 \\ \sin(\alpha_0) & \cos(\alpha_0) & 0 & 0 \\ 0 & 0 & 1 & 0 \\ 0 & 0 & 0 & 1 \end{bmatrix} \end{array} \right. \qquad (1.121)$$

$$\left\{ \begin{array}{l} \left. M_{i-1 \ i} \right|_{t=t_0+\Delta t} = \begin{bmatrix} \cos(\alpha_{0i}+\delta\alpha_i) & -\sin(\alpha_{0i}+\delta\alpha_i) & 0 & 0 \\ \sin(\alpha_{0i}+\delta\alpha_i) & \cos(\alpha_{0i}+\delta\alpha_i) & 0 & 0 \\ 0 & 0 & 1 & 0 \\ 0 & 0 & 0 & 1 \end{bmatrix}; \quad \delta\alpha \equiv \sum_{i=1}^{n} \delta\alpha_i \\[6mm] \left. M_{0n} \right|_{t=t_0+\Delta t} = \prod_{i=1}^{n} \left. M_{i-1 \ i} \right|_{t=t_0+\Delta t} = \begin{bmatrix} \cos(\alpha_0+\delta\alpha) & -\sin(\alpha_0+\delta\alpha) & 0 & 0 \\ \sin(\alpha_0+\delta\alpha) & \cos(\alpha_0+\delta\alpha) & 0 & 0 \\ 0 & 0 & 1 & 0 \\ 0 & 0 & 0 & 1 \end{bmatrix} \end{array} \right. \qquad (1.122)$$

Starting from (1.121 and 1.122), the matrix $Q_{(0)}$ in (1.123) is constructed to represent the helicoidal transformation likewise in (1.81).

$$
Q_{(0)} = M_{0n}\big|_{t=t_0+\Delta t} \cdot \left(M_{0n}\big|_{t=t_0} \right)^{-1}
$$

$$
= \prod_{i=1}^{n} M_{i-1 \ i}\big|_{t=t_0+\Delta t} \cdot \left(\prod_{i=1}^{n} M_{i-1 \ i}\big|_{t=t_0} \right)^{-1}
$$

$$
= \prod_{i=1}^{n} \left[M_{i-1 \ i}\big|_{t=t_0+\Delta t} \cdot \left(M_{n-i \ n-i+1}\big|_{t=t_0} \right)^{-1} \right] \tag{1.123}
$$

$$
= \begin{bmatrix} \cos(\delta\alpha) & -\sin(\delta\alpha) & 0 & 0 \\ \sin(\delta\alpha) & \cos(\delta\alpha) & 0 & 0 \\ 0 & 0 & 1 & 0 \\ 0 & 0 & 0 & 1 \end{bmatrix}
$$

According to (1.82–1.85), the properties of the helicoidal movement are extracted from (1.123) to gain ϑ, h, and u in (1.124), obtaining a considerable analogy with the results for kinematic pair described in Figure 1.15.

$$
\vartheta = \sum_{i=1}^{n} \delta\alpha_i; \quad h=0; \quad u = \begin{bmatrix} 0 & 0 & 1 & 0 \end{bmatrix}^T \tag{1.124}
$$

As a different example, we consider a surrogate of the prismatic pair indicated in Figure 1.16, which is generated starting from a fixed joint with n structural compliances in series to remove the proper degree of constraint. Therefore, the position matrix is represented by (1.125), which is analogous to (1.87) and the matrix L_i, associated to the generic i-th degree of freedom, takes the form of (1.88), which also represents the matrix L that is associated to the summary parameter z. Thus, the speed and acceleration matrixes can be represented by (1.126). Then, starting from the two position matrixes evaluated at the instants t_0 and the $t_0+\Delta t$ (1.127 and 1.128), the helicoidal rototranslation matrix (1.129) can be achieved.

$$
\begin{cases} M_{i-1 \ i} = \begin{bmatrix} 1 & 0 & 0 & 0 \\ 0 & 1 & 0 & 0 \\ 0 & 0 & 1 & z_i \\ 0 & 0 & 0 & 1 \end{bmatrix}; \quad z \equiv \sum_{i=1}^{n} z_i \\[4em] M_{0n} = \prod_{i=1}^{n} M_{i-1 \ i} = \begin{bmatrix} 1 & 0 & 0 & 0 \\ 0 & 1 & 0 & 0 \\ 0 & 0 & 1 & z \\ 0 & 0 & 0 & 1 \end{bmatrix} \end{cases} \tag{1.125}
$$

$$W_{0n} = \sum_{i=1}^{n} L_i \cdot \dot{z}_i = L \cdot \sum_{i=1}^{n} \dot{z}_i = L \cdot \dot{z}$$

$$H_{0n} = \sum_{i=1}^{n} L_i \cdot \ddot{z}_i = L \cdot \sum_{i=1}^{n} \ddot{z}_i = L \cdot \ddot{z}$$

(1.126)

$$\left\{ \begin{aligned} \boldsymbol{M}_{i-1 \ i}\Big|_{t=t_0} &= \begin{bmatrix} 1 & 0 & 0 & 0 \\ 0 & 1 & 0 & 0 \\ 0 & 0 & 1 & z_{0i} \\ 0 & 0 & 0 & 1 \end{bmatrix}; \quad z_0 \equiv \sum_{i=1}^{n} z_{0i} \\ \boldsymbol{M}_{0n}\Big|_{t=t_0} &= \prod_{i=1}^{n} \boldsymbol{M}_{i-1 \ i}\Big|_{t=t_0} = \begin{bmatrix} 1 & 0 & 0 & 0 \\ 0 & 1 & 0 & 0 \\ 0 & 0 & 1 & z_0 \\ 0 & 0 & 0 & 1 \end{bmatrix} \end{aligned} \right.$$

(1.127)

$$\left\{ \begin{aligned} \boldsymbol{M}_{i-1 \ i}\Big|_{t=t_0+\Delta t} &= \begin{bmatrix} 1 & 0 & 0 & 0 \\ 0 & 1 & 0 & 0 \\ 0 & 0 & 1 & z_{0i}+\delta z_i \\ 0 & 0 & 0 & 1 \end{bmatrix}; \quad \delta z \equiv \sum_{i=1}^{n} \delta z_i \\ \boldsymbol{M}_{0n}\Big|_{t=t_0+\Delta t} &= \prod_{i=1}^{n} \boldsymbol{M}_{i-1 \ i}\Big|_{t=t_0+\Delta t} = \begin{bmatrix} 1 & 0 & 0 & 0 \\ 0 & 1 & 0 & 0 \\ 0 & 0 & 1 & z_0+\delta z \\ 0 & 0 & 0 & 1 \end{bmatrix} \end{aligned} \right.$$

(1.128)

$$\boldsymbol{Q}_{(0)} = \begin{bmatrix} 1 & 0 & 0 & 0 \\ 0 & 1 & 0 & 0 \\ 0 & 0 & 1 & \sum_{i=1}^{n} \delta z_i \\ 0 & 0 & 0 & 1 \end{bmatrix} = \begin{bmatrix} 1 & 0 & 0 & 0 \\ 0 & 1 & 0 & 0 \\ 0 & 0 & 1 & \delta z \\ 0 & 0 & 0 & 1 \end{bmatrix}$$

(1.129)

The ratio between the summary parameter z and the parameters z_i, which is associated with their respective compliances, is analogous to the ratio between α and α_i in revolute pairs. Therefore, the algorithm used to obtain (1.129) is similar to the one to obtain (1.123), and then the parameters of the helicoidal motion can be extracted from (1.129) and listed in (1.130).

$$\vartheta = 0; \quad h = \sum_{i=1}^{n} \delta z_i; \quad u = \begin{bmatrix} 0 & 0 & 1 & 0 \end{bmatrix}^T \tag{1.130}$$

Another interesting kinematic pair is the surrogate of a cylindrical pair (Figure 1.17), which is generated starting from a fixed joint with n structural compliances to remove a revolute constraint, and with m structural compliances to remove the corresponding prismatic constraint. Hence, the position matrix M of the pair is obtained in (1.131) by multiplying in the desired order the matrixes $M_{i-1,i}$, which are indicated in (1.118) and (1.125).

$$M = \begin{bmatrix} \cos\left(\sum_{i=1}^{n} \alpha_i\right) & -\sin\left(\sum_{i=1}^{n} \alpha_i\right) & 0 & 0 \\ \sin\left(\sum_{i=1}^{n} \alpha_i\right) & \cos\left(\sum_{i=1}^{n} \alpha_i\right) & 0 & 0 \\ 0 & 0 & 1 & \sum_{i=1}^{m} z_i \\ 0 & 0 & 0 & 1 \end{bmatrix} = \begin{bmatrix} \cos(\alpha) & -\sin(\alpha) & 0 & 0 \\ \sin(\alpha) & \cos(\alpha) & 0 & 0 \\ 0 & 0 & 1 & z \\ 0 & 0 & 0 & 1 \end{bmatrix} \tag{1.131}$$

The matrixes $L_{\alpha i}$ and L_{zj}, which are associated with the parameters α_i and z_j, are similar, respectively, to the matrixes L_α and L_z of the summary parameters α and z indicated in (1.93). Then the speed matrixes (1.132) and the acceleration matrixes (1.133) can be deduced with results analogous to (1.94), on account of the annulment of product between $L_{\alpha i}$ and L_{zj}; proceeding as for the previous kinematic pairs, then the helicoidal rototranslation matrix can be derived (1.134) and, from it, the properties of the helicoidal motion can be extracted, as indicated in (1.135).

$$W = \sum_{i=1}^{n} L_{\alpha i} \cdot \dot{\alpha}_i + \sum_{i=1}^{m} L_{zi} \cdot \dot{z}_i = L_\alpha \cdot \dot{\alpha} + L_z \cdot \dot{z} \tag{1.132}$$

$$H = \sum_{i=1}^{n} \left(L_{\alpha i} \ddot{\alpha}_i + L_{\alpha i}^2 \dot{\alpha}_i^2 \right) + \sum_{j=1}^{m} \left(L_{zj} \ddot{z}_j + L_{zj}^2 \dot{z}_j^2 \right) + \sum_{i=1}^{n} \sum_{j=1}^{m} \left(L_{\alpha i} L_{zj} \dot{\alpha}_i \dot{z}_j \right)$$

$$+ 2 \cdot \sum_{i=2}^{n} \sum_{k=1}^{i-1} \left(L_{\alpha k} L_{\alpha i} \dot{\alpha}_k \dot{\alpha}_i \right) + 2 \cdot \sum_{j=2}^{m} \sum_{k=1}^{j-1} \left(L_{zk} L_{zj} \dot{z}_k \dot{z}_j \right) \tag{1.133}$$

$$= L_\alpha^2 \cdot \dot{\alpha}^2 + L_\alpha \cdot \ddot{\alpha} + L_z \cdot \ddot{z}$$

$$
Q = \begin{bmatrix} \cos\left(\sum_{i=1}^{n} \delta\alpha_i\right) & -\sin\left(\sum_{i=1}^{n} \delta\alpha_i\right) & 0 & 0 \\ \sin\left(\sum_{i=1}^{n} \delta\alpha_i\right) & \cos\left(\sum_{i=1}^{n} \delta\alpha_i\right) & 0 & 0 \\ 0 & 0 & 1 & \sum_{j=1}^{m} \delta z_j \\ 0 & 0 & 0 & 1 \end{bmatrix} \tag{1.134}
$$

$$
= \begin{bmatrix} \cos(\delta\alpha) & -\sin(\delta\alpha) & 0 & 0 \\ \sin(\delta\alpha) & \cos(\delta\alpha) & 0 & 0 \\ 0 & 0 & 1 & \delta z \\ 0 & 0 & 0 & 1 \end{bmatrix}
$$

$$
\vartheta = \sum_{i=1}^{n} \delta\alpha_i; \quad h = \sum_{j=1}^{m} \delta z_j; \quad u = \begin{bmatrix} 0 & 0 & 1 & 0 \end{bmatrix}^{T} \tag{1.135}
$$

The next kinematic pair is the surrogate of the spherical one (Figure 1.18), obtained from a fixed joint removing three revolute constraints with, respectively, n_1, n_2, and n_3 structural compliances. Therefore the position matrix M of the kinematic pair (1.137) is achieved by multiplying the matrixes $M_{j\ i\text{-}1\ i}$ that are indicated in (1.136).

$$
M_{1\ i\text{-}1\ i} = \begin{bmatrix} 1 & 0 & 0 & 0 \\ 0 & c_{1i} & -s_{1i} & 0 \\ 0 & s_{1i} & c_{1i} & 0 \\ 0 & 0 & 0 & 1 \end{bmatrix}; \quad M_{2\ i\text{-}1\ i} = \begin{bmatrix} c_{2i} & 0 & s_{2i} & 0 \\ 0 & 1 & 0 & 0 \\ -s_{2i} & 0 & c_{2i} & 0 \\ 0 & 0 & 0 & 1 \end{bmatrix};
$$

$$
M_{3\ i\text{-}1\ i} = \begin{bmatrix} c_{3i} & -s_{3i} & 0 & 0 \\ s_{3i} & c_{3i} & 0 & 0 \\ 0 & 0 & 1 & 0 \\ 0 & 0 & 0 & 1 \end{bmatrix} \tag{1.136}
$$

where c_{ji} and s_{ji} are, respectively, the cosine and the sine of α_{ji}.

$$M = \prod_{j=1}^{3} \prod_{i=1}^{n_j} M_{j \ i-1 \ i} = \begin{bmatrix} c_2 c_3 & -c_2 s_3 & s_2 & 0 \\ s_1 s_2 c_3 + c_1 s_3 & -s_1 s_2 s_3 + c_1 s_3 & -s_1 c_2 & 0 \\ -c_1 s_2 c_3 + s_1 s_3 & c_1 s_2 s_3 + s_1 c_3 & c_1 c_2 & 0 \\ 0 & 0 & 0 & 1 \end{bmatrix} \quad (1.137)$$

where c_j and s_j are, respectively, the cosine and the sine of the synthetic parameter α_j, which is indicated in (1.138).

$$\alpha_j = \sum_{i=1}^{n_j} \alpha_{ji} \quad (1.138)$$

The matrixes L_{ij}, which are associated with every degree of freedom, are equal to the matrixes L of the summary degrees of freedom and are indicated in (1.99). These are used to calculate the speed matrix (1.139) and the acceleration matrix (1.140). Then we point out two position matrixes, evaluated simplifying (1.137), as for the other kinematic pairs, to obtain (1.141 and 1.142), which are associated with the time instants t_0 and $t_0 + \Delta t$ to calculate the helicoidal rototranslation matrix and to extract some parameters of the spherical pair.

The computation of the rototranslation matrix, starting from (1.141 and 1.142), as for the previous kinematic pairs, leads to results that are analogous to the ones indicated in the C.1 appendix; these results represent a concise form that can be associated with the use of the summary parameters of the spherical pair. We omit, just to avoid useless long formulas, the expressions of the components of $Q_{(0)}$ in terms of every single degree of freedom associated with every structural compliance. The reader will be able to verify personally the equivalence of the extensive approach with respect to the concise one; then the rotation ϑ (1.143) around the helicoidal axis can be achieved with the direction of the rototranslation, which is described by the vector u (1.144).

$$W = \sum_{i=1}^{3} \sum_{j=1}^{n_j} L_{ij} \cdot \dot{\alpha}_{ij} = \begin{bmatrix} 0 & -\dot{\alpha}_3 & \dot{\alpha}_2 & 0 \\ \dot{\alpha}_3 & 0 & -\dot{\alpha}_1 & 0 \\ -\dot{\alpha}_2 & \dot{\alpha}_1 & 0 & 0 \\ 0 & 0 & 0 & 0 \end{bmatrix} \quad (1.139)$$

$$H = \sum_{i=1}^{3} \sum_{j=1}^{n_j} \left(L_{ij} \ddot{\alpha}_{ij} \right) + \sum_{h=1}^{3} \sum_{k=1}^{3} \sum_{i=1}^{n_h} \sum_{j=1}^{n_k} \left(L_{hi} L_{kj} \dot{\alpha}_{hi} \dot{\alpha}_{kj} \right)$$

$$= \sum_{i=1}^{3} L_i \left(\sum_{j=1}^{n_j} \ddot{\alpha}_{ij} \right) + \sum_{h=1}^{3} \sum_{k=1}^{3} L_h L_k \left(\sum_{i=1}^{n_j} \dot{\alpha}_{hi} \right) \left(\sum_{j=1}^{n_j} \dot{\alpha}_{kj} \right)$$

$$= \sum_{i=1}^{3} L_i \ddot{\alpha}_i + \sum_{h=1}^{3} \sum_{k=1}^{3} L_h L_k \dot{\alpha}_h \dot{\alpha}_k \qquad (1.140)$$

$$= \begin{bmatrix} -\dot{\alpha}_2^2 - \dot{\alpha}_3^2 & \dot{\alpha}_2 \dot{\alpha}_1 - \ddot{\alpha}_3 & \dot{\alpha}_3 \dot{\alpha}_1 + \ddot{\alpha}_2 & 0 \\ \dot{\alpha}_1 \dot{\alpha}_2 + \ddot{\alpha}_3 & -\dot{\alpha}_1^2 - \dot{\alpha}_3^2 & \dot{\alpha}_3 \dot{\alpha}_2 - \ddot{\alpha}_1 & 0 \\ \dot{\alpha}_1 \dot{\alpha}_3 - \ddot{\alpha}_2 & \dot{\alpha}_2 \dot{\alpha}_3 + \ddot{\alpha}_1 & -\dot{\alpha}_1^2 - \dot{\alpha}_2^2 & 0 \\ 0 & 0 & 0 & 0 \end{bmatrix}$$

$$\begin{cases} M_1 {}_{i-1\ i}\big|_{t=t_0} = \begin{bmatrix} 1 & 0 & 0 & 0 \\ 0 & c_{01i} & -s_{01i} & 0 \\ 0 & s_{01i} & c_{01i} & 0 \\ 0 & 0 & 0 & 1 \end{bmatrix}; \quad M_2 {}_{i-1\ i}\big|_{t=t_0} = \begin{bmatrix} c_{02i} & 0 & s_{02i} & 0 \\ 0 & 1 & 0 & 0 \\ -s_{02i} & 0 & c_{02i} & 0 \\ 0 & 0 & 0 & 1 \end{bmatrix} \\[24pt] M_3 {}_{i-1\ i}\big|_{t=t_0} = \begin{bmatrix} c_{03i} & -s_{03i} & 0 & 0 \\ s_{03i} & c_{03i} & 0 & 0 \\ 0 & 0 & 1 & 0 \\ 0 & 0 & 0 & 1 \end{bmatrix}; \quad \alpha_{0j} = \sum_{i=1}^{n_j} \alpha_{0ji} \qquad (1.141) \\[24pt] M\big|_{t=t_0} = \\[12pt] \prod_{j=1}^{3} \prod_{i=1}^{n_j} M_j {}_{i-1\ i}\big|_{t=t_0} = \begin{bmatrix} c_{02}c_{03} & -c_{02}s_{03} & s_{02} & 0 \\ s_{01}s_{02}c_{03} + c_{01}s_{03} & -s_{01}s_{02}s_{03} + c_{01}s_{03} & -s_{01}c_{02} & 0 \\ -c_{01}s_{02}c_{03} + s_{01}s_{03} & c_{01}s_{02}s_{03} + s_{01}c_{03} & c_{01}c_{02} & 0 \\ 0 & 0 & 0 & 1 \end{bmatrix} \end{cases}$$

where c_{0ji} and s_{0ji} are, respectively, the cosine and sine of α_{0ji}, that is, the parameter that describes the displacement of the i-th compliance used to remove the j-th degree of constraint to turn a fixed joint into a spherical kinematic pair.

$$
\left\{
\begin{array}{l}
\left.M_1{}_{i-1}{}_i\right|_{t=t_0+\Delta t} =
\begin{bmatrix}
1 & 0 & 0 & 0 \\
0 & c_{01i+\delta1i} & -s_{01i+\delta1i} & 0 \\
0 & s_{01i+\delta1i} & c_{01i+\delta1i} & 0 \\
0 & 0 & 0 & 1
\end{bmatrix} \quad\quad (1.142)
\\[3em]
\left.M_2{}_{i-1}{}_i\right|_{t=t_0+\Delta t} =
\begin{bmatrix}
c_{02i+\delta2i} & 0 & s_{02i+\delta2i} & 0 \\
0 & 1 & 0 & 0 \\
-s_{02i+\delta2i} & 0 & c_{02i+\delta2i} & 0 \\
0 & 0 & 0 & 1
\end{bmatrix}
\\[3em]
\left.M_3{}_{i-1}{}_i\right|_{t=t_0+\Delta t} =
\begin{bmatrix}
c_{03i+\delta3i} & -s_{03i+\delta3i} & 0 & 0 \\
s_{03i+\delta3i} & c_{03i+\delta3i} & 0 & 0 \\
0 & 0 & 1 & 0 \\
0 & 0 & 0 & 1
\end{bmatrix} ; \quad \delta\alpha_j = \sum_{i=1}^{n_j} \delta\alpha_{ji}
\end{array}
\right.
$$

$$
\left.M\right|_{t=t_0+\Delta t} = \prod_{j=1}^{3}\prod_{i=1}^{n_j} \left.M_j{}_{i-1}{}_i\right|_{t=t_0+\Delta t}
$$

$$
=
\begin{bmatrix}
c_{02+\delta2}c_{03+\delta3} & -c_{02+\delta2}s_{03+\delta3} \\
s_{01+\delta1}s_{02+\delta2}c_{03+\delta3} + c_{01+\delta1}s_{03+\delta3} & -s_{01+\delta1}s_{02+\delta2}s_{03+\delta3} + c_{01+\delta1}c_{03+\delta3} \\
-c_{01+\delta1}s_{02+\delta2}c_{03+\delta3} + s_{01+\delta1}s_{03+\delta3} & c_{01+\delta1}s_{02+\delta2}s_{03+\delta3} + s_{01+\delta1}c_{03+\delta3} \\
0 & 0
\end{bmatrix}
$$

$$
\left.
\begin{array}{cc}
s_{02+\delta2} & 0 \\
-s_{01+\delta1}c_{02+\delta2} & 0 \\
c_{01+\delta1}c_{02+\delta2} & 0 \\
0 & 1
\end{array}
\right]
$$

where $c_{0ji+\delta ji}$ and $s_{0ji+\delta ji}$ are, respectively, the cosine and sine of $\alpha_{0ji} + \delta\alpha_{ji}$, which are the parameters α_{0ji} and which are increased by a value $\delta\alpha_{ji}$, as a consequence of the time increment Δt.

$$
c = 1 - \delta\alpha_1\delta\alpha_3 s_{02} - \delta\alpha_1\delta\alpha_2\delta\alpha_3 c_{02}/2
$$

$$
s = \pm\Big(\delta\alpha_1^2\delta\alpha_2^2 - 2\delta\alpha_1\delta\alpha_2^2\delta\alpha_3 s_{02} + \delta\alpha_1^2\delta\alpha_3^2 c_{02}^2 + 8\delta\alpha_1\delta\alpha_3 s_{02}
$$

$$
+ 4\delta\alpha_2^2 + 4\delta\alpha_3^2 + 4\delta\alpha_1^2 + \delta\alpha_2^2\delta\alpha_3^2 + 4\delta\alpha_1\delta\alpha_2\delta\alpha_3 c_{02}\Big)^{1/2}
$$

$$
(1.143)
$$

$$
\vartheta = \arctan(s, c)
$$

$$u_1 = \left(2\delta\alpha_3 s_{02} + 2\delta\alpha_1 + \delta\alpha_2\delta\alpha_3 c_{02}\right)\Big/\Big(\delta\alpha_1^2\delta\alpha_2^2 - 2\delta\alpha_1\delta\alpha_2^2\delta\alpha_3 s_{02} + \delta\alpha_1^2\delta\alpha_3^2 c_{02}^2$$

$$+ 8\delta\alpha_1\delta\alpha_3 s_{02} + 4\delta\alpha_1^2 + 4\delta\alpha_2^2 + 4\delta\alpha_3^2 + \delta\alpha_2^2\delta\alpha_3^2 + 4\delta\alpha_1\delta\alpha_2\delta\alpha_3 c_{02}\Big)^{1/2}$$

$$u_2 = -\left(\delta\alpha_1\delta\alpha_3 c_{02}c_{01} - 2\delta\alpha_2 c_{01} + \delta\alpha_1\delta\alpha_2 s_{01} + 2\delta\alpha_3 s_{01}c_{02} - \delta\alpha_2\delta\alpha_3 s_{01} s_{02}\right)\Big/$$

$$\left(\delta\alpha_1^2\delta\alpha_2^2 - 2\delta\alpha_1\delta\alpha_2^2\delta\alpha_3 s_{02} + \delta\alpha_1^2\delta\alpha_3^2 c_{02}^2 + 8\delta\alpha_1\delta\alpha_3 s_{02} + 4\delta\alpha_1^2 + 4\delta\alpha_2^2\right.$$

$$\left. + 4\delta\alpha_3^2 + \delta\alpha_2^2\delta\alpha_3^2 + 4\delta\alpha_1\delta\alpha_2\delta\alpha_3 c_{02}\right)^{1/2}$$

$$u_3 = -\left(-2\delta\alpha_2 s_{01} - 2\delta\alpha_3 c_{01}c_{02} - \delta\alpha_1\delta\alpha_2 c_{01} + \delta\alpha_1\delta\alpha_3 c_{02} s_{01} + \delta\alpha_2\delta\alpha_3 c_{01} s_{02}\right)\Big/$$

$$\left(\delta\alpha_1^2\delta\alpha_2^2 - 2\delta\alpha_1\delta\alpha_2^2\delta\alpha_3 s_{02} + \delta\alpha_1^2\delta\alpha_3^2 c_{02}^2 + 8\delta\alpha_1\delta\alpha_3 s_{02} + 4\delta\alpha_1^2 + 4\delta\alpha_2^2\right.$$

$$\left. + 4\delta\alpha_3^2 + \delta\alpha_2^2\delta\alpha_3^2 + 4\delta\alpha_1\delta\alpha_2\delta\alpha_3 c_{02}\right)^{1/2} \tag{1.144}$$

Examining a spherical kinematic pair, we consider a surrogate of the planar pair (Figure 1.19), which is generated, starting from a fixed joint through the introduction of suitably concentrated compliances, then the corresponding position matrix M can be expressed with (1.145).

$$\begin{cases} M_1 {}_{i-1} {}_{i} = \begin{bmatrix} 1 & 0 & 0 & \xi_{1i} \\ 0 & 1 & 0 & 0 \\ 0 & 0 & 1 & 0 \\ 0 & 0 & 0 & 1 \end{bmatrix}; \quad M_2 {}_{i-1} {}_{i} = \begin{bmatrix} 1 & 0 & 0 & 0 \\ 0 & 1 & 0 & \xi_{2i} \\ 0 & 0 & 1 & 0 \\ 0 & 0 & 0 & 1 \end{bmatrix} \\[20pt] M_3 {}_{i-1} {}_{i} = \begin{bmatrix} c_{3i} & -s_{3i} & 0 & 0 \\ s_{3i} & c_{3i} & 0 & 0 \\ 0 & 0 & 1 & 0 \\ 0 & 0 & 0 & 1 \end{bmatrix} \\[20pt] M = \prod_{j=1}^{3}\prod_{i=1}^{n_j} M_j {}_{i-1} {}_{i} \end{cases} \tag{1.145}$$

The matrixes L, which are associated with the degrees of freedom described by the parameters ξ_{ji}, are represented by (1.146), and they allow the calculation of the speed matrix W (1.147) and the acceleration matrix H (1.148).

$$\left\{ L_{1i} = \begin{bmatrix} 0 & 0 & 0 & 1 \\ 0 & 0 & 0 & 0 \\ 0 & 0 & 0 & 0 \\ 0 & 0 & 0 & 0 \end{bmatrix}; \quad L_{2i} = \begin{bmatrix} 0 & 0 & 0 & 0 \\ 0 & 0 & 0 & 1 \\ 0 & 0 & 0 & 0 \\ 0 & 0 & 0 & 0 \end{bmatrix} \right.$$

$$\left. L_{3i} = \begin{bmatrix} 0 & -1 & 0 & 0 \\ 1 & 0 & 0 & 0 \\ 0 & 0 & 0 & 0 \\ 0 & 0 & 0 & 0 \end{bmatrix} \right\}$$

(1.146)

$$W = \sum_{j=1}^{3} \sum_{i=1}^{n_j} L_{ji} \dot{\xi}_{ji} = \begin{bmatrix} 0 & -\sum_{i=1}^{n_3} \dot{\xi}_{3i} & 0 & \sum_{i=1}^{n_1} \dot{\xi}_{1i} \\ \sum_{i=1}^{n_3} \dot{\xi}_{3i} & 0 & 0 & \sum_{i=1}^{n_2} \dot{\xi}_{2i} \\ 0 & 0 & 0 & 0 \\ 0 & 0 & 0 & 0 \end{bmatrix}$$

(1.147)

$$H = \sum_{j=1}^{3} \sum_{i=1}^{n_j} \left(L_{ji} \ddot{\xi}_{ji} + L_{ji}^2 \dot{\xi}_{ji}^2 \right) = \begin{bmatrix} -\sum_{i=1}^{n_3} \dot{\xi}_{3i}^2 & -\sum_{i=1}^{n_3} \ddot{\xi}_{3i} & 0 & \sum_{i=1}^{n_1} \ddot{\xi}_{1i} \\ \sum_{i=1}^{n_3} \ddot{\xi}_{3i} & \sum_{i=1}^{n_3} \dot{\xi}_{3i}^2 & 0 & \sum_{i=1}^{n_2} \ddot{\xi}_{2i} \\ 0 & 0 & 0 & 0 \\ 0 & 0 & 0 & 0 \end{bmatrix}$$

(1.148)

We consider the two usual distinct spatial configurations that are associated with the two time instants t_0 and $t_0+\Delta t$ (1.149 and 1.150) to calculate the rototranslation matrix that turns one configuration into the other (1.151).

$$\left\{ \begin{array}{l} \mathbf{M}_1{}_{i-1i}\Big|_{t=t_0} = \begin{bmatrix} 1 & 0 & 0 & \xi_{10i} \\ 0 & 1 & 0 & 0 \\ 0 & 0 & 1 & 0 \\ 0 & 0 & 0 & 1 \end{bmatrix}; \quad \mathbf{M}_2{}_{i-1i}\Big|_{t=t_0} = \begin{bmatrix} 1 & 0 & 0 & 0 \\ 0 & 1 & 0 & \xi_{20i} \\ 0 & 0 & 1 & 0 \\ 0 & 0 & 0 & 1 \end{bmatrix} \\[3em] \mathbf{M}_3{}_{i-1i}\Big|_{t=t_0} = \begin{bmatrix} c_{30i} & -s_{30i} & 0 & 0 \\ s_{30i} & c_{30i} & 0 & 0 \\ 0 & 0 & 1 & 0 \\ 0 & 0 & 0 & 1 \end{bmatrix} \\[3em] \mathbf{M}\Big|_{t=t_0} = \prod_{j=1}^{3}\prod_{i=1}^{n_j} \mathbf{M}_j{}_{i-1i}\Big|_{t=t_0} \end{array} \right. \tag{1.149}$$

$$\left\{ \begin{array}{l} \mathbf{M}_1{}_{i-1i}\Big|_{t=t_0+\Delta t} = \begin{bmatrix} 1 & 0 & 0 & \xi_{10i}+\delta\xi_{1i} \\ 0 & 1 & 0 & 0 \\ 0 & 0 & 1 & 0 \\ 0 & 0 & 0 & 1 \end{bmatrix} \\[3em] \mathbf{M}_2{}_{i-1i}\Big|_{t=t_0+\Delta t} = \begin{bmatrix} 1 & 0 & 0 & 0 \\ 0 & 1 & 0 & \xi_{20i}+\delta\xi_{2i} \\ 0 & 0 & 1 & 0 \\ 0 & 0 & 0 & 1 \end{bmatrix} \\[3em] \mathbf{M}_3{}_{i-1i}\Big|_{t=t_0+\Delta t} = \begin{bmatrix} c_{30i+\delta 3i} & -s_{30i+\delta 3i} & 0 & 0 \\ s_{30i+\delta 3i} & c_{30i+\delta 3i} & 0 & 0 \\ 0 & 0 & 1 & 0 \\ 0 & 0 & 0 & 1 \end{bmatrix} \\[3em] \mathbf{M}\Big|_{t=t_0+\Delta t} = \prod_{j=1}^{3}\prod_{i=1}^{n_j} \mathbf{M}_j{}_{i-1i}\Big|_{t=t_0+\Delta t} \end{array} \right. \tag{1.150}$$

$$\left\{ \begin{array}{l} \xi_{j0} = \sum_{i=1}^{n_j} \xi_{j0i}; \quad \delta_j = \sum_{i=1}^{n_j} \delta_{ji} \\[2em] Q = \begin{bmatrix} c_{\delta 3} & -s_{\delta 3} & 0 & s_{\delta 3}\xi_{20} - c_{\delta 3}\xi_{10} + \xi_{10} + \delta_{10} \\ s_{\delta 3} & c_{\delta 3} & 0 & -c_{\delta 3}\xi_{20} - s_{\delta 3}\xi_{10} + \xi_{20} + \delta_{20} \\ 0 & 0 & 1 & 0 \\ 0 & 0 & 0 & 1 \end{bmatrix} \end{array} \right. \tag{1.151}$$

Starting from (1.151), some results can be deduced to concisely describe the helicoidal motion associated with the planar compliant pair, and these results are analogous to the observations just outlined for the theoretical (without compliances) planar motion.

Finally, the last kinematic pair examined is the surrogate of the helicoidal pair generated from a fixed joint removing the proper constraints with some structural compliance. This kind of kinematic pair looks like the cylindrical pair, but the relation (1.152) is introduced between the parameters α_i and z_i, that are shown in (1.131); consequently, the position matrix M can be represented by (1.153).

$$z_i = p_i \frac{\alpha_i}{2\pi} \tag{1.152}$$

where p_i is a constant parameter that represents the pitch of the helix and is associated with the i-th compliance. The matrix L_i, which is associated with the i-th degree of freedom, is then identified by (1.154), and it allows a fast calculation of the speed matrix W (1.155) and the acceleration matrix H (1.156).

$$
\left\{
\begin{aligned}
M_i &= \begin{bmatrix} \cos(\alpha_i) & -\sin(\alpha_i) & 0 & 0 \\ \sin(\alpha_i) & \cos(\alpha_i) & 0 & 0 \\ 0 & 0 & 1 & p_i \frac{\alpha_i}{2\pi} \\ 0 & 0 & 0 & 1 \end{bmatrix} \\[2em]
M &= \prod_{i=1}^{n} M_i = \begin{bmatrix} \cos\left(\sum_{i=1}^{n}\alpha_i\right) & -\sin\left(\sum_{i=1}^{n}\alpha_i\right) & 0 & 0 \\ \sin\left(\sum_{i=1}^{n}\alpha_i\right) & \cos\left(\sum_{i=1}^{n}\alpha_i\right) & 0 & 0 \\ 0 & 0 & 1 & \dfrac{\sum_{i=1}^{n} p_i\alpha_i}{2\pi} \\ 0 & 0 & 0 & 1 \end{bmatrix}
\end{aligned}
\right.
\tag{1.153}
$$

$$
L_i = \begin{bmatrix} 0 & -1 & 0 & 0 \\ 1 & 0 & 0 & 0 \\ 0 & 0 & 1 & \dfrac{p_i}{2\pi} \\ 0 & 0 & 0 & 1 \end{bmatrix} \tag{1.154}
$$

$$
W = \sum_{i=1}^{n} L_i \dot{\alpha}_i =
\begin{bmatrix}
0 & -\sum_{i=1}^{n} \dot{\alpha}_i & 0 & 0 \\
\sum_{i=1}^{n} \dot{\alpha}_i & 0 & 0 & 0 \\
0 & 0 & 0 & \dfrac{\sum_{i=1}^{n} p_i \dot{\alpha}_i}{2\pi} \\
0 & 0 & 0 & 0
\end{bmatrix}
\tag{1.155}
$$

$$
H = \sum_{i=1}^{n} L_i \ddot{\alpha}_i + \sum_{j=1}^{n}\sum_{i=1}^{n} L_i L_j \dot{\alpha}_i \dot{\alpha}_j
$$

$$
= \sum_{i=1}^{n} L_i \ddot{\alpha}_i + L^2 \left(\sum_{i=1}^{n} \dot{\alpha}_i^2 + 2\sum_{i=2}^{n}\sum_{j=1}^{i-1} \dot{\alpha}_i \dot{\alpha}_j \right)
$$

$$
= \sum_{i=1}^{n} L_i \ddot{\alpha}_i + L^2 \left(\sum_{i=1}^{n} \dot{\alpha}_i \right)^2
$$

$$
=
\begin{bmatrix}
-\left(\sum_{i=1}^{n} \dot{\alpha}_i\right)^2 & -\sum_{i=1}^{n} \ddot{\alpha}_i & 0 & 0 \\
\sum_{i=1}^{n} \ddot{\alpha}_i & -\left(\sum_{i=1}^{n} \dot{\alpha}_i\right)^2 & 0 & 0 \\
0 & 0 & 0 & \dfrac{\sum_{i=1}^{n} p_i \ddot{\alpha}_i}{2\pi} \\
0 & 0 & 0 & 0
\end{bmatrix}
\tag{1.156}
$$

Even for this kinematic pair, we consider two positions that are associated to the instants t_0 (1.157) and $t_0 + \Delta t$ (1.158) to deduce the matrix $Q_{(0)}$ (1.159), which turns a spatial configuration into the other.

$$
\begin{cases}
\left. M_i \right|_{t=t_0} =
\begin{bmatrix}
\cos(\alpha_{0i}) & -\sin(\alpha_{0i}) & 0 & 0 \\
\sin(\alpha_{0i}) & \cos(\alpha_{0i}) & 0 & 0 \\
0 & 0 & 1 & p_i \dfrac{\alpha_{0i}}{2\pi} \\
0 & 0 & 0 & 1
\end{bmatrix} \\[4em]
\left. M \right|_{t=t_0} = \prod_{i=1}^{n} \left. M_i \right|_{t=t_0} =
\begin{bmatrix}
\cos\left(\sum_{i=1}^{n} \alpha_{0i}\right) & -\sin\left(\sum_{i=1}^{n} \alpha_{0i}\right) & 0 & 0 \\
\sin\left(\sum_{i=1}^{n} \alpha_{0i}\right) & \cos\left(\sum_{i=1}^{n} \alpha_{0i}\right) & 0 & 0 \\
0 & 0 & 1 & \dfrac{\sum_{i=1}^{n} p_i \alpha_{0i}}{2\pi} \\
0 & 0 & 0 & 1
\end{bmatrix}
\end{cases} \tag{1.157}
$$

$$
\begin{cases}
\left. M_i \right|_{t=t_0+\Delta t} =
\begin{bmatrix}
\cos(\alpha_{0i}+\delta\alpha_i) & -\sin(\alpha_{0i}+\delta\alpha_i) & 0 & 0 \\
\sin(\alpha_{0i}+\delta\alpha_i) & \cos(\alpha_{0i}+\delta\alpha_i) & 0 & 0 \\
0 & 0 & 1 & p_i \dfrac{\alpha_{0i}+\delta\alpha_i}{2\pi} \\
0 & 0 & 0 & 1
\end{bmatrix} \tag{1.158}\\[4em]
\left. M \right|_{t=t_0+\Delta t} = \\[2em]
\prod_{i=1}^{n} \left. M_i \right|_{t=t_0+\Delta t} =
\begin{bmatrix}
\cos\left(\sum_{i=1}^{n} \alpha_{0i}+\delta\alpha_i\right) & -\sin\left(\sum_{i=1}^{n} \alpha_{0i}+\delta\alpha_i\right) & 0 & 0 \\
\sin\left(\sum_{i=1}^{n} \alpha_{0i}+\delta\alpha_i\right) & \cos\left(\sum_{i=1}^{n} \alpha_{0i}+\delta\alpha_i\right) & 0 & 0 \\
0 & 0 & 1 & \dfrac{\sum_{i=1}^{n} p_i\left(\alpha_{0i}+\delta\alpha_i\right)}{2\pi} \\
0 & 0 & 0 & 1
\end{bmatrix}
\end{cases}
$$

$$Q_{(0)} = M\big|_{t=t_0+\Delta t}\left(M\big|_{t=t_0}\right)^{-1} = \begin{bmatrix} \cos\left(\displaystyle\sum_{i=1}^{n}\delta\alpha_i\right) & -\sin\left(\displaystyle\sum_{i=1}^{n}\delta\alpha_i\right) & 0 & 0 \\[4mm] \sin\left(\displaystyle\sum_{i=1}^{n}\delta\alpha_i\right) & \cos\left(\displaystyle\sum_{i=1}^{n}\delta\alpha_i\right) & 0 & 0 \\[4mm] 0 & 0 & 1 & \dfrac{\displaystyle\sum_{i=1}^{n}p_i\delta\alpha_i}{2\pi} \\[4mm] 0 & 0 & 0 & 1 \end{bmatrix} \quad (1.159)$$

Using (1.82), we obtain the rotation ϑ around the helicoidal axis (1.160), and with (1.83), we identify the axis of the motion. From this, with (1.84), we obtain the h pitch of the helicoidal motion (1.160).

$$\vartheta = \sum_{i=1}^{n}\delta\alpha_i; \qquad h = \frac{\displaystyle\sum_{i=1}^{n}p_i\delta\alpha_i}{2\pi} \qquad (1.160)$$

After an examination of the kinematics of six classes of kinematic pairs that are generated from a fixed joint, with the introduction of appropriate concentrated compliances, we propose a table of comparison (Table 1.2), which considers some properties of these kinematic pairs.

For the kinematic pairs proposed in Table 1.2, proper structural compliances are introduced in a fixed joint to obtain the desired level of mobility; a similar approach can be used starting from nonfixed joints. Consequently, we consider, as an example, the revolute pair that is represented in Figure 1.12; we remove the prismatic constraint, which is normal to the planes of rotation, with proper structural compliances to obtain a surrogate of the cylindrical pair whose position matrix M is expressed by (1.161). The matrixes L of the kinematic pair are as in (1.93), and allow the computation of the speed matrix W (1.162) and the acceleration matrix H (1.163). Then, as for previous kinematic pairs, two spatial configurations are associated to the time instants t_0 and $t_0+\Delta t$ to build the rototranslation matrix $Q_{(0)}$, which turns one configuration into the other (1.164).

TABLE 1.2

Comparison of Kinematic Properties for the Kinematic Pairs Under Examination

Joint	L	Degree of Freedom	Degree of Constraint	u	h	ϑ
Revolute	L^{ri}	n_1	5	$u_i = 1$ $u_j = 0$ $i \neq j$	0	$\Sigma_a \delta\theta_{ia}$
Prismatic	L^{ti}	n_1	5	$u_i = 1$ $u_j = 0$ $i \neq j$	$\Sigma_a \delta x_{ia}$	0
Cylindrical	L^{ri}, L^{ti}	$n_1 + n_2$	4	$u_i = 1$ $u_j = 0$ $i \neq j$	$\Sigma_a \delta x_{ia}$	$\Sigma_a \delta\theta_{ia}$
Spherical	L^{r1}, L^{r2}, L^{r3}	$n_1 + n_2 + n_3$	3	(1.144)	0	(1.102)
Planar	L^{ti}, L^{tj}, L^{rk}	$n_1 + n_2 + n_3$	3	$u_k = 1,$ composed	$\Sigma_a \delta x_{ia}$ $\Sigma_a \delta x_{ja}$	$\Sigma_a \delta\theta_{ka}$
Helicoidal	$L^{ri} + (p_a\alpha_a/2\pi)L^{ti}$	n_1	5	$u_i = 1$ $u_j = 0$ $i \neq j$	$\Sigma_a p_a \delta\theta_{ia}/2\pi$	$\Sigma_a \delta\theta_{ia}$

$$\left\{\begin{array}{l} M_i = \begin{bmatrix} \cos(\alpha) & -\sin(\alpha) & 0 & 0 \\ \sin(\alpha) & \cos(\alpha) & 0 & 0 \\ 0 & 0 & 1 & z_i \\ 0 & 0 & 0 & 1 \end{bmatrix} \\ M = \prod_{i=1}^{n} M_i = \begin{bmatrix} \cos(\alpha) & -\sin(\alpha) & 0 & 0 \\ \sin(\alpha) & \cos(\alpha) & 0 & 0 \\ 0 & 0 & 1 & \sum_{i=1}^{m} z_i \\ 0 & 0 & 0 & 1 \end{bmatrix} \end{array}\right. \tag{1.161}$$

$$W = L_\alpha \dot\alpha + \sum_{i=1}^{n} L_{zi}\dot z_i = \begin{bmatrix} 0 & -\dot\alpha & 0 & 0 \\ \dot\alpha & 0 & 0 & 0 \\ 0 & 0 & 0 & \sum_{i=1}^{n}\dot z_i \\ 0 & 0 & 0 & 0 \end{bmatrix} \tag{1.162}$$

$$H = L_\alpha^2 \dot{\alpha}^2 + L_\alpha \ddot{\alpha} + \sum_{i=1}^{n} L_{zi} \ddot{z}_i = \begin{bmatrix} -\dot{\alpha}^2 & -\ddot{\alpha} & 0 & 0 \\ \ddot{\alpha} & -\dot{\alpha}^2 & 0 & 0 \\ 0 & 0 & 0 & \sum_{i=1}^{n} \ddot{z}_i \\ 0 & 0 & 0 & 0 \end{bmatrix} \quad (1.163)$$

$$\begin{cases} M\Big|_{t=t_0} = \begin{bmatrix} \cos(\alpha_0) & -\sin(\alpha_0) & 0 & 0 \\ \sin(\alpha_0) & \cos(\alpha_0) & 0 & 0 \\ 0 & 0 & 1 & \sum_{i=1}^{n} z_{0i} \\ 0 & 0 & 0 & 1 \end{bmatrix} \\[4em] M\Big|_{t=t_0+\Delta t} = \begin{bmatrix} \cos(\alpha_0+\Delta\alpha) & -\sin(\alpha_0+\Delta\alpha) & 0 & 0 \\ \sin(\alpha_0+\Delta\alpha) & \cos(\alpha_0+\Delta\alpha) & 0 & 0 \\ 0 & 0 & 1 & \sum_{i=1}^{n}(z_{0i}+\delta z_i) \\ 0 & 0 & 0 & 1 \end{bmatrix} \quad (1.164) \\[4em] Q_{(0)} = M\Big|_{t=t_0+\Delta t}\left(M\Big|_{t=t_0}\right)^{-1} = \begin{bmatrix} \cos(\Delta\alpha) & -\sin(\Delta\alpha) & 0 & 0 \\ \sin(\Delta\alpha) & \cos(\Delta\alpha) & 0 & 0 \\ 0 & 0 & 1 & \sum_{i=1}^{n}(\delta z_i) \\ 0 & 0 & 0 & 1 \end{bmatrix} \end{cases}$$

From (1.164) and with (1.82–1.85), helicoidal parameters of the motion can be extracted from the rototranslation matrix $Q_{(0)}$: the axis of the motion is x_3, and the rotation angle ϑ and the pitch of the motion h are represented by (1.165).

$$\vartheta = \Delta\alpha; \quad h = \sum_{i=1}^{n} \delta z_i \quad (1.165)$$

Now, it is interesting to compare some kinematic properties of the surrogate of the cylindrical pair obtained starting from a revolute pair with the homologous properties of a theoretical cylindrical pair (without compliances) and of a surrogate of the cylindrical pair that is generated starting from a fixed joint (Table 1.3).

As a further example, starting from a revolute pair (Figure 1.22), we remove, with two classes of structural compliances, two prismatic constraints that are parallel to the plane of the motion, obtaining a surrogate of the planar pair whose position matrix M is expressed by (1.166). Then, with the L matrixes

TABLE 1.3

Comparison of Kinematic Properties of a Cylindrical Pair
and Its Surrogate

Joint	L	Degree of Freedom	Degree of Constraint	u	H	ϑ
Theoretical	L^{ri}, L^{ti}	2	4	$u_i = 1$ $u_j = 0$ $i \neq j$	Δx_i	$\Delta \theta_i$
From fixed	L^{ri}, L^{ti}	$n_1 + n_2$	4	$u_i = 1$ $u_j = 0$ $i \neq j$	$\Sigma_a \delta x_{ia}$	$\Sigma_a \delta \theta_{ka}$
From revolute	L^{ri}, L^{ti}	$n_1 + 1$	4	$u_i = 1$ $u_j = 0$ $i \neq j$	$\Sigma_a \delta x_{ia}$	$\Delta \theta_i$

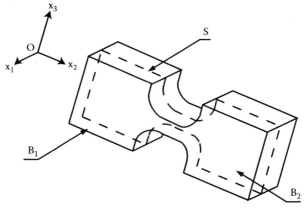

FIGURE 1.22
Example of a compliant revolute pair.

shown in (1.146) and in (1.115), we can calculate the speed matrix **W** (1.167) and the acceleration matrix **H** (1.168). After that, two spatial configurations are associated to the time instants t_0 and $t_0 + \Delta t$ to compute the rototranslation matrix $Q_{(0)}$, which turns one configuration into the other (1.169).

$$
\begin{cases}
M_{1 \ i-1 \ i} = \begin{bmatrix} 1 & 0 & 0 & x_i \\ 0 & 1 & 0 & 0 \\ 0 & 0 & 1 & 0 \\ 0 & 0 & 0 & 1 \end{bmatrix}; \quad M_{2 \ i-1 \ i} = \begin{bmatrix} 1 & 0 & 0 & 0 \\ 0 & 1 & 0 & y_i \\ 0 & 0 & 1 & 0 \\ 0 & 0 & 0 & 1 \end{bmatrix} \\
M_3 = \begin{bmatrix} \cos(\alpha) & -\sin(\alpha) & 0 & 0 \\ \sin(\alpha) & \cos(\alpha) & 0 & 0 \\ 0 & 0 & 1 & 0 \\ 0 & 0 & 0 & 1 \end{bmatrix}; \quad M = \prod_{j=1}^{2} \prod_{i=1}^{n_j} M_{j \ i-1 \ i} M_3
\end{cases}
$$

$$(1.166)$$

$$W = \sum_{i=1}^{n_1} L_{1i}\dot{x}_i + \sum_{i=1}^{n_2} L_{2i}\dot{y}_i + L_3\dot{\alpha} = \begin{bmatrix} 0 & -\dot{\alpha} & 0 & \sum_{i=1}^{n_1}\dot{x}_i \\ \dot{\alpha} & 0 & 0 & \sum_{i=1}^{n_2}\dot{y}_i \\ 0 & 0 & 0 & 0 \\ 0 & 0 & 0 & 0 \end{bmatrix} \qquad (1.167)$$

$$H = \sum_{i=1}^{n_1} L_{1i}\ddot{x}_i + \sum_{i=1}^{n_2} L_{2i}\ddot{y}_i + L_3\ddot{\alpha} + L_3^2\dot{\alpha}^2 = \begin{bmatrix} -\dot{\alpha}^2 & -\ddot{\alpha} & 0 & \sum_{i=1}^{n_1}\ddot{x}_i \\ \ddot{\alpha} & \dot{\alpha}^2 & 0 & \sum_{i=1}^{n_2}\ddot{y}_i \\ 0 & 0 & 0 & 0 \\ 0 & 0 & 0 & 0 \end{bmatrix} \qquad (1.168)$$

$$Q_{(0)} = \begin{bmatrix} \cos(\Delta\alpha) & -\sin(\Delta\alpha) & 0 & \sin(\Delta\alpha)\sum_{i=1}^{n_1}y_i + \left[1-\cos(\Delta\alpha)\right]\sum_{i=1}^{n_1}x_i + \sum_{i=1}^{n_1}\delta x_i \\ \sin(\Delta\alpha) & \cos(\Delta\alpha) & 0 & -\sin(\Delta\alpha)\sum_{i=1}^{n_1}x_i + \left[1-\cos(\Delta\alpha)\right]\sum_{i=1}^{n_1}y_i + \sum_{i=1}^{n_1}\delta y_i \\ 0 & 0 & 1 & 0 \\ 0 & 0 & 0 & 1 \end{bmatrix}$$

$$(1.169)$$

Consequently, the helicoidal parameters of the pair (1.82–1.85) can be deduced from (1.169); these parameters, together with other properties of the kinematic pair, are compared with those of a simple theoretical pair, as well as with those of a surrogate of the planar pair that is generated starting from a fixed joint (Table 1.4).

Compliances in the kinematic pair can be due also to the presence of continuous compliant properties; this way of modeling allows us to represent, with a good degree of precision, kinematic pairs realized in micro- and meso-actuators through compliant parts. These compliances are realized with inhomogeneous materials characterized by variable properties or, more commonly, with neckings, that is, the reduction in the section of the element that has to become compliant. To introduce a revolute mobility between two

TABLE 1.4

Comparison of Kinematic Properties of a Planar Pair and Its Surrogate

Joint	L	Degree of Freedom	Degree of Constraint	u	h	ϑ
Theoretical	L^{ti}, L^{tj}, L^{rk}	3	3	$u_k = 1$, composed	$\Delta x_i\, \Delta x_j$	$\Delta\theta_k$
From fixed	L^{ti}, L^{tj}, L^{rk}	$n_1 + n_2 + n_3$	3	$u_k = 1$, composed	$\Sigma_a \delta x_{ia}$ $\Sigma_a \delta x_{ja}$	$\Sigma_a \delta x_{ka}$
From revolute	L^{ti}, L^{tj}, L^{rk}	$n_1 + n_2 + 1$	3	$u_k = 1$, composed	$\Sigma_a \delta x_{ia}$ $\Sigma_a \delta x_{ja}$	$\Delta\theta_k$

parts B_1 and B_2 of the body B in the x_1-direction obtaining a surrogate of a theoretical revolute pair, we can produce a necking in every section S of B with its orthogonal exiting univector, which is also parallel to x_1 (Figure 1.22).

Similarly, we can realize a surrogate of a spherical pair between two parts B_1 and B_2 of the body B with a necking in every section S of B, with its perpendicular univector, which is also orthogonal to x_2 (Figure 1.23).

Furthermore, to distribute properly the level of flexibility, we can act on the shape of the necking produced on the section S, optimizing the relative movement of the parts B_1 and B_2, or, differently, we can produce a series of neckings characterized by the same shape and by identical or diverse size, according to the design needs and technological opportunities. Before detailing the study of compliant kinematic pairs realized with continuous flexibilities, it is useful to introduce some concept of dynamics. In fact, with these conceptual and computational tools, the concept of flexibility can be defined more precisely and, consequently, the dynamic and kinetostatic behavior of the mechanical pair can be described in a more accurate way.

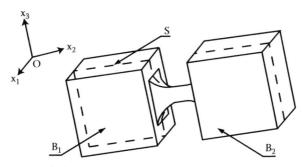

FIGURE 1.23
Example of a compliant spherical pair.

1.3   Dynamics

In the kinematic domain, the evolution of a mechanical system is studied from a merely descriptive viewpoint. In this case, the laws of classical dynamics add a way to determine motion through a causal connection between the concepts of force and of motion; in particular, a unique motion corresponds to a given force when the position and the initial act of motion are assigned. This causal relation constitutes the fundamental principle of mechanics and allows us to establish a link between the processes of motion and the processes of force, and it has a strictly deterministic nature in classical mechanics, but it can assume a statistical character or other types of forms in other modern mechanics. An observer who is invariable with time is really useful to examine these processes because it is characterized by the fact that, once a reference system is fixed on this observer, the motions of every isolated point can be unambiguously determined with proper initial conditions. With respect to this reference system, the observer examines the movements of a mechanical system that is represented by a material system (a point, a body, or a set of them), by constitutive forces, i.e., internal forces characterizing the fundamental constitution of the body, and by constraints connected to the material system. Furthermore, external forces can be impressed to obtain a dynamic effect. Remaining in the domain of classical mechanics, the causal principle is expressed by the possibility of a description of the evolution of a mechanical system in an ordered temporal domain, where Γ_{t2} depends on the states Γ_{t1} and not on the states Γ_{t3}, with t_3 subsequent to t_2, which, in its turn, is subsequent to t_1. With this choice, also remaining in the domain of classical mechanics, we admit, however, the possibility of alternative descriptions that sometimes can be really elegant or useful in control theory. In different situations, the dynamic relations that represent the mechanical system can be described with differential equations, but this possibility is allowed only if the mechanical system is a dynamic system with sufficient level of regularity. The previous hints on dynamics of systems can be deeply analyzed; therefore, we refer to a proper bibliographic monograph (Mesarovic, M. D., Takahara, Y. 1975) and we pass on to a brief comment on the dynamics of a material point.

1.3.1   Particle Dynamics

A material point P, as in section 1.2.1, is assumed with a motion as depicted in Figure 1.3, and a physical property is associated with this point P, which is related to the quantity of matter of P and identified by its mass. Furthermore, the set of forces acting on P is indicated by the resulting term F, which, generally, depends on the time, position, and speed of P, and then, under proper hypotheses, (1.170) expresses the connection between the forces acting on P and the motion of P, through the inertial term m.

$$m\frac{d^2x}{dt^2} = F\left(x, \frac{dx}{dt}, t\right) \tag{1.170}$$

The relation in (1.170) is a second-order differential equation in normal form with respect to the variable x; therefore, under wide hypotheses on its regularity and with proper initial conditions, it admits a solution that allows us to associate a temporal evolution of P with the set of causes F that produces it. An easier, but with applicative interest, procedure consists in the inverse operation permitting us to obtain a resultant force that is able to generate the desired motion, with the solution of a simple linear equation. It is interesting to observe that the mechanical system here, represented as a simple point P, can reveal one or more internal variables α_i, and consequently, the expression (1.171) describes a wider class of dynamic situations.

$$m\frac{d^2x}{dt^2} = F\left(x, \frac{dx}{dt}, \alpha, t\right)$$
$$\frac{d\alpha}{dt} = G\left(x, \frac{dx}{dt}, \alpha\right) \tag{1.171}$$

where α is the n-dimensional vector of the internal variables of the system. The dynamics of the material point can be studied and analyzed properly to face interesting applicative problems, as shown in specialist monographs (i.e., Bhat R. B., Lopez-Gomez A., 2001).

1.3.2 Dynamics of a Rigid Body

In the dynamics of a rigid body, some complications are introduced in addition to the case of the material point; further, there is a slightly deeper analysis than that in the previous section. Particularly, instead of the force F acting on the material point P, the concept of matrix of actions is described by $\Phi_{(i)}$, which represents the resulting system of forces and torques acting on the body B with respect to the reference frame (i), according to the definition (1.172).

$$\Phi_{(i)} = \begin{bmatrix} 0 & -t_3 & t_2 & f_1 \\ t_3 & 0 & -t_1 & f_2 \\ -t_2 & t_1 & 0 & f_3 \\ -f_1 & -f_2 & -f_3 & 0 \end{bmatrix} = \left[\begin{array}{c|c} T_{(i)} & {}^3F_{V(i)} \\ \hline -{}^3F_{V(i)}^T & 0 \end{array} \right] \tag{1.172}$$

where ${}^3F_{V(i)}$ identifies the first three components of $F_{(i)}$, and $F_{(i)}$ and $T_{(i)}$ represent, respectively, the resultant force and torque computed with respect to the origin of the reference system, which is considered, conventionally, as the pole. Furthermore, we observe that $\Phi_{(i)}$ is hemisymmetric, i.e., it is equal to the opposite of its transpose, and, as indicated in (1.172), it is obtained applying a force $F_{V(i)}$ to an infinitesimal volume dV, which is located in the geometrical point P identified by the position p, and successively integrating volume V of the body B, as shown in (1.173).

$$
\begin{aligned}
\Phi_{(i)} &= \int_V \left(F_{V(i)} P_{(i)}^T - P_{(i)} F_{V(i)}^T \right) \cdot dV \\[2mm]
&= \int_V \left[\begin{array}{c|c} {}^3F_{V(i)}\,{}^3P_{(i)}^T - {}^3P_{(i)}\,{}^3F_{V(i)}^T & {}^3F_{V(i)} \\[1mm] -{}^3F_{V(i)}^T & 0 \end{array} \right] dV = \left[\begin{array}{c|c} T_{(i)} & {}^3F_{V(i)} \\[1mm] -{}^3F_{V(i)}^T & 0 \end{array} \right]
\end{aligned}
\tag{1.173}
$$

where ${}^3F_{V(i)}$ and ${}^3p_{(i)}$ identify, respectively, the first three components of the vectors $F_{V(i)}$ and $p_{(i)}$, and $T_{(i)}$ is the torque, written in matricial form using the components of the vectorial product between ${}^3F_{V(i)}$ and ${}^3p_{(i)}$. Analogously, the matrix $\Gamma_{(i)}$ of the momentum of the body B is introduced with respect to the reference system, describing in a unique matrix the linear and angular momenta of the body B (1.174).

$$
\Gamma_{(i)} = \left[\begin{array}{ccc|c} 0 & -\gamma_3 & \gamma_2 & \rho_1 \\ \gamma_3 & 0 & -\gamma_1 & \rho_2 \\ -\gamma_2 & \gamma_1 & 0 & \rho_3 \\ \hline -\rho_1 & -\rho_2 & -\rho_3 & 0 \end{array} \right] = \left[\begin{array}{c|c} \gamma_{(i)} & {}^3\rho_{(i)} \\ \hline -{}^3\rho_{(i)}^T & 0 \end{array} \right]
\tag{1.174}
$$

where ${}^3\rho_{(i)}$ identifies the first three components of $\rho_{(i)}$, which is the momentum, and $\gamma_{(i)}$ is the angular momentum computed considering, conventionally, as pole the origin of the reference system. For this matrix also we observe the presence of hemisymmetry. The definition of the matrix $\Gamma_{(i)}$ (1.174) is obtained assuming an infinitesimal neighborhood of the point P and providing it with a mass dm; then the position and speed of the particle P are described, respectively, by $p_{(i)}$ and its time derivative, and are used to generate a differential relation that is integrated over the whole mass m of the body B, as shown in (1.175).

$$
\begin{aligned}
\Gamma_{(i)} &= \int_m \left(\frac{dp_{(i)}}{dt} p_{(i)}^T - p_{(i)} \frac{dp_{(i)}^T}{dt} \right) \cdot dm \\[2mm]
&= \int_m \left[\begin{array}{c|c} \dfrac{{}^3dp_{(i)}}{dt}\,{}^3p_{(i)}^T - {}^3p_{(i)}\dfrac{{}^3dp_{(i)}^T}{dt} & \dfrac{{}^3dp_{(i)}}{dt} \\[3mm] -\dfrac{{}^3dp_{(i)}^T}{dt} & 0 \end{array} \right] dm = \left[\begin{array}{c|c} \gamma_{(i)} & {}^3\rho_{(i)} \\ \hline -{}^3\rho_{(i)}^T & 0 \end{array} \right]
\end{aligned}
\tag{1.175}
$$

where $^3\boldsymbol{p}_{(i)}$ identifies the first three components of the vector $\boldsymbol{p}_{(i)}$, and $\boldsymbol{\gamma}_{(i)}$ is the angular momentum written in matricial form using the components of the vectorial product between $^3\boldsymbol{p}_{(i)}$ and its temporal derivative.

The inertia of a rigid body can be represented by a matricial operator, which is more complex than a scalar mass (1.171), and it is identifiable with the inertia matrix or pseudo-tensor of inertia, $\boldsymbol{J}_{(i)}$, referred to in the frame (i) in (1.176), in which m is the mass of the body, and the elements I_{ij} of the submatrix $^3\boldsymbol{J}$ are indicated in (1.177), together with the coordinates x_{gi} of the barycenter of the body.

$$J_{(i)} = \begin{bmatrix} I_{11} & I_{21} & I_{31} & m \cdot x_{g1} \\ I_{12} & I_{22} & I_{32} & m \cdot x_{g2} \\ I_{13} & I_{23} & I_{33} & m \cdot x_{g3} \\ m \cdot x_{g1} & m \cdot x_{g2} & m \cdot x_{g3} & m \end{bmatrix} = \begin{bmatrix} {}^3J & {}^3G \\ {}^3G^T & m \end{bmatrix} \tag{1.176}$$

$$I_{ij} = \int_m x_i x_j \cdot dm; \quad x_{gi} = \frac{1}{m} \int_m x_i \cdot dm \tag{1.177}$$

We observe that the terms of the submatrix $^3\boldsymbol{J}$ are intentionally denoted with the symbol I_{ij} to distinguish them from the usual inertial terms J_{ij} expressed in (1.178) and related to the terms I_{ij} by (1.179).

$$J_{ii} = \int_m \left(x_j^2 + x_i^2 \right) \cdot dm; \quad J_{ij} = -\int_m x_i x_j \cdot dm; \quad i \neq j \neq k \tag{1.178}$$

$$I_{ii} = \frac{-J_{ii} + J_{jj} + J_{kk}}{2}; \quad I_{ij} = -J_{ij}; \quad J_{ii} = I_{jj} + I_{kk}; \quad i \neq j \neq k \tag{1.179}$$

The definition (1.176) of the pseudo-tensor of inertia $\boldsymbol{J}$ is obtained assuming an infinitesimal neighborhood of the point P provided with an infinitesimal mass dm and belonging to B, and integrating over the whole body, as shown in (1.180), where $^3\boldsymbol{p}$ is the vector of the first three components of $\boldsymbol{p}$.

$$J_{(i)} = \int_m \boldsymbol{p} \cdot \boldsymbol{p}^T dm = \int_m \begin{bmatrix} {}^3\boldsymbol{p} \cdot {}^3\boldsymbol{p}^T & {}^3\boldsymbol{p} \\ {}^3\boldsymbol{p}^T & 1 \end{bmatrix} dm = \begin{bmatrix} {}^3J & G \\ G & m \end{bmatrix} \tag{1.180}$$

Further, the matrixes $\boldsymbol{\Phi}$, $\boldsymbol{\Gamma}$, and $\boldsymbol{J}$ can be expressed with respect to a reference frame (i) or to a reference frame (j), and the transformation formulas indicated in (1.181) can be adopted to pass from one reference to another using the matrix $\boldsymbol{M}_{ij}^T$ instead of the matrix $\boldsymbol{M}_{ij}^{-1}$, which appears in the relation used in kinematics for the reference change.

$$\Phi_{(j)} = M_{ji}\Phi_{(i)}M_{ji}^T; \quad \Gamma_{(j)} = M_{ji}\Gamma_{(i)}M_{ji}^T; \quad J_{(j)} = M_{ji}J_{(i)}M_{ji}^T \qquad (1.181)$$

The transformation rules in (1.181) can be easily achieved from the definitions in (1.173, 1.175, and 1.180) applying a reference change to the vectors $F_{(j)}$ and $p_{(j)}$ and, then, bringing the position matrixes out of the integral, as shown in (1.182–1.184).

$$
\begin{aligned}
\Phi_{(j)} &= \int_V \left(F_{V(j)}p_{(j)}^T - p_{(j)}F_{V(j)}^T \right) \cdot dV \\
&= \int_V \left(M_{ji}F_{V(i)}p_{(i)}^T M_{ji}^T - M_{ji}p_{(i)}F_{V(i)}^T M_{ji}^T \right) \cdot dV \\
&= M_{ji} \int_V \left(F_{V(i)}p_{(i)}^T - p_{(i)}F_{V(i)}^T \right) \cdot dV \cdot M_{ji}^T \\
&= M_{ji}\Phi_{(i)}M_{ji}^T
\end{aligned}
\qquad (1.182)
$$

$$
\begin{aligned}
\Gamma_{(j)} &= \int_V \left(\dot{p}_{(j)}p_{(j)}^T - p_{(j)}\dot{p}_{(j)}^T \right) \cdot dV \\
&= \int_V \left(M_{ji}\dot{p}_{(i)}p_{(i)}^T M_{ji}^T - M_{ji}p_{(i)}\dot{p}_{(i)}^T M_{ji}^T \right) \cdot dV \\
&= M_{ji} \int_V \left(\dot{p}_{(i)}p_{(i)}^T - p_{(i)}\dot{p}_{(i)}^T \right) \cdot dV \cdot M_{ji}^T \\
&= M_{ji}\Gamma_{(i)}M_{ji}^T
\end{aligned}
\qquad (1.183)
$$

$$
\begin{aligned}
J_{(j)} &= \int_m \left(p_{(j)}p_{(j)}^T \right) \cdot dm \\
&= \int_m \left(M_{ji}p_{(i)}p_{(i)}^T M_{ji}^T \right) \cdot dm \\
&= M_{ji} \int_m \left(p_{(i)}p_{(i)}^T \right) \cdot dm \cdot M_{ji}^T \\
&= M_{ji}J_{(i)}M_{ji}^T
\end{aligned}
\qquad (1.184)
$$

The operators Φ, Γ, and J can be related to the kinematic entities H and W of a body B_k with respect to an inertial reference frame (0), projecting the dynamic relation on any frame (i), as shown in (1.185), which is enough to illustrate the dynamics of a rigid body.

$$\Phi_{(i)} = H_{0k(i)}J_{k(i)} - J_{k(i)}H_{0k(i)}^T; \quad \Gamma_{k(i)} = W_{0k(i)}J_{k(i)} - J_{k(i)}W_{0k(i)}^T \qquad (1.185)$$

Further investigations can be performed (Legnani, 2003), or energetic entities can be proposed, to obtain an alternative description, which can be very useful in multiphysical problems; therefore, in the following text, the concepts of kinetic energy E_k and potential U of the conservative forces acting on the body are applied, sometimes substituted with the generalized potential of conservative forces and some nonconservative actions (Marion J. B. 1965). These forces are used to gain a sufficiently wide-ranging formulation to achieve an acceptable level of applicability (1.186), where L is the Lagrangian of the body B_j, and f_{qi} is the generalized force associated with the degree of freedom identified by the free variable q_i and computed from the virtual work of the actions Φ on B_j due to a virtual displacement, recognizing that this virtual work is equal to the work of the generalized force f_{qi}. More general and in-depth formulations than the previous one can certainly be proposed; therefore, for further investigations, we suggest that specific monographs, such as (Readman, 1994) or (West, 1993) be consulted.

$$E_k = \frac{1}{2}\left(W_{0j}J_jW_{0j}^T\right); \quad L = E_k + U$$

$$\frac{d}{dt}\frac{\partial L}{\partial \dot{q}_i} - \frac{\partial L}{\partial q_i} = f_{q_i} \qquad (1.186)$$

1.3.3 Dynamics of a Continuum Body

As anticipated in section 1.2.4, the kinematics and dynamics of a continuum in classical mechanics are often handled under Cauchy's hypothesis, which introduces some simplifying characteristics on forces and couples acting on the deformable continuum, particularly on the resulting forces and torques acting on a finite volume ΔV extracted from the inside, and also on a finite portion ΔS_f of the loaded surface of the deformable continuum. In fact, referring to Figure 1.24, where the portion ΔV and ΔS_f appear in a schematic representation together with the actions Δw and Δp applied to them, Cauchy's

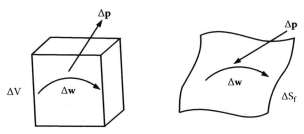

FIGURE 1.24
Schema of actions on a portion of volume ΔV and surface ΔS_f.

hypothesis can be summarized with (1.187), which represents a way to define the body forces b and the surface forces t, proposing an additive hypothesis on the neglectability of actions due to the torques Δw.

$$
\begin{cases}
\lim_{\Delta V \to 0} \dfrac{\Delta w}{\Delta V} = 0; & \lim_{\Delta V \to 0} \dfrac{\Delta p}{\Delta V} = b; & x \in V \\[2mm]
\lim_{\Delta S_f \to 0} \dfrac{\Delta w}{\Delta S_f} = 0; & \lim_{\Delta S_f \to 0} \dfrac{\Delta p}{\Delta S_f} = t; & x \in S_f
\end{cases}
\tag{1.187}
$$

Further, a simplifying hypothesis on the absence of body inertia forces in the continuum can be useful for a wide class of physical problems, but it can be too restrictive for some situations involving MEMS (Micro Electro Mechanical System). In fact, it prescribes that strain accelerations are negligible with respect to macroscopic movements of the whole body, and often it can be considered a valid hypothesis except for the situations where relative movements of a part of the body become important. Therefore, in future paragraphs, we can remove it by observing interesting inertial and vibrational phenomena. Another additive hypothesis that is often confirmed by daily practice consists in the conservation of the actions b and t during the deformation process of the continuum. This hypothesis can be maintained for a wide class of situations, but it can also be removed to examine particularly variable actions. Moreover, in continuum theories, time and space concentrate loads are not generally avoided because they can produce discontinuities that are mathematically difficult to be deal with. Actually, an analytical description of deformable continuum with proper mathematical tools, eventually drawing from distribution theories, allows a tractability of discontinuities stressing some properties in extreme cases, and these can be interesting from an academic or technical viewpoint. Notwithstanding this possibility, the continuum theory is usually introduced as a forerunner of a numerical discretization, permitting the use of proper algorithms to provide approximate solutions. These algorithms, in the presence of discontinuities, can diverge, or they can produce too inaccurate results. Therefore, a set of analyses should be performed to observe asymptotic phenomena, and the physical meaning of numerical results should be properly verified.

After this brief introduction, we consider a deformable body B, as in Figure 1.25, endowed with a volume V, separated from the external space by a surface S that can be subdivided in a part S_f loaded with actions $t(x)$ and in a part S_u provided with constraint reactions $r(x)$. Then, two temporal configurations are considered for the body B (initial configuration 0 and final configuration 1); the first is just previous to, whereas the latter is just subsequent to the deformation process, and both are in state of equilibrium.

Examining a point P in the deformed configuration, a vector of Cauchy's stress σ_α can be defined in its neighborhood with its direction exiting from a surface $\Delta \Sigma$ that belongs to the neighborhood of P respecting the hypothesis (1.188), equilibrating and transferring the actions from the rest of the body B.

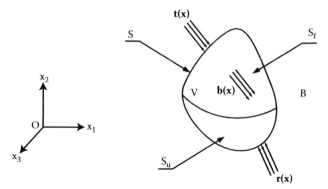

FIGURE 1.25
Deformable continuum B subject to external actions and constraint reactions.

$$\begin{cases} \lim_{\Delta\Sigma\to 0} \dfrac{\Delta p}{\Delta\Sigma} = \boldsymbol{\sigma}_\alpha \\[3mm] \lim_{\Delta\Sigma\to 0} \dfrac{\Delta w}{\Delta\Sigma} = \mathbf{0} \end{cases} ; \quad x_1 \in V \qquad (1.188)$$

The set of all possible vectors of stress $\boldsymbol{\sigma}_\alpha$ exiting from $\Delta\Sigma$ represents the stress state of B in the point P; it can be demonstrated that every vector $\boldsymbol{\sigma}_\alpha$ can be represented with a linear combination of three mutually independent vectors $\boldsymbol{\sigma}_i$, i.e., the stress vectors with directions of the three normals to the planes of the adopted Cartesian reference system. Therefore, a stress matrix $\boldsymbol{\sigma}$ can be defined as a column composition of the stress vectors $\boldsymbol{\sigma}_i$ with directions of the axes x_i of the reference system adopted to describe the deformed configuration of the body B. Furthermore, it can be demonstrated that the stress matrix $\boldsymbol{\sigma}$ is a double symmetric tensor, which can be represented using the components of the vectors $\boldsymbol{\sigma}_i$ as in (1.189), where the engineering notation is shown distinguishing between tangential and normal components.

$$\boldsymbol{\sigma} = \begin{bmatrix} \sigma_{11} & \sigma_{12} & \sigma_{13} \\ \sigma_{21} & \sigma_{22} & \sigma_{23} \\ \sigma_{31} & \sigma_{32} & \sigma_{33} \end{bmatrix} = \begin{bmatrix} \sigma_1 & \tau_{12} & \tau_{13} \\ \tau_{21} & \sigma_2 & \tau_{23} \\ \tau_{31} & \tau_{32} & \sigma_3 \end{bmatrix} \qquad (1.189)$$

As for the tensor of small strain in section 1.2.4, in the case of stress tensors also, we can search for the principal stress and principal directions of stress pertaining to the computational problem of the diagonalization of the symmetric matrix $\boldsymbol{\sigma}$ to obtain its eigenvalues and eigenvectors corresponding to the stress vectors exiting from the surfaces, which are parallel to the coordinated planes of the principal stress-reference system. The stress tensor can be used to write the indefinite equilibrium equations in differential form (1.190), where n_i are the components of the normal vector exiting from S and calculated in the point with coordinates x.

$$\begin{cases} \dfrac{\partial \sigma_{ij}}{\partial x_i} + b_j = 0; & x \in V \\ \sigma_{ij} n_i = t_j; & x \in S_f \\ \sigma_{ij} n_i = r_j; & x \in S_u \end{cases} \tag{1.190}$$

The condition of small deformation can be recognized in a wide range of situations where the deformed configuration x_1 can be almost superimposed on the undeformed configuration x_0, allowing a considerable computational and notational reduction, i.e., we can avoid the lower index to distinguish between deformed and undeformed configurations.

Then, the constitutive relation is introduced, examining the property of the material that constitutes the body B to connect the stress field defined on the body B with its kinematic field through a proper mathematical relation. Particularly, some hypotheses on the constitutive relation are supposed to produce a model with a good level of applicability, maintaining the possibility of removing them when it is necessary. The first one consists of the linearity of the constitutive relation, and then a reversible behavior is assumed for the material, avoiding hysteretic phenomena. Furthermore, the dependences of the constitutive relation on time and load speed are neglected, and finally, crack and damage phenomena are excluded, resulting, however, in a wide class of mechanical applications. Consequent to these hypotheses, the constitutive expression can be represented in indicial and matricial forms directly in (1.191) and inversely in (1.192).

$$\sigma_{ij}(x) = D_{ijhk}(x) \cdot \varphi_{hk}(x)$$
$$\sigma(x) = D(x) : \varphi(x) \tag{1.191}$$

$$\varphi_{ij}(x) = C_{ijhk}(x) \cdot \sigma_{hk}(x)$$
$$\varphi(x) = C(x) : \sigma(x) \tag{1.192}$$

$$D_{ijhk} = D_{hkij}; \quad C_{ijhk} = C_{hkij} \tag{1.193}$$

With some observations on the potential related to a deformation and under proper hypotheses, the reversibility (or elasticity) of the material can be associated with the symmetry of the tensors D and C, as indicated in (1.193). Further, assuming the isotropy and homogeneity of the material and applying the symmetry of the tensors D, C, σ and φ, a further simplified expression of high engineering interest can be reorganized, which transform the quadruple tensors D and C in double tensors and convert the double symmetric tensors σ and φ of stress and strain in column vectors, obtaining, respectively, the direct and inverse elastic constitutive relation (1.194 and 1.195). In (1.194) and (1.195), the engineering notation is used for the stress and strain vectors; moreover, E is the Young's modulus, and v, the Poisson's coefficient is the first Lamè's constant and it is related to the second Lamè's constant (the tangential elastic modulus G) by the expression in (1.196).

$$
\begin{bmatrix} \sigma_{11} \\ \sigma_{22} \\ \sigma_{33} \\ \tau_{12} \\ \tau_{13} \\ \tau_{23} \end{bmatrix} = \frac{E}{1+v} \begin{bmatrix} \dfrac{1-v}{1-2v} & \dfrac{v}{1-2v} & \dfrac{v}{1-2v} & 0 & 0 & 0 \\[2mm] \dfrac{v}{1-2v} & \dfrac{1-v}{1-2v} & \dfrac{v}{1-2v} & 0 & 0 & 0 \\[2mm] \dfrac{v}{1-2v} & \dfrac{v}{1-2v} & \dfrac{1-v}{1-2v} & 0 & 0 & 0 \\[2mm] 0 & 0 & 0 & \dfrac{1}{2} & 0 & 0 \\[2mm] 0 & 0 & 0 & 0 & \dfrac{1}{2} & 0 \\[2mm] 0 & 0 & 0 & 0 & 0 & \dfrac{1}{2} \end{bmatrix} \begin{bmatrix} \varepsilon_{11} \\ \varepsilon_{22} \\ \varepsilon_{33} \\ \gamma_{12} \\ \gamma_{13} \\ \gamma_{23} \end{bmatrix} \tag{1.194}
$$

$$
\begin{bmatrix} \varepsilon_{11} \\ \varepsilon_{22} \\ \varepsilon_{33} \\ \gamma_{12} \\ \gamma_{13} \\ \gamma_{23} \end{bmatrix} = \frac{1}{E} \begin{bmatrix} 1 & -v & -v & 0 & 0 & 0 \\ -v & 1 & -v & 0 & 0 & 0 \\ -v & -v & 1 & 0 & 0 & 0 \\ 0 & 0 & 0 & 2(1+v) & 0 & 0 \\ 0 & 0 & 0 & 0 & 2(1+v) & 0 \\ 0 & 0 & 0 & 0 & 0 & 2(1+v) \end{bmatrix} \begin{bmatrix} \sigma_{11} \\ \sigma_{22} \\ \sigma_{33} \\ \tau_{12} \\ \tau_{13} \\ \tau_{23} \end{bmatrix} \tag{1.195}
$$

$$
G = \frac{E}{2(1+v)} \tag{1.196}
$$

1.3.4 Dynamics of a Flexible Beam

Flexible beams can be represented with continuous dynamic systems characterized by an infinite number of degrees of freedom and governed by nonlinear coupled partial differential equations, and approximations with discrete models can be used for practical analysis and synthesis. Among the most common approaches, we can mention the assumed mode, finite element, and lumped parameters methods; therefore, we can use different approaches to study the kinematics and dynamics of free flexible beams and constrained beams, especially for interesting simplified constrains, i.e., fixed revolute and prismatic ends.

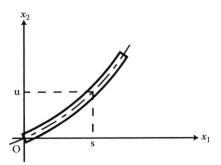

FIGURE 1.26
Bending beam described in a local reference system.

The movement of a flexible beam can be decomposed in a rigid and deforming movement. Because rigid motions were discussed in section 1.3.2, we concentrate on the deforming action and, among the modalities of deformation, privileging flexional vibrations for the geometric nature of the beam. Some reasonable hypotheses are assumed to limit the field of analysis to a class of interesting problems from an applicative viewpoint; particularly, the bending of the flexible beam is supposed to be described using a linear elastic theory. Thus, every flexible beam considered is constituted by elastic homogeneous and isotropic material, the transversal plane sections remain plane after the deformation, the transversal sections have two symmetry axes, and therefore the center of shear coincides with the centroid of the section, and the transversal actions do not produce torsion. These hypotheses are reasonable for compliant systems with small speed and negligible centrifugal actions, and the effect of shear should be considered for high vibrational speed and, in particular cases, some other assumptions can be removed to improve the beam's ability for modelization. Furthermore, the maximum bending should be an order of magnitude less then the length of the beam (Cannon, Schmitz, 1984), and the section rotation can be neglected when the number of assumed modes or the number of finite elements is not too high. Therefore, under the proposed premises, the dynamic Euler–Bernoulli equation can be introduced for a compliant beam with length l (Figure 1.26) in a point on its neutral axis (Meirovitch, 1986), describing the free transversal vibrations in the reference (0) with (1.197), where $u(s,t)$ represents the transversal displacement of a material point with a coordinate s on the neutral axis at the time instant t, and the bending stiffness $EI(s)$ is composed of the product of the elasticity modulus E and the moment of inertia $I(s)$ of the transversal section, ρ is the density of the material, and $A(s)$ is the area of the transversal section.

$$\frac{\partial^2}{\partial s^2}\left(EI(s)\frac{\partial^2 u(s,t)}{\partial s^2}\right)+\rho A(s)\frac{\partial^2 u(s,t)}{\partial t^2}=0 \qquad (1.197)$$

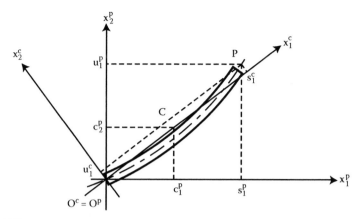

FIGURE 1.27
Local reference systems (p) and (c), respectively, for pinned and clamped boundary conditions, where C is the center of mass and P is a generic point on the neutral axis of the beam.

Proper boundary conditions, where s is equal to zero and l, and proper initial conditions, when t is equal to zero, should be set to obtain a unique solution for the motion equation (1.197) in classical mechanics, specifically, two initial conditions should be set involving the transversal displacement and its first-order time derivative because the vibrational motion is governed by a second-order differential equation with respect to the time variable t. Further, four boundary conditions should be set involving the transversal displacement and its space derivatives to the third order because the same equation is of the fourth order with respect to the space variable s. The boundary conditions can be classified in two categories, depending on whether they impose geometrical constraints or torque and force balances, and are called, respectively, geometrical (or imposed) and natural (or dynamic) constraints. Geometrical boundary conditions produce relations that regard the bending $u(s,t)$ or its tangent, represented by the partial derivative of $u(s,t)$ with respect to s, whereas dynamic boundary conditions affect the moment or the shearing action indicated in (1.198). Commonly, in models of compliant kinematic chain, the boundary conditions at the base of the beam, which is the powered side, are chosen between pinned and clamped ends. The pinned end is described by (1.199), and its reference frame (0^c) is characterized by an axis x_0 always tangential to the compliant beam (Figure 1.27), whereas the clamped end is represented by (1.200), where J_a is the total inertia of the actuating system at the base of the beam, and the reference system (0^p) has the x_0 axis always passing through the center of mass (Figure 1.27). At the other end of the beam, often the proposed boundary conditions are free end and mass end; in the first case, the moment and the shearing action are set to zero, obtaining (1.201), whereas in the latter case the boundary constraint is described by (1.202), where M_L and J_L are the mass and the inertia moment, respectively, of the load on the end of the beam.

$$EI(s)\frac{\partial^2 u(s,t)}{\partial s^2}; \quad \frac{\partial}{\partial s}\left[EI(s)\frac{\partial^2 u(s,t)}{\partial s^2}\right] \tag{1.198}$$

$$\left[u(s,t)\right]_{s=0} = 0; \quad \left[\frac{\partial u(s,t)}{\partial s}\right]_{s=0} = 0 \tag{1.199}$$

$$\left[u(s,t)\right]_{s=0} = 0; \quad \left[EI(s)\frac{\partial^2 u(s,t)}{\partial s^2}\right]_{s=0} = J_a\left[\frac{\partial^2}{\partial t^2}\left(\frac{\partial u(s,t)}{\partial s}\right)\right]_{s=0} \tag{1.200}$$

$$\left[EI(s)\frac{\partial^2 u(s,t)}{\partial s^2}\right]_{s=l} = 0; \quad \left[\frac{\partial}{\partial s}\left(EI(s)\frac{\partial^2 u(s,t)}{\partial s^2}\right)\right]_{s=l} = 0 \tag{1.201}$$

$$\left[\left[EI(s)\frac{\partial^2 u(s,t)}{\partial s^2}\right]_{s=l} = -J_L\left[\frac{\partial^2}{\partial t^2}\left(\frac{\partial u(s,t)}{\partial s}\right)\right]_{s=l}\right.$$
$$\left[\left[\frac{\partial}{\partial s}\left(EI(s)\frac{\partial^2 u(s,t)}{\partial s^2}\right)\right]_{s=l} = M_L\left[\frac{\partial^2 u(s,t)}{\partial t^2}\right]_{s=l}\right. \tag{1.202}$$

Some approximations can be introduced in the continuum model pro-posed to obtain numerical results, i.e., the separation of temporal and spatial variables in the expression of $u(s,t)$ (Meirovitch, 1967). In fact, the Cauchy problem can be decomposed into a boundary problem and an initial value problem using this property.

As a first example, we can a consider a compliant beam with length l, con-stant mass density ρ, uniform section A, and uniform bending stiffness EI, connected to a revolute joint animated by an actuation system and a load on the other end with mass M_L and moment of inertia J, as shown in Figure 1.28.

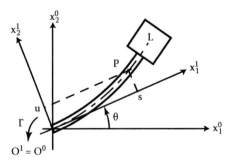

FIGURE 1.28
Scheme of a rotating compliant beam.

To describe the dynamics of the beam, a fixed reference frame (0) is selected with the axis x_3 coinciding with the axis of the revolute joint and with the axes x_1 and x_2 in the motion plane of the beam. On the other hand, a mobile reference frame (1) is set on the beam with the axis x_3^1 coinciding with the axis x_3 and having an axis x_1^1 always tangential to the beam. Finally, $\theta(t)$ is the angle describing the relative motion of (1) with respect to (0), and $\Gamma(t)$ is the torque applied by the actuator system of the revolute joint to obtain a rotation $\theta(t)$. Therefore, the equation of bending vibrations $u(s,t)$ can be described by (1.203) in the spatial domain limited by the length of the bar with camped end/mass end, or, with dimensionless variables (1.204). The bending equation becomes (1.205), which can be further simplified with a free-end boundary condition (avoiding the mass end) only when the inertia of the load is negligible compared with the inertia of the beam.

$$
\begin{cases}
EI\dfrac{\partial^4 u(s,t)}{\partial s^4} + \rho A \dfrac{\partial^2 u(s,t)}{\partial t^2} = 0 \\[3mm]
\left[u(s,t)\right]_{s=0} = 0; \qquad \left[EI(s)\dfrac{\partial^2 u(s,t)}{\partial s^2}\right]_{s=l} = -J_L\left[\dfrac{\partial^2}{\partial t^2}\left(\dfrac{\partial u(s,t)}{\partial s}\right)\right]_{s=l} \\[3mm]
\left[\dfrac{\partial u(s,t)}{\partial s}\right]_{s=0} = 0; \qquad \left[\dfrac{\partial}{\partial s}\left(EI(s)\dfrac{\partial^2 u(s,t)}{\partial s^2}\right)\right]_{s=l} = M_L\left[\dfrac{\partial^2 u(s,t)}{\partial t^2}\right]_{s=l}
\end{cases}
\tag{1.203}
$$

$$
\tilde{u}(s,t) = \frac{u(s,t)}{l} \qquad \eta = \frac{s}{l} \qquad \tau = \frac{tU_g}{l} \qquad U_g = \frac{1}{l}\sqrt{\frac{EI}{\rho A}}
\tag{1.204}
$$

$$
\begin{cases}
\dfrac{\partial^4 \tilde{u}(\eta,\tau)}{\partial \eta^4} + \dfrac{\partial^2 \tilde{u}(\eta,\tau)}{\partial \tau^2} = 0 \qquad 0 < \eta < 1 \\[3mm]
\left[\tilde{u}(\eta,\tau)\right]_{\eta=0} = 0; \qquad \left[\dfrac{\partial^2 \tilde{u}(\eta,\tau)}{\partial \eta^2}\right]_{\eta=1} = -\dfrac{J_L}{\rho A l^3}\left[\dfrac{\partial^2}{\partial \tau^2}\left(\dfrac{\partial \tilde{u}(\eta,\tau)}{\partial \eta}\right)\right]_{\eta=1} \\[3mm]
\left[\dfrac{\partial \tilde{u}(\eta,\tau)}{\partial \eta}\right]_{\eta=0} = 0; \qquad \left[\dfrac{\partial^3 \tilde{u}(\eta,\tau)}{\partial \eta^3}\right]_{\eta=1} = \dfrac{M_L}{\rho A l}\left[\dfrac{\partial^2 \tilde{u}(\eta,\tau)}{\partial \tau^2}\right]_{\eta=1}
\end{cases}
\tag{1.205}
$$

Starting from expression (1.205), some numerical results can be obtained with a proper discretization, and therefore, the solution is assumed to be variable separable (1.206), and the partial differential equation takes the form of (1.207), where the shape functions $\Psi(\eta)$ depend only on the spatial variable, and the generalized coordinates $q_i(\eta)$ depend only on the temporal variable; furthermore, both the members of (1.207) should be a real negative constant (Meirovitch, 1967), as shown in (1.208 and 1.209) and the boundary conditions are reduced to (1.210).

$$\tilde{u}(\eta, \tau) = \psi(\eta)q_f(\tau) \tag{1.206}$$

$$\frac{1}{q_f(\tau)}\frac{d^2 q_f(\tau)}{d\tau^2} = -\frac{1}{\psi(\eta)}\frac{d^4 \psi(\eta)}{d\eta^4} \tag{1.207}$$

$$\frac{d^2 q_f(\tau)}{d\tau^2} + \omega^2 q_f(\tau) = 0 \tag{1.208}$$

$$\frac{d^4 \psi(\eta)}{d\eta^4} - \omega^2 \psi(\eta) = 0; \quad 0 < \eta < 1 \tag{1.209}$$

The problem of searching for values of the parameter ω^2, avoiding trivial solutions of (1.209) and being subjected to boundary conditions (1.210), produces a transcendental equation in ω^2 that is recognized to be an eigenvalues/eigenvectors problem, and the $\Psi(\eta)$ associated with these values are eigenfunctions of the system (Meirovitch, 1967). The result of this characteristic, or frequency, equation is an infinite set of characteristic values, and their square roots are the natural frequencies ω_i, with the associated natural modes $\Psi_i(\eta)$.

$$\begin{aligned}
\left[\psi(\eta)\right]_{\eta=0} &= 0; & \left[\frac{\partial^2 \psi(\eta)}{\partial \eta^2}\right]_{\eta=1} &= \frac{J_L \omega^2}{\rho A l^3}\left[\frac{\partial \psi(\eta)}{\partial \eta}\right]_{\eta=1} \\
\left[\frac{\partial \psi(\eta)}{\partial \eta}\right]_{\eta=0} &= 0; & \left[\frac{\partial^3 \psi(\eta)}{\partial \eta^3}\right]_{\eta=1} &= -\frac{M_L \omega^2}{\rho A l}\left[\psi(\eta)\right]_{\eta=1}
\end{aligned} \tag{1.210}$$

One of the four constants of integration cannot be determined unambiguously; consequently, the shape of an eigenfunction is recognized but not its amplitude, which can be conventionally determined with a normalization process, obtaining the normal natural modes of bending vibration.

A similar discussion can be held on another interesting class of joints—prismatic joints—and, referring to Figure 1.29, a translating compliant beam composed of homogeneous and uniform material is examined. Then, a fixed-reference frame (0) is set with the axes x_3 and x_1 on the motion plane of the beam, and with x_3 along the axis of the prismatic joint, analogously a mobile frame (1) is connected to the beam with the origin on the internal end and with the axes parallel to the corresponding axes of the fixed frame (0). The beam has a drift translational speed equal to $U(t)$ along the direction x_3, the length of the external part of the beam is equal to $l(t)$ and changes from a value between zero and l_0, and the internal part of the beam is supposed to be not subject to bending vibrations during the motion. Finally, $F(t)$ is the

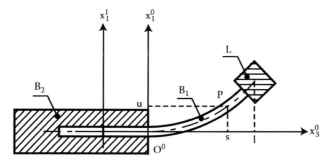

FIGURE 1.29
Scheme of a compliant beam B_1 endowed with a load L and translating with respect to a frame B_2, with clamped end mass and boundary conditions; P is a generic point on the neutral axis and its position with respect to the fixed frame (0) on the body B_2 is identified by the couple (s, u).

moving force acting on the beam along the axis x_3, and M_L and J_L are, respectively, the mass and the moment of inertia of the load on the external end of the beam, and therefore the Euler–Bernoulli equation of the free vibration can be represented by (1.211) on the external part of the translating beam.

$$\begin{vmatrix} EI\dfrac{\partial^4 u(s,t)}{\partial s^4}+\rho A\left(\dfrac{\partial^2 u(s,t)}{\partial t^2}+2U\dfrac{\partial^2 u(s,t)}{\partial s\partial t}+U^2\dfrac{\partial^2 u(s,t)}{\partial s^2}+\dfrac{dU}{dt}\dfrac{\partial u(s,t)}{\partial t}\right)=0 \\[3mm] \left[u(s,t)\right]_{s=0}=0;\quad \left[EI\dfrac{\partial^2 u(s,t)}{\partial s^2}\right]_{s=l(t)}=-J_L\left[\dfrac{\partial^2}{\partial t^2}\left(\dfrac{\partial u(s,t)}{\partial s}\right)\right]_{s=l(t)} \\[3mm] \left[\dfrac{\partial u(s,t)}{\partial s}\right]_{s=0}=0;\quad \left[EI\dfrac{\partial^3 u(s,t)}{\partial s^3}\right]_{s=l(t)}=M_L\left[\dfrac{\partial^2 u(s,t)}{\partial t^2}\right]_{s=l(t)} \end{vmatrix} \tag{1.211}$$

The expression (1.211) describes an interesting and complex moving boundary problem because the length $l(t)$ of the external part of the beam is time dependent without a known general analytical solution. In the literature there are some finite difference numerical algorithms on variable or fixed grids, whereas, in the following text, we expose a well-known approximate approach with a reduced computational strength, after introducing dimensionless variables (1.212), to obtain a dimensionless form of the proposed differential equation problem as in (1.213), and then the assumption of separable variables points to a simplified problem.

$$\begin{cases} \tilde{u}(s,t)=u(s,t)/l;\quad \eta=s/l^0 \\[2mm] \tau=t/(l^0/U_g);\quad U_g=(l/l_0)\sqrt{(EI)/(\rho A)} \end{cases} \tag{1.212}$$

$$
\begin{cases}
\dfrac{\partial^4 \tilde{u}(\eta,\tau)}{\partial \eta^4} + \dfrac{\partial^2 \tilde{u}(\eta,\tau)}{\partial \tau^2} + 2\dfrac{U}{U_g}\dfrac{\partial^2 \tilde{u}(\eta,\tau)}{\partial \eta \partial \tau} + \left(\dfrac{U}{U_g}\right)^2 \dfrac{\partial^2 \tilde{u}(\eta,\tau)}{\partial \eta^2} + \left(\dfrac{d}{d\tau}\left(\dfrac{U}{U_g}\right)\right)\dfrac{\partial \tilde{u}(\eta,\tau)}{\partial t} = 0 \\[4mm]
\left[\tilde{u}(\eta,\tau)\right]_{\eta=0} = 0; \qquad \left[\dfrac{\partial^2 \tilde{u}(\eta,\tau)}{\partial \eta^2}\right]_{\eta=\frac{l(t)}{l^0}} = -\dfrac{J_L}{\rho A l^3}\left[\dfrac{\partial^2}{\partial \tau^2}\left(\dfrac{\partial \tilde{u}(\eta,\tau)}{\partial \eta}\right)\right]_{\eta=\frac{l(t)}{l^0}} \\[4mm]
\left[\dfrac{\partial \tilde{u}(\eta,\tau)}{\partial \eta}\right]_{\eta=0} = 0; \qquad \left[\dfrac{\partial^3 \tilde{u}(\eta,\tau)}{\partial \eta^3}\right]_{\eta=\frac{l(t)}{l^0}} = \dfrac{M_L}{\rho A l}\left[\dfrac{\partial^2 \tilde{u}(\eta,\tau)}{\partial \tau^2}\right]_{\eta=\frac{l(t)}{l^0}}
\end{cases}
\tag{1.213}
$$

$$
\psi(\eta)\dfrac{d^2 q_f(\tau)}{d\tau^2} + 2\dfrac{U}{U_g}\dfrac{d\psi(\eta)}{d\eta}\dfrac{dq_f(\tau)}{d\tau} =
$$

$$
-\left(\dfrac{d^4\psi(\eta)}{d\eta^4} + \left(\dfrac{U}{U_g}\right)^2 \dfrac{d^2\psi(\eta)}{d\eta^2} + \dfrac{d}{d\tau}\left(\dfrac{U}{U_g}\right)\dfrac{d\psi(\eta)}{d\eta}\right) q_f(\tau)
\tag{1.214}
$$

The equation (1.214) is separable only when speed U of the beam is negligible compared with the group speed U_g of the beam and the translational acceleration of the beam is negligible. In these conditions, expression (1.214) is approximated to (1.207) and can be solved as just shown.

The expounded results can be generalized to obtain models of multibody compliant systems, adopting a matricial treatment analogous to the approach just proposed for the kinematics and dynamics of multibody systems connected with noncompliant prismatic and revolute lower pairs. We do not elaborate on this subject, which has been widely examined in several monographic works (Lobontiu, 2003), (Howell, 2001).

1.3.5 Dynamics with Compliant Constraints

In this section we resume the topic discussed in section 1.2.6 suggesting a dynamic viewpoint, which allows us to point out some useful properties for the evaluation of compliant constraints. For this purpose, proper indicators can be proposed with dynamic characteristics that allow us to compare different types of couplings with each other, facilitating the design and realizing activities of micro/mini devices for actuation. In the scientific literature, a considerable effort was made to identify interesting indicators, and in the strictly technical domain, mechanical stiffness is often emphasized as the unique characteristic parameter.

A particularly consistent and complete work can be identified in (Lobontiu, 2003) as a referral guidance for this section, and therefore, this reference, together with other monographs—i.e., (Howell, 2001), (Smith, 2000)—can be used for the purpose of a comprehensive investigation. As just proposed in

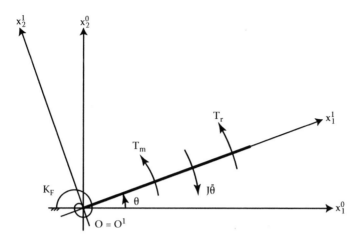

FIGURE 1.30
Beam *B* rotating around *O* by means of a concentrated compliance.

section 1.2.5, a fixed body *T* (the frame) is taken with a body *B*, which is mobile with respect to *T*, and the dynamic pair *B-T* is examined, observing the relative movements of its members; specifically we are interested in the dynamic pairs produced through a localized compliance, which is able to transform a mechanical force into a displacement by means of a strain. For the sake of simplicity and without an excessive loss of applicability, we refer only to the linear elastic strains of isotropic and homogeneous materials, as just proposed in section 1.3.3. Thus, a compliance can be represented as a ratio between a relative displacement of *B* and the originating force. Moreover, in the absence of hysteretic phenomena, the stiffness of the pair can be described exactly by its reciprocal (i.e., the ratio between applied force and resulting displacement). This simple definition can be properly extended with the principle of virtual work, introducing the concepts of virtual force and virtual displacement, which allow the expansion of the domain for stiffness and compliance until it includes the whole pair *B-T* and not only the local application point of a force.

As a first illustrative example, a rigid beam *B* (Figure 1.30) is observed while it moves with respect to the frame *T*, and its mobility is allowed by a revolute constraint produced with a compliance C_F concentrated in the hinge and identified in the figure by its reciprocal, the stiffness K_F. The subscript *F* points out the functional aspect held by these parameters as they allow the desired revolute mobility. Therefore, the dynamics of this beam can be represented by (1.215) with respect to a fixed reference frame, where *J* is an indicator of the inertia of the beam, θ describes the revolute degree of freedom, T_r is the resistance torque reduced to the motor axis, and T_m is the motor torque.

$$T_m - T_r - K_F\theta - J\ddot{\theta} = 0 \qquad (1.215)$$

$$T_{\text{eff}} - K_F \theta = 0 \tag{1.216}$$

The expression (1.215) is written in a scalar form for the sake of simplicity, and it points out the planar dynamics of the observed revolute joint. Then, the equation (1.216) is obtained by gathering the torques that are able to generate rotations, resist to rotations, and accumulate potential energy in such a way as to highlight the effective component T_{eff} for the rotation. In fact, equation (1.216) shows how the compliance C_F represents an indicator of ability to produce rotations θ caused by effective torques T_{eff}. Therefore, the functional compliance C_F is an indicator of the functionality of the hinge constraint realized with a compliance, and it is, for this reason, the principal indicator of its performance. Furthermore, some secondary compliances can be present at the same time with the functional compliance C_F, and hence, remaining in the plain for expository simplicity, they produce relative movements of the position of the real center of the hinge O^1 with respect of the theoretical center O, contributing to inaccuracies in the movement of the beam B; in fact, it results in a movement of the rotation axis. Then, the functional compliance is assumed to act between the beam B and a fictitious configuration (Figure 1.31) to maintain uncoupled secondary and functional movements as just proposed in section 1.2.6 for kinematic pairs with different concentrated compliances.

With the same sign convention suggested in (1.215), the motor actions have a versus concordant with the axes of the reference frame, whereas the resistance actions have an opposite versus. Further, the versus of the functional movement θ and secondary motions x_1 and x_2 is taken as positive when concordant with the axes of the reference frame, and therefore the expression (1.217) can represent the dynamics of the beam B, where the secondary

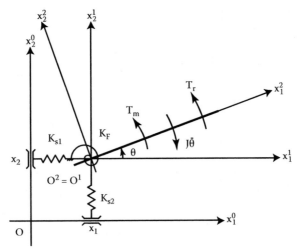

FIGURE 1.31
Rotating beam B by means of a concentrated compliance in the presence of secondary compliances.

motions are produced by the resistance forces F_{r1} and F_{r2} reduced at the hinge O^2. Equation (1.217) can be expressed in a simplified form by introducing the concept of effective action in this situation. Therefore, the functional aspects of C_F are highlighted as just shown in (1.215), and the ability of K_{s1} and K_{s2} to represent the accuracy of the movement are pointed out, correlating them to the insensibility of secondary motions in the presence of effective mechanical loads (1.219).

$$\begin{cases} T_m - T_r - K_F\theta - J\ddot{\theta} = 0 \\ -F_{r1} - K_{s1}x_1 - m\ddot{x}_1 = 0 \\ -F_{r2} - K_{s2}x_2 - m\ddot{x}_2 = 0 \end{cases} \tag{1.217}$$

$$T_{eff} - K_F\theta = 0; \quad F_{eff1} - K_{s1}x_1 = 0; \quad F_{eff2} - K_{s2}x_2 = 0 \tag{1.218}$$

Summarizing, in the dynamics of compliant joints, a high functional compliance is required to assure the desired mobility, and a high secondary stiffness is essential to respect kinematic constraints and to allow a sufficient trajectory precision. These constraints should be substantially provided with selective compliances. The study of techniques for the realization of these special connections opened the way to authentic and specific research possibilities, producing interesting results for applications to meso- and micro-actuators.

The relations between displacements and actions can be coupled, hence focusing only on aspects associated with stiffness and compliance. Under wide hypotheses (Lobontiu, 2003), the dynamics of the moving body B can be represented with a symmetric matrix (1.219) in which K_{pi} are the parasitic stiffness of the coupling that correlates functional and secondary movements.

$$\begin{bmatrix} T_{eff} \\ F_{eff1} \\ F_{eff2} \end{bmatrix} = \begin{bmatrix} K_F & K_{p1} & K_{p2} \\ K_{p1} & K_{s1} & K_{ps} \\ K_{p2} & K_{ps} & K_{s2} \end{bmatrix} \begin{bmatrix} \theta \\ x_1 \\ x_2 \end{bmatrix} \tag{1.219}$$

The observations expounded up to this point can be generalized to a three-dimensional space in which the bodies B and T of the dynamic pair are rigid, and consequently their relative position in the space can be identified by at least six degree of freedom with six associated, effective mechanical loads. These consist of three independent relative rotations around the axes of a reference frame (0) on T with three independent relative translations along the same axes, obtaining, in the case of a revolute pair, a relation between effective actions and functional and secondary movements as shown in (1.220), where K_F is the functional stiffness, K_p is the column vector of the parasitic stiffness, K_s is the symmetric matrix of the secondary stiffness, L_{effs} are the effective secondary loads, and u_s are the secondary displacements.

$$\left[\frac{T_{\mathit{eff}}}{L_{\mathit{effS}}}\right]=\left[\begin{array}{c|c}K_F & K_p^T \\ \hline K_p & K_s\end{array}\right]\left[\frac{\theta}{u_s}\right]$$
(1.220)

With regard to the physical meaning of the relation between effective loads and movements, we prefer to use the inverse relation of (1.220) shown in (1.221), as proposed also in Lobontiu (2003), where the matrix C of the compliance is pointed out.

$$\left[\frac{\theta}{u_s}\right]=\left[\begin{array}{c|c}C_F & C_p^T \\ \hline C_p & C_s\end{array}\right]\left[\frac{T_{\mathit{eff}}}{L_{\mathit{effS}}}\right]$$
(1.221)

In (1.221), the characteristics of the compliance matrix are indicated by proper components, such as C_F, which represents functional compliance and should assume a high value to allow the desired mobility; the submatrix C_S, which describes the sensitivity of secondary movements to effective secondary actions and should be composed of elements with a low absolute value to assure the respect of the desired kinematic constraints and thus the precision of the realized trajectory; the submatrix C_P and its transpose C_P^T, which identify, respectively, the correlation between secondary movements and functional mechanical action and the correlation between functional movement and secondary effective actions, and therefore should be composed by elements with a low absolute value to ensure the respect of the desired motion profile, i.e., the desired temporal evolution for the functional parameter θ.

It is interesting to remember that the symmetry of the matrix K, and of the matrix C when K is invertible, assure their diagonalizability, and this allows us to obtain a set of independent coordinates that describe unrelated (through the stiffness matrix) functional and secondary movements. Unfortunately, the connection persists because the new set of effective mechanical loads, with inertial actions present, bring about internally the correlation between secondary and functional movements. On the other hand, when the inertial effects are negligible either because of reduced inertias or reduced accelerations, the secondary movements can always be separated from the functional movements. A practical consequence of a similar situation consists of the possibility of compensating systematic production and assembly errors with a proper dynamic calibration, obtaining considerable increments in kinematic and dynamic performances. Therefore, a trade-off between production precision and effort in the calibration phase should be appraised to gain satisfactory results without excessive efforts. When functional and secondary movements are uncoupled, as in (1.217), the kinematics of the system can be represented as in section 1.2.5 with a corresponding matricial description of the dynamics of the joint to allow the employment of action matrix Φ, as proposed in section 1.3.2. Thus, the expression (1.222) can represent the dynamic relation in a form that allows the use of powerful computational algorithms, but it introduces difficulties in understanding the physical meaning of each scalar term of the matrix $\underline{K}$.

$$\phi = \underline{K} \cdot M \tag{1.222}$$

A simple observation on the matrix $\underline{K}$ can be made by examining the structure of the matrixes M and Φ (1.223) and trying to obtain the structure of the matrix $\underline{K}$ with a simple inversion of the matrix M and with a successive block product between Φ and M^{-1} (1.224). The result is that $\underline{K}$ is not generally symmetric.

$$\left[\begin{array}{c|c} T & {}^3F \\ \hline -{}^3F^T & 0 \end{array}\right] = \underline{K} \left[\begin{array}{c|c} R & T_M \\ \hline 0 & 1 \end{array}\right] \tag{1.223}$$

$$\underline{K} = \left[\begin{array}{c|c} T & {}^3F \\ \hline -{}^3F^T & 0 \end{array}\right] \left[\begin{array}{c|c} R^T & -R^T T_M \\ \hline 0 & 1 \end{array}\right] = \left[\begin{array}{c|c} TR^T & -T\left(R^T T_M\right) + {}^3F \\ \hline -{}^3F^T R^T & {}^3F^T\left(R^T T_M\right) \end{array}\right] \tag{1.224}$$

As just mentioned, a method to realize compliant constraints in an approximate way can consist of producing compliances on the beam near the coupling joint or even on the whole beam. The compliances realized near the coupling joint are usually executed by means of proper profiles of the beam obtained with section narrowings to gain the desired functional mobility, minimizing the presence of secondary motions, whereas compliances extended to the whole beam are generated using adequate materials that are able to achieve a suitable level of structural strain sagging against the presence of mechanical load. In the latter situation, the beam section can be constant along the whole extension of the neutral axis carrying out the task of containment for secondary movements. The study of the behavior of a compliant mechanism with compliant hinges realized in this way can be subdivided into three sequential phases: the definition of a pseudo-rigid-body model (Figure 1.32), i.e., a model of the compliant mechanism based on the substitution of compliant elements with ideal hinges and associated twisting springs connected to each other with rigid links (Howell, 2001); the determination of the hinge compliances; and analysis of the pseudo-rigid-body model obtained.

The study of a mini compliant mechanism is proposed in the following text, using a schematic representation formed by two articulated quadrilateral mechanisms provided with hinges realized from a fixed constraint introducing proper functional compliances. These mechanisms are composed of a three-dimensional linear guide used for experimental investigations to improve the design and prototype of the whole system (Table 1.5).

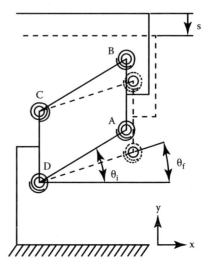

FIGURE 1.32
Quadrilateral mechanism for static experiments.

TABLE 1.5

Geometrical Characteristics of the First
Quadrilateral Mechanism

Geometrical Parameter	Assumed Value
AB	13.0 mm
BC	30.3 mm
θ_i	30°

The first trial (Figure 1.32) is conducted by means of a specimen compression until the yield point, and the descent of the test bench bar is kept quasi-static to preserve the equilibrium of the system during the whole process. A first geometrical examination of this experimental setup shows a correlation between the final position of the bar AD, identified by the angular variable θ_f and the imposed vertical displacement s (1.225). Therefore, fixing the other geometrical parameters, the angular displacement $\Delta\theta$ of the bar AD can be computed as a difference between its initial and final positions θ_i and θ_f. Further, the particular parallelogram structure of the articulated mechanism implies the same relative angular displacement for every hinge. The same geometrical analysis also shows that the beam AB is subjected to a vertical displacement, and thus a film of lubricant was used to reduce the metal–metal sliding friction.

$$\theta_f = \arcsin\left(\sin\theta_i - \frac{s}{d}\right) \qquad (1.225)$$

After a geometrical definition of the mechanical model, the determination of the equivalent spring stiffness used in the pseudo-rigid-body model allows a description of the dynamic behavior for the four bar mechanisms; this stiffness is strictly related to the geometry of the proposed bending hinge, which consists, as shown in Figure 1.33, of a non-symmetric circular hinge with

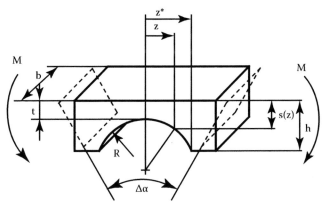

FIGURE 1.33
Nonsymmetric circular hinge.

TABLE 1.6

Geometrical Characteristics of the Proposed
Compliant Hinge

Geometrical Parameter	Assumed Value [mm]
t	0.4
b	50
R	20
h	4°

the parametric values listed in Table 1.6. Under this hypothesis, assuming a beam model with a variable section for the flexural hinge to allow the concentration of its properties on the neutral axis, the expression (1.226), or more explicitly, the expression (1.227), can represent the compliance of each hinge. Further, in many applicative situations, β and γ have negligible absolute values, i.e., the thickness t of the hinge and thickness h of the connected bodies are frequently negligible, compared with the radius R of the hinge. Hence, the simplified expression (1.228) can approximately represent the compliance described in (1.227) and, avoiding secondary motions, the associated stiffness can be calculated as the reciprocal of the estimated compliance.

The analyzed flexural hinges are also the zones of the mechanism with highest mechanical stress. Therefore, an evaluation of the maximal stress σ_{MAX} can be done by computing the bending torque M (1.226) and using the simplified expression (1.229), under proper hypotheses, i.e., linear and elastic material, De Saint Venant's assumptions, homogeneous and isotropic material, and so on.

$$\frac{\Delta\alpha}{M} = c = \int_{-z^*}^{+z^*} \frac{1}{EJ(z)} dz; \quad J(z) = \frac{b \cdot s(z)^3}{12} \qquad (1.226)$$

$$
\begin{cases}
c = \dfrac{12}{EbR^2} \cdot \dfrac{1}{\beta^{\frac{5}{2}}(\beta+2)^{\frac{5}{2}}} \cdot \left\{ \dfrac{\sqrt{\beta(\beta+2)}\sqrt{1-(1+\beta-\gamma)^2}}{2\gamma} \right. \\[4ex]
\qquad\qquad \cdot \left[\dfrac{\beta(\beta+1)(2+\beta-\gamma)}{\gamma} + 3\beta + 3 + 2\beta^2 \right] \\[4ex]
\left. \qquad\qquad + 3(\beta+1)\arctan\left[\sqrt{\dfrac{\beta+2}{\beta}} \, \dfrac{\gamma-\beta}{\sqrt{1-(1+\beta-\gamma)^2}} \right] \right\} \\[4ex]
\beta \equiv \dfrac{t}{2R}; \quad \gamma \equiv \dfrac{h}{2R}
\end{cases}
\tag{1.227}
$$

$$
c = \dfrac{9\pi\sqrt{R}}{\sqrt{2}Ebt^{\frac{5}{2}}}
\tag{1.228}
$$

$$
\sigma_{MAX} = \dfrac{2\sqrt{2}}{3\pi}\Delta\alpha E \sqrt{\dfrac{t}{r}}
\tag{1.229}
$$

Within a stress range that does not produce material yield, the pseudo-rigid-body model obtained can be examined with an energetic approach. The whole system can be dynamically compared with a prismatic joint consisting of a vertical spring characterized by a stiffness k_{vert}. Observing that the potential energy E_c stored in a single hinge is a quarter of the energy E_{mec} stored in the equivalent prismatic joint, the expression of k_{vert} can be deduced from (1.230), and the dynamics of the system can be described by the approximate dynamics of the prismatic joint (1.231).

$$
k_{vert} = \dfrac{4k \cdot \left[\theta_i - \arcsin\left(\sin\theta_i - \dfrac{s}{d}\right) \right]^2}{s^2}; \quad E_c = 0.5 \cdot k \cdot \Delta\theta^2; \quad E_{mec} = 0.5 k_{vert} s^2
\tag{1.230}
$$

$$
P = k_{vert} \cdot s
\tag{1.231}
$$

The model proposed in (1.231) should be validated and calibrated by means of experimental tests. Thus, a specimen was realized in Al 7075 T651 (Ergal), an aluminum alloy with optimal mechanical properties, using wire erosion technology in two passes: the first is a rough one that releases the residual

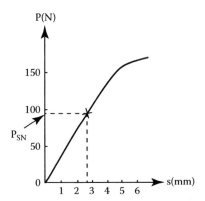

FIGURE 1.34
Experimental curve obtained with a quasi-static load process.

stresses in the material, whereas the latter is a finishing; finally the specimen was anodized for protection against corrosion.

The experimental curve represented in Figure 1.34 can be superimposed, in the linear domain, on the diagram of expression (1.231). Therefore, the proposed approximations do not limit the validity of the model. Further, the comparison, in the linear domain, with a least-squares line and finite element analyses shows a good correspondence between numerical and experimental models. Finally, the calculation of σ_{MAX} with (1.229) appears too conservative; in fact, the forecasted yield point is really lower than the experimental results; therefore, a wider measurement campaign should be performed on a congruous number of specimens to clear this incongruity.

Besides just described static, or quasi-static, tests, dynamic and fatigue analyses also were carried out with a four-bar mechanism analogous to the previous one provided with a vertical mass adequately dimensioned to be submitted to mechanical vibrations by means of an electrodynamic shaker rigidly connected to the link CD of the sample. Initially, the specimens are subjected to a frequency sweep to identify the dynamic characterization of the system, and then they are stressed with oligocyclic fatigue loads according to the forecasted applicative situations. Due to the displacement Δs_{in} imposed by the shaker to the link CD, a movement Δs_{out} of the link AB can be observed; assuming a Fourier expansion for Δs_{in}, the resulting output motion, Δs_{out} is not generally phased with the input motion, and hence the relative motion between CD and AB constitutes the dynamic load for the four-bar mechanism. The scheme of this second quadrilateral mechanism (Figure 1.35), with the geometrical characteristics listed in Table 1.7, is different from the previous one (Figure 1.34) that limits transversal accelerations, i.e., secondary motions, of the calibrated mass. In fact, the morphology of the second mechanism offers transversal motions with a lower order of magnitude than the previous sample, assuming the same relative displacement between AB and CD.

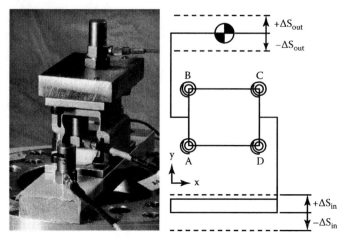

FIGURE 1.35
Four-bar mechanism for dynamic loads.

TABLE 1.7

Geometrical Characteristics of the Four-Bar
Mechanism for Dynamic Experiments

Geometrical Parameter	Assumed Value [mm]
AB	15
BC	26

The proposed system can be dynamicly associated with a simple prismatic joint realized, starting from a fixed connection and removing it with
a compliance; therefore, it can be described by a one-degree-of-freedom
model with a relative natural angular frequency ω_0 (1.232), or its natural frequency f_0. Then the equivalent mass can be valued as the sum between the
calibrated mass and the mass of the link AB, and the determination of the
equivalent stiffness k_{vert} can be achieved with a procedure similar to the one
used for quasi-static analysis. Knowing the stiffness of a single compliant
hinge (1.230), the dependence of k_{vert} on the displacement s is neglected for
the sake of simplicity by means of a proper choice of the displacement s as
a mean value of 1.5 mm in the presence of a maximum value of 2.5 mm.
Therefore, the behavior of the mechanism can be described by equations
(1.233) and (1.234), which represent, respectively, the modulus and the phase
of the transfer function of the system. The graphical representation of (1.233)
is depicted in Figure 1.36.

$$\omega_0 = \sqrt{\frac{k_{vert}}{m_e}} \qquad (1.232)$$

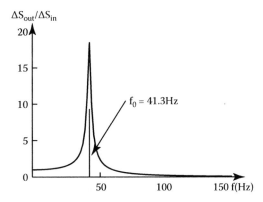

FIGURE 1.36
Modulus of the theoretical transfer function.

$$\frac{\Delta S_{out}}{\Delta S_{in}} = \sqrt{\frac{k_{vert}^2 + \left(2\pi f c\right)^2}{\left(k_{vert}^2 - 4m_e\pi^2 f^2\right)^2 + \left(2\pi f c\right)^2}} \qquad (1.233)$$

$$tg\left(\phi\right) = \frac{2\pi m_e c f}{k_{vert}^2 - 4\pi^2 k_{vert}m_e f^2 + 4\pi^2 c f^2} \qquad (1.234)$$

The experimental sweep was carried out within a 0–1000 frequency range with a maximum acceleration equal to 0.5 g. The transfer function, measured from the accelerometer on the translating plane of the shaker to the other accelerometer on the test mass on the specimen (Figure 1.35), is depicted in Figure 1.37 and compared with the forecasted diagram in the figure. The result is a good correspondence between the experimental and theoretical approaches, particularly in the peak at the frequency f_0 equal to 41.3 Hz, and

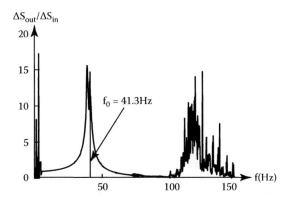

FIGURE 1.37
Modulus of the experimental transfer function.

therefore the model can be validated. The sample is realized in a metallic material, and hence the damping coefficient assumes a low value. In fact, calibration of the model should be implemented to further reduce the value of the coefficient c in (1.233); moreover, the presence of a low-frequency peak in Figure 1.37 can be attributed to the motion controller, whereas the high-frequency peak is associated with the second natural frequency of the system. However, a further study of this second peak was not performed because the application range for the four-bar compliant mechanism is expected to be at a lower frequency.

Oligocyclic fatigue tests were executed on the same specimens used for the frequency analysis, exposing them to a sinusoidal excitation of input and output bars with an amplitude of the movement equal to 2 mm and a frequency of 30 Hz over the expected lifetime of 12000 cycles; the theoretical expected results should be negligible according to the ASM Specialty Handbook for aluminium 7075 T651.

The two samples A and B were subjected, as listed in Table 1.8, to different cyclic movements with semiamplitudes Δs at a frequency equal to 30 Hz; specimen B showed a residual plasticization at the end of the test with a visible yield of the material, also confirmed by an additive analysis of both samples A and B. Each specimen supported the number of cyclic excitations, but they did not meet the forecasted damage, probably because of the electroerosion process in the presence of thin sections. However, further studies should be performed to confirm this intuition.

TABLE 1.8

Oligocyclic Fatigue Test on Two Specimens

Δs [mm]	A	B
1.0	12,000	16,800
1.3	12,000	—
1.6	—	9,900

1.4   Fundamental Notions of Interacting Fields

1.4.1   Electromagnetic Fields

Although the research on electromagnetic phenomena and their ability to generate mechanical interactions is really ancient, the first scientific results were from William Gilbert (Gilbert, 1600), who, in 1600, published *De Magnete*, a treatise on the principal properties of a magnet—the presence of two poles and the attraction of opposite poles. In 1750, John Michell (Michell, 1751), and then in 1785, Charles Coulomb (Coulomb, 1785–1789) developed a quantitative model for these attraction forces discovered by Gilbert. In 1820, Oersted (Oersted, 1820) and, independently, Biot and Savart (1820), discovered the mechanical interaction between an electric current and a magnet. In 1821, Michael Faraday (Faraday, 1821) discovered the moment of the magnetic force, and Ampere (Ampère, 1820) observed a magnetic equivalence of an electric circuit. In 1876, Rowland (Rowland, 1876) demonstrated that the magnetic effects due to moving electric charges are equivalent to the effects due to electric currents. In 1831, Faraday (Faraday, 1832) and, independently, Joseph Henry (Henry, 1831) discovered the possibility of generating an electric current with a variable magnetic field. In 1865, James Clerk Maxwell (Maxwell, 1865) developed the first comprehensive theory on electromagnetic fields, introducing the modern concepts of electromagnetic waves. Though not considered so by his contemporaries, his theory was revolutionary as it did not require the presence of a media, the ether, to propagate the electromagnetic field. Later, in 1887, Hertz (Hertz, 1887) experimentally demonstrated the existence of electromagnetic waves. It opened up the possibility of neglecting the ether and, in 1905, aided the formulation of the theory of relativity by Albert Einstein (Einstein, 1905). While electromagnetic physics was being studied, new advances in electromagnetic motion systems were developed: the earliest experiments were undertaken by M. H. Jacobi (Jacobi, 1835) in 1834 (moving a boat); the first complete electric motor was built by Antonio Pacinotti (Pacinotti, 1865) in 1860, the first induction motor was invented and analyzed by G. Ferraris in 1885 (Ferraris, 1888) and, later, independently, by N. Tesla (Tesla, 1888), who registered a patent in the United States in 1888. Many other macro electromagnetic motors were later developed, and research in this field remained very active.

Research in the field of electric micro-actuators conventionally started in 1960 with W. McLellan, who developed a 1/64th-in. cubed micromotor in answer to a challenge by R. Feynman. Since then, an indefinite number of inventions and prototypes have been presented to the scientific community, patented, and marketed. Thus, outlining the history of microelectromagnetic actuators is an almost impossible task. However, by observing the new technologies produced, we are able to trace the key inventions and ideas in the formation of actual components: i.e. the isotropic and anisotropic etching techniques that were developed in the 1960s generated bulk micromachining in 1982, and sacrificial layer techniques, also developed in the 1960s, generated surface micromachining in 1985. Some more recent technologies include silicon fusion bonding, LIGA technology, and microelectro and discharge machining.

1.4.2 Piezoelectric Materials

The discovery of piezoelectric phenomena is due to Pierre and Jacque Curie, who experimentally demonstrated the connection between crystallographic structure and macroscopic piezoelectric phenomena and published their results in 1880 on the direct piezoelectric effect (from mechanical energy to electric energy). The next year, Lippman demonstrated, theoretically, the existence of an inverse piezoelectric effect (from electric energy to mechanical energy). The Curie brothers then added value to Lippman's theory with new experimental data, opening the way for piezoelectric actuators. After tenacious theoretical and experimental work in the scientific community,

Voigt synthesized all the knowledge in the field using a properly tensorial approach, and in 1910 published a comprehensive study on piezoelectricity. The first application of a piezoelectric system, following the historical discovery of the Curie brothers, was a sensor (direct effect), a submarine ultrasonic detector developed by Lengevin and French in 1917. Between the first and second World Wars, many applications of natural piezoelectric crystals appeared, the most important being ultrasonic transducers, microphones, accelerometers, bender element actuators, signal filters, and phonograph pick-ups. During the Second World War, research was stimulated in the United States, Japan, and Soviet Union, resulting in the discovery of piezoelectric properties of piezoceramic materials exhibiting dielectric constants up to 100 times higher than common-cut crystals. The research on new piezoelectric materials continued during the second half of the twentieth century with the development of barium titanate and lead zirconate titanate piezoceramics. Knowledge was also gained on the mechanisms of piezoelectricity and the possibilities of using additives to improve the piezoceramic properties. These new results allowed high-performance and low-cost applications, and allowed the exploitation of a new design approach (piezocomposite structures, polymeric materials, new geometries, etc.) to develop new classes of sensors and, especially interesting for us, new classes of actuators.

A piezoelectric material is characterized by the ability to convert electrical power to mechanical power (inverse piezoelectric effect) by a crystallographic deformation; when piezoelectric crystals are polarized by an electric tension on two opposite surfaces, they change their structure, causing an elongation or a shortening, according to the electric field polarity. The electric charge is converted to a mechanical strain, enabling a relative movement between two material points on the actuator and, if an external force or moment is applied to one of the two selected points opposing a resistance to the movement, this "conceptual actuator" is able to overcome the force/moment, resulting in mechanical power generation (Figure 1.38). The most used piezoelectric materials are piezoceramics such as PZT, a polycrystalline ferroelectric material with a tetragonal/rhombahedral structure; these materials are generally composed of large divalent metal ions, such as lead, tetravalent metal ions, such as titanium or zirconium (Figure 1.39), and oxygen ions. Under the Curie temperature, these materials exhibit a structure without a center of symmetry; but, when the piezoceramics are exposed to temperatures higher than Curie point, their structure is subjected to a transformation becoming symmetric and losing their piezoelectric ability.

Common piezoelectric materials are piezoceramics such as lead zirconate titanate (PZT) and piezoelectric polymers such as polyvinylidene fluoride (PVDF). Further, the research proposed new formulations to improve the performance of piezoceramics, such as PZN-PT and PMN-PT, expanding the strain range from 0.1 to 0.2% (for PZT), to 1% (for the new formulations), and enabling a power density five times higher than that of PZT. On the other hand, piezoelectric polymers are usually configured in film structures

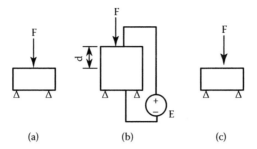

(a)   (b)   (c)

FIGURE 1.38
Inverse piezoelectric effect: an external force *F* is applied to a piezoelectric parallelepiped as in configuration (a); when an electric tension generator applies power to the actuator, it results in a displacement *d*, as in configuration (b); if the electric power is disconnected, the piezoelectric parallelepiped returns to its initial condition, as in configuration (c).

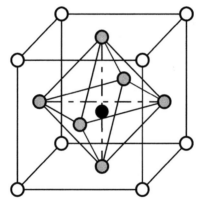

FIGURE 1.39
Structure of a PZT cell under Curie temperature: white particles are large divalent metal ions, gray particles are oxygen ions, and the black particle is a tetravalent metal ion.

and exhibit high voltage limits, but have low stiffness and electromechanical coupling coefficients; therefore, they are not usually chosen as actuators.

Constitutive equations can be used to describe the relationship between the electric field and mechanical strain in the piezoelectric media (1.235), where *D* is the three-dimensional vector of the electric displacement, *E* is the three-dimensional vector of the electric field density, *S* is the second-order tensor of mechanical strain, *T* is the second-order tensor of mechanical stress, s^E is the fourth-order tensor of elastic compliance, ε^T is the second-order tensor of permeability, and *d* is the third-order tensor of piezoelectric strain. Note that $\underline{d}$ is related to *d* through a transposition as shown in the matrix form of the constitutive equation (1.236).

$$
\begin{aligned}
S_{ij} &= s^E_{ijkl} T_{kl} + \underline{d}_{ijk} E_k \\
D_i &= d_{ijk} T_{jk} + \varepsilon^T_{ij} E_j
\end{aligned}
\quad , \quad i, j, k, l \in 1, 2, 3
\tag{1.235}
$$

$$S_I = s_{I,J}^E T_J + d_{I,j} E_j$$
$$D_i = d_{i,J} T_J + \varepsilon_{i,j}^T E_j$$
$$, \quad i,j \in 1,2,3; I,J \in 1,2\ldots6 \qquad (1.236)$$

The first version of the constitutive equation is useful to express the dimensions of the tensors, whereas the second is more concise; however, there are different ways of writing these equations. Another interesting version (1.237) gives the strain in terms of stress and electric displacement, introducing the voltage matrix g and the matrix β; further, if the tensors/vectors D, S, E, and T are rearranged into nine-dimensional column vectors, the constitutive equation can then take the form of (1.237–1.241), according to the selection of dependent and independent variables.

$$S_I = s_{I,J}^D T_J + g_{I,j} E_j$$
$$E_i = -g_{i,J} T_J + \beta_{i,j}^T D_j$$
$$, \quad i,j \in 1,2,3; I,J \in 1,2\ldots6 \qquad (1.237)$$

$$\begin{cases} D = \varepsilon^T E + d : T \\ S = d^t E + s^E : T \end{cases} \qquad (1.238)$$

$$\begin{cases} D = \varepsilon^S E + e : S \\ T = -e^t E + c^E : S \end{cases} \qquad (1.239)$$

$$\begin{cases} E = \beta^T D - g : T \\ S = g^t D + s^D : T \end{cases} \qquad (1.240)$$

$$\begin{cases} E = \beta^S D - h : S \\ T = -h^t D + c^D : S \end{cases} \qquad (1.241)$$

The preceding constitutive equations exhibit linear relationships between the applied field and the resulting strain. As an example, we can consider the tensor d; the experimental values of its components are obtained by an approximation as it depends upon the strain and the applied electric field. This approximation consists of the hypothesis of low variation of applied voltage and resulting reduced strain. If the observed domain is out of the field of linearity, then new values should be used to estimate the tensor d (the constitutive equations are linear, but the value of d is different in each small considered region). On the contrary, a unique nonlinear constitutive equation could be used with the piezoelectric strain tensor d dependent on S and E, resulting in a theoretically correct but really complex approach. Another interesting consideration is based on the "aging effect" of piezoceramic materials represented by a logarithmic decay of their properties with time; therefore, over time, a new value of *d* should be estimated to obtain a correct model for the piezoceramic material.

The hypothesis of low strain on the linear constitutive equations for each small region of the adopted variables allows the expression in equation 1.242, where *v* is the speed of a basic element of piezoceramic material.

$$\nabla_s v = \frac{\partial S}{\partial t}, \qquad \nabla_s (\bullet) \equiv \frac{1}{2}\left(\nabla(\bullet) + \nabla(\bullet^T)\right) \qquad (1.242)$$

Using equation 1.242, the equation of motion (1.243), the Maxwell equations (1.244), and the previously mentioned constitutive equations 1.239 and 1.240 (the latter used only to find the time derivative of *D*), we can obtain the general Christoffel equations of motion (1.245).

$$\nabla \circ T = \rho \frac{\partial v}{\partial t} - F \qquad (1.243)$$

where ρ is the density of the material and *F* is the resulting internal force reduced to a surface force (with the divergence theorem).

$$\begin{cases} -\nabla \times E = \mu_0 \dfrac{\partial H}{\partial t} \\ \nabla H = \dfrac{\partial D}{\partial t} + J \end{cases} \qquad (1.244)$$

where μ_0 is the permeability constant, *H* is the electromagnetic induction, and *J* is current density.

$$\begin{cases} \nabla \circ c^E : \nabla_s v = \rho \dfrac{\partial^2 v}{\partial t^2} - \dfrac{\partial F}{\partial t} + \nabla \circ e \dfrac{\partial E}{\partial t} \\ -\nabla \times \nabla \times E = \mu_0 \varepsilon^S \dfrac{\partial^2 E}{\partial t^2} + \mu_0 e : \nabla_s \dfrac{\partial v}{\partial t} + \mu_0 \dfrac{\partial J}{\partial t} \end{cases} \qquad (1.245)$$

To obtain a simple set of equations from equation 1.245, we can neglect the presence of force, current density, and the rotational term of *E*. Then the Fourier theorem allows us to transform, under reasonable hypotheses, a periodical function (or in general, a function defined into a finite time frame) into a sum of trigonometric functions such that we can consider only a single wave propagating through the media. The usual geometries of piezoelectric actuators are planar; then, planar waves can be expressed (1.246), where ω is the angular pulsation, $\underline{I}$ is the direction of the wave, and the constant *k* should be determined. Under these hypotheses, a simplified set of Christoffel equations can be obtained (1.247), where V is the electric potential and l are the matrixes of the directional cosines.

$$f(r,t) = e^{j(\omega \cdot t - k \underline{I} \circ r)} \qquad (1.246)$$

$$\begin{cases} -k^2 \left(l_{iK} c_{KL}^E l_{Lj} \right) \cdot v_j + \rho \omega^2 v_i = -j\omega \cdot k^2 \left(l_{iK} e_{Kj} l_j \right) \cdot V \\ \omega^2 k^2 \left(l_i \varepsilon_{ij}^S l_j \right) \cdot V = -j\omega \cdot k^2 \left(l_i e_{iL} l_{Lj} \right) \cdot v_j \end{cases} \qquad (1.247)$$

The resulting equation can be solved to calculate the value of the potential energy for a desired speed (under the limitations of the used technology). Knowing the model for the actuating principle, this equation can be implemented in a more complex model of a complete actuator, or it can be used as a simplified model for the behavior of the actuator. The second approach allows the definition of a "virtual" piezoelectric object implementing, through a proper calibration, corrections of the piezoelectric matrix to take into account some unconsidered phenomena.

1.4.3 Shape Memory Alloys

The first known study on shape memory effect is credited to Olander (Olander, 1932) who observed that an object composed of an Au–Cd alloy, if plastically deformed and then heated, is able to compensate the plastic deformation and to recover its original shape. In 1938, Greninger and Mooradian (Greninger and Mooradian 1938), changing the temperature of a Cu–Zn alloy, observed the formation of a crystalline phase. Then, in 1950, Chang and Read (Chang and Read, 1951) used an x-ray analysis to understand the phenomenon of crystalline phase formation in shape memory alloys and, in 1958, showed the first shape memory Au–Cd actuator. An important milestone in the development of shape memory alloys was reached by researchers at the Naval Ordinance Laboratory in 1961, led by Buehler (Buehler and Wiley, 1965), who observed the same shape memory effect in an Ni–Ti alloy that is cheaper, easier to work with, and less dangerous than Au–Cd alloys; consequently, many applications appeared on the market during the 1970s, initially with a static nature and then with dynamic capabilities.

Control techniques were later developed to improve performances, with a particular focus on reducing cooling and heating time, and further data was acquired in reference to the properties of different shape memory alloys. These new results, along with the ability to mass produce NiTi microelements, will be a key to opening the door to a much wider use of smart materials. During the last two decades of the twentieth century, many advances have been made in the modeling of SMA (shape memory alloys) behavior, including the control in regard to the change of shape; new design approaches to improve performances, particularly response time and reliability; and allowing for low-cost mass production. For instance, the response time of the system was significantly reduced by observing the relationship between the geometry of the system, its thermal inertia, and using some interesting material combinations to generate a Peltier cell.

The reliability of an SMA is based on the knowledge of the physical properties of the alloy. This knowledge is useful to prevent irreversible damages and to guarantee a repeatable behavior. The aims of preventing damage and gaining repeatable behavior were attained through experimental works with new simple and complex models, varying design techniques, and with the development of precise inputs and control of heating and cooling. Advances in the control area are some of the most important aspects of SMA systems in order to enhance repeatable high-speed and high-precision performances of micro-SMA machines. Further, the consequential cost reduction of SMA objects followed shortly, thus allowing for the mass production of industrial and medical applications.

The behavior of a shape memory alloy that deforms at a low temperature and regains its original undeformed shape when heated to a higher temperature is due to the thermoelastic martensitic phase transformation and its reversal (Bo and Lagoudas, 1999a). Many alloys exhibit shape memory effect, and the level of commercial interest of each and every alloy is correlated with their ability to easily recover their initial position or to exert a significantly high force. When an NiTi alloy is subjected to an external force, the various planes slide without a break in the crystallographic connections; the atoms of the structure are subjected only to a reduced movement and, through a subsequent heating, it reverts the structure to its initial position, resulting in the production of a significantly high mechanical force.

The martensitic phase consists of a thick arrangement of crystallographic planes, characterized by a high relative mobility, and can appear in two different forms, depending on the history of the material: twinned martensite is derived by cooling the austenitic phase and has a herringbone pattern, whereas detwinned martensite is due to a sliding of the crystallographic planes. A scheme of a simple cycle of transformations able to exhibit the shape memory effect is shown in Figure 1.40: a NiTi bar with a fixed end and a free end without any external load is observed. The first condition of the material (a) is the austenitic phase; after an external cooling (A), the material is converted into twinned martensite (b); however, from a macroscopic viewpoint, we cannot observe any deformation of the material or any external force exerted by the bar. Then, a mechanical traction is applied to the bar (B) and some portions of the bar are converted, during its elongation, into detwinned martensite (c). Finally, after an external heating (C), the material is converted again into austenite, and the bar recovers its original shape.

It should be observed that the martensitic generation process does not require an introduction of external substances and depends only on the achievement of a critical temperature (M_s, or martensite start). The second important aspect of this transformation is the heat production, and finally a hysteretic phenomenon is observed when, at the same temperature, an austenitic phase and a martensitic phase coexist. The austenite-to-martensite transformation is

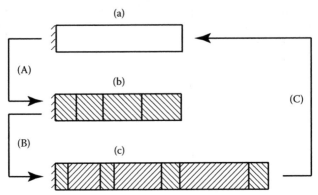

FIGURE 1.40
Shape memory effect.

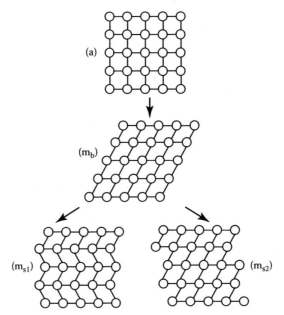

FIGURE 1.41
Martensite formation.

subdivided into two subsequent stages: bain strain and lattice-invariant shear (Figure 1.41). Bain strain generates a martensitic structure (m_b) from austenite (a) and through a movement of the plane interface, and then the lattice-invariant transformation can generate a reversible deformation with a twinning (m_{s1}) or a permanent deformation with a slip (m_{s2}). A reversible deformation is needed to exhibit a shape memory effect, so it is important to avoid slipped martensite (m_{s2}) and generate twinned martensite (m_{s1}).

From a macroscopic viewpoint, the martensitic formation can be represented in the temperature diagram (Figure 1.42). An interesting characteristic

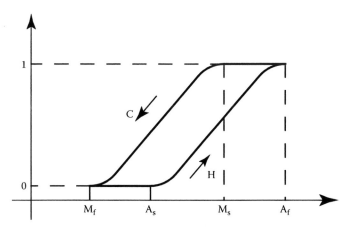

FIGURE 1.42
Martensitic fraction versus temperature: M_s, M_f, A_s, and A_f are, respectively, martensite start, martensite finish, austenite start, and austenite finish critical temperatures. H and C are, respectively, heating and cooling transformations.

of this diagram is the hysteresis, which can occur between 20 and 40°C and is associated with the microfrictions within the structure. Even though austenitic and martensitic macroscopic properties are slightly different, the martensitic fraction can be used as the dependent variable to define the behavior of the shape memory material.

The classical shape memory effect appears in the cycle of Figure 1.40 and is usually called one-way shape memory effect (OWSM) because the shape transformation can be thermally commanded during the martensite-to-austenite conversion. This is because the material is able to "remember" only the austenitic shape. A more complete shape memory effect is the two-way memory effect (TWSM), which consists in the ability to recover an austenitic as well as a martensitic shape; however, it is able to reach only reduced forces and deformations, so it is not very common in commercial applications. TWSM can be obtained from a shape memory alloy after a thermomechanical training, which generates a new microstructure (stress-biased martensite) and reduces the mechanical properties of the material.

The microscopic and macroscopic properties of the NiTi alloys can be changed opportunely, varying the percentage of Ni or introducing additive components (Table 1.9): an increase in the percentage of Ni can decrease critical temperatures and increase the austenitic breaking stress. Fe and Cr are used to reduce critical temperatures, Cu reduces the hysteretic cycle and decreases the necessary stress to deform the martensite, and O and C are usually avoided because they diminish the mechanical performance. Some thermomechanical treatments are necessary to induce a proper shape memory effect; however, because shape memory alloys are commercialized after intermediate treatments, a classification such as in Table 1.10 can be used to select the correct SMA components.

TABLE 1.9

Standard NiTi Alloys

Alloy Code	A_s [°C]	A_f [°C]	Composition [%]
S	–5 to 15	10 to 20	~55.8 Ni
N	–20 to –5	0 to 20	~56.0 Ni
C	–20 to –5	0 to 10	~55.8 Ni, 0.25 Cr
B	15 to 45	20 to 40	~55.6 Ni
M	45 to 95	45 to 95	55.1 to 55.5 Ni
H	> 95	95 to 115	<55.0 Ni

TABLE 1.10

Some Common Commercial SMA Components

Treatment	Description	Characteristics
Cold processed	The client executes thermal treatment	Do not exhibit SME
Straight annealed	Thermal treatment is executed by producer	Wire exhibiting SME
Flat annealed	Thermal treatment is executed by producer	Plate exhibiting SME
Special annealed	Thermal treatment is executed by producer	Special shape exhibiting SME

An austenitic shape must be "saved" to induce the shape memory effect; this can be obtained through a thermal treatment that consists in a heating of the cold processed specimen in a hot mold to over 400°C for several minutes, and hardening will then result after a rapid cooling. The TWSM can also be obtained through a more complex approach due to a cyclic repetition of cold and hot deformations, which allow the "saving" of an austenitic shape as well as a martensitic shape. Unfortunately, the TWSM process will lead to a degradation of mechanical performances and, if compared with OWSM, the recoverable deformation is only 2% (versus 5–8%). In addition, its maximum stress is significantly reduced, SME disappears if the temperature of the material exceeds 250°C, and it is time-unstable, and consequently, the TWSM mechanical performance cannot be guaranteed after a high number of load cycles. It should be noted that not only NiTi exhibits SMA effect; some Cu alloys can be used, such as CuZnAl or CuAlNi and others with Mn, as well as some Fe alloys (FePt, FePd, and FeNiCoTi).

1.5   Hints at Microcontinua

Microcontinuum field theories assign to physical continuum a structure more complex than that established by classical mechanics. They examine material bodies with microscopic internal structure that can be buckled and that can interact with mechanical and electromagnetic fields. The object of this approach is to describe some continuum properties that are not explained by the classical theory. Particularly, three structures (3M) are studied; the more general is called micromorphic, the second and third, respectively called microstretch and micropolar, are subclasses of the first.

The micromorphic theory (Mm) constitutes a systematic and logical extension of the classical field theory and considers material bodies made up of infinitesimal non-point-shaped deformable particles characterized by an internal continuous structure (subcontinuum); the macroscopic behavior of the material body is caused by the combination of its microscopic behaviors. A generic particle of a microcontinuum is a set of material points and represents a microsystem with properties constructed to comply with the continuity hypothesis on the order of micrometers. This approach permits us to assign additive degrees of freedom, not seen by the classic theory, to the continuum according to new field equations. As schematically shown in Figure 1.43, the classical field theory is a particular case of the microcontinuum field theories; further, microstretch, micropolar, and micromorphics are constituted by deformable microparticles subjected to different kind of load. The microstretchs are not subjected to shear deformation, the micropolars represent the simplest class of microcontinua because its microparticles are not deformable but are be able to rotate and translate rigidly, and the classical continua derive from micropolars annulling microparticles' spatial dimensions. The difference between the classic theory and the microcontinuum field theory is clear: the former studies the evolution of every single point; the latter studies the evolution of a point with its neighborhood that represents the microparticle. The second approach allows an improved description of some microphysical phenomena to the detriment of

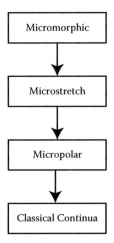

FIGURE 1.43
Subclasses of microcontinuum field theories; classic continua represent the simplest class.

a complex mathematical treatment. Different substances and materials can be described with microcontinuum field theories such as blood flow, liquid crystals, porous materials, polymeric substances, and solids with microfracture or dislocations; these materials demand consideration of the movements of their microconstituents such suspension particles, blood cells, and grains. Examples of micromorphic substances are polymers with flexible molecules, liquid crystals with lateral chains, animal blood with deformable cells, and suspensions with deformable particles. Examples of microstretch continua are liquid cement, elastic suspensions and porous materials, and examples of micropolar materials are liquid crystals with rigid molecules, suspensions with deformable particles, magnetic fluids, and muddy fluids.

To simplify the mathematical expressions, all formulas of this section are written in Einstein compact form: the summation notation is implicit and identified by repeated indexes, and spatial derivatives are denoted with a comma followed by some indexes associated with the spatial derivatives in a three-dimensional space.

1.5.1   Kinematics

A microcontinuum is a continuous collection of deformable point particles (Eringen, 1999). Physically, the particles are point particles, i.e., they are infinitesimal in size, they do not violate continuity of matter, and yet, they are deformable. A single point particle is represented in this way: a P_0 point called centroid coinciding with the particle barycenter and located by the position vector $x_0 = (x_{01}; x_{02}; x_{03})$, and some vectors ξ_{0i} are associated with P_0 and locate the positions of some points of the same particle with regard to P_0. This kinematic schematization is related to a reference microcontinuum configuration usually coinciding with the undeformed configuration. The centroid, initially in P_0 localized by x_0, changes its position during particle deformation, assuming the position P localized by the position vector x, whereas the vectors ξ_{0i}, with origin in P_0, are transformed into the vectors ξ_i, with origin in P. In the easiest case, with only one vector ξ_{0i} indicated by ξ_0 for the sake of simplicity, the microcontinuum is called microcontinuum of grade one; in this book the treatment of microcontinuum field theory is concerned only with this case. During particle deformation in a microcontinuum of grade one, P_0 moves to P, and x_0 and ξ_0 move, respectively, to x and ξ, as shown in (1.248) and (1.249), where t is the time. Then, a generic point A of a microcontinuum is localized by the position vector x_0' in the reference configuration and by the position vector x' in the deformed configuration (1.250).

$$x_k = x_k(x_{K0}, t); \quad k = 1, 2, 3 \quad K = 1, 2, 3 \tag{1.248}$$

$$\xi_k = \xi_k(x_{K0}, \xi_{K0}, t); \quad k = 1, 2, 3 \quad K = 1, 2, 3 \tag{1.249}$$

$$x_0' = x_0 + \xi_0; \quad x' = x + \xi \tag{1.250}$$

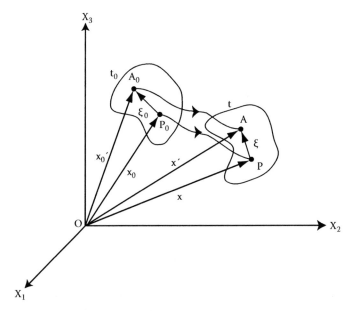

FIGURE 1.44
Deformation of a microcontinuum point particle of grade one.

The vectors x_0 and ξ_0 do not depend on time because they are related to the reference configuration; further, x and ξ are characterized by sufficient continuity and derivability to comply with the continuity hypothesis of the matter. Some kinematic properties of a point particle of grade one are represented in Figure 1.44, describing its movement from the reference configuration at the starting time t_0 to the deformed configuration at the time t'.

The function (1.248) is the macromotion (or simply, the motion); on the other hand, the expression in (1.249) represents the micromotion. Then, the function (1.249) can be approximated with its Taylor series calculated with respect to ξ with center ξ_0 to the first order because the components of ξ and ξ_0 are small; further, when ξ_0 is null, (1.251) can be written in the McLaurin form. The derivatives of the components ξ_k with respect to the components ξ_{0K} describe the components of χ: a second-order tensor is not present in the classical theory, and is called the microdeformation tensor. Thus, according to definition (1.252), the expression (1.251) assumes the form (1.253), representing the linear approximation of the components ξ_k with respect to the components ξ_{0K}.

$$\xi_k = \left(\frac{\partial \xi_k}{\partial \xi_{0K}} \right)_{\xi_{0K}=0} \xi_{0K}(x_{0K},t) \tag{1.251}$$

$$\chi_{kK} \equiv \left(\frac{\partial \xi_k}{\partial \xi_{0K}} \right)_{\xi_{0K}=0} \qquad k, K = 1, 2, 3 \tag{1.252}$$

$$\xi_k = \chi_{kK}(x_{0K}, t)\xi_{0K}(x_{0K}, t) \tag{1.253}$$

A material body is called micromorphic continuum of grade one, or simply micromorphic continuum, if its movement is described by (1.248) and (1.253) or by their inverted expressions (1.254), where the functions χ_{Kk} are components of the second-order tensor χ, called the inverse microdeformation tensor, representing the inverse of the microdeformation tensor χ.

$$x_{K0} = x_{K0}(x_k, t); \quad \xi_{0K} = \chi_{Kk}(x_k, t)\xi_k(x_k, t); \quad K = 1,2,3; \quad k = 1,2,3 \tag{1.254}$$

To describe the strain tensors of microcontinua, the strain tensor concept of the classical continua is kept; thus, the points P and P' of a classic continuum are depicted in Figure 1.45 with their respective coordinates x and x', where x' is equal to x increased by the vectorial quantity Δx.

For a generic deformation, the P^* point with x^* coordinates is the final position of the material point from the position P. Further, P'^* with coordinate x'^* is the final position associated with initial position P', as shown in (1.255 and 1.256), where u and u' are the displacement vectors, respectively, associated with P and P'. The vector Δx is defined by the difference between the coordinates of P' and P, whereas Δx^* is described by the difference between the coordinates of P'^* and P^*, and for that, Δx^* is the deformation of Δx (1.257). Furthermore, the components of $u'(x')$ (1.258) can be approximated with their first-order McLaurin series with respect to $(\Delta x_1; \Delta x_2; \Delta x_3)$; thus, relations (1.259–1.261) can be deduced because $u'(x)$ is equal to $u(x)$.

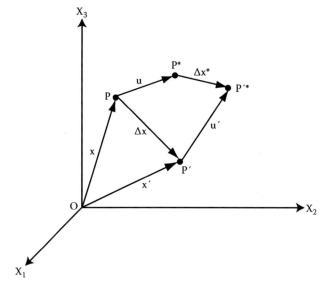

FIGURE 1.45
Deformation of two points in a classical continuum.

$$x^* = x + u(x) \tag{1.255}$$

$$x'^* = x' + u'(x') \tag{1.256}$$

$$x'^* - x^* = \Delta x + u'(x') - u(x) \tag{1.257}$$

$$u_i'\left(x_1', x_2', x_3'\right) = u_i'\left(x_1 + \Delta x_1, x_2 + \Delta x_2, x_3 + \Delta x_3\right); \quad i = 1,\ 2,\ 3 \tag{1.258}$$

$$u_i'(x') = u_i(x) + \frac{\partial u_i'}{\partial x_j}\Delta x_j + O\left(\Delta x_j^{\ 2}\right); \quad i = 1,\ 2,\ 3 \tag{1.259}$$

$$\Delta u_i(x) = u_i'(x') - u_i(x) = \frac{\partial u_i'}{\partial x_j}\Delta x_j = du_i \quad i = 1,\ 2,\ 3 \tag{1.260}$$

$$d\bar{x}_i = dx_i + du_i \quad i = 1,\ 2,\ 3 \tag{1.261}$$

The square ds^2 of the displacement can be written as in (1.262), that is, a quadratic form with respect to the infinitesimals $dx_i dx_j$, where δ_{kl} is the Kronecker symbol. A similar expression can be written also for the deformed configuration (1.263), or as in (1.264) with the definition of du_i given by $u_{i,k}dx_k$. Expressions (1.262) and (1.264) differ for the E_{ik} tensor defined by (1.265), describing the local deformation of a classical continuum, whereas its nonlinear term, given by $u_{m,i}u_{m,k}$, is usually small and can be neglected. In this case, E_{ik} is called Cauchy's infinitesimal strain tensor and is indicated by φ_{ik}, (1.266).

$$ds^2 = \delta_{ij}dx_i dx_j \tag{1.262}$$

$$\overline{ds}^2 = \delta_{ij}d\bar{x}_i d\bar{x}_j = \delta_{ij}\left(dx_i + du_i\right)\left(dx_j + du_j\right) \tag{1.263}$$

$$\overline{ds}^2 = \left(\delta_{ik} + u_{i,k} + u_{k,i} + u_{m,i}u_{m,k}\right)dx_i dx_k \tag{1.264}$$

$$E_{ik} \equiv \frac{1}{2}\left(u_{i,k} + u_{k,i} + u_{m,i}u_{m,k}\right) \tag{1.265}$$

$$\varphi_{ik} \equiv \frac{1}{2}\left(u_{i,k} + u_{k,i}\right) \tag{1.266}$$

The microcontinuum kinematics can be described with an approach similar to that of the classical theory with some additional considerations. In the reference configuration, the points P and P_1 are localized, respectively, with the coordinates x' and x_1' (1.267); then, according to (1.253), the first relation of (1.267) can be expressed as in (1.268), where X_K is the K-th column vector of χ; further, after differentiating with respect to space variables (x_{01}, x_{02}, x_{03}) and $(\xi_{01}, \xi_{02}, \xi_{03})$, (1.268) can be written like (1.269). Finally, the relation (1.271) is obtained substituting (1.269) in (1.270).

$$x' = x + \xi; \qquad x_1' = x_1 + \xi_1; \qquad x_1' = x' + \Delta x'; \tag{1.267}$$

$$x' = x\left(x_0, t\right) + X_K\left(x_0, t\right)\xi_{0K}\left(x_0, t\right) \tag{1.268}$$

$$dx' = dx_{,K}\, dx_{0K} + X_{K,L}\xi_{0K}dx_{0L} + X_K d\xi_{0K} \tag{1.269}$$

$$\overline{ds}^{\,2} = dx' \cdot dx' = \left(dx'\right)^2 \tag{1.270}$$

$$
\begin{aligned}
\overline{ds}^{\,2} &= x_{,K}\, x_{,L}\, dx_{0K}dx_{0L} + X_K X_L d\xi_{0K}d\xi_{0L} + X_{M,K} X_{N,L}\xi_{0M}\xi_{0N}dx_{0K}dx_{0L} \\
&\quad + 2x_{,K}\, X_L dx_{0K}d\xi_{0L} + 2X_L X_{M,K}\xi_{0M}dx_{0K}d\xi_{0L} + 2x_{,L}\, X_{M,K}\xi_{0M}dx_{0K}dx_{0L}
\end{aligned}
\tag{1.271}
$$

The first and third summations in (1.269) are single summations, whereas the second is a double summation. Further, the first three implicit summations of the relation (1.271) are their squares; in fact, the single summations become double summations, and the double summations become quadruple summations, whereas the double products are due to triple and double summations. The relation (1.271) is a quadratic form with respect to (dx_{0k}, dx_{0l}), $(dx_{0k}, d\xi_{0l})$, $(d\xi_{0k}, d\xi_{0l})$ infinitesimals and can be written in the canonical form (1.272). F_{KL}, G_{KL}, and H_{KL} are tensors and are defined in (1.273) with the associated strain tensors C_{KL}, C_{KL}^*, Γ_{KLM}^* as in (1.275) with the definitions in (1.274).

$$\overline{ds}^{\,2} = F_{KL}dx_{0K}dx_{0L} + G_{KL}dx_{0K}d\xi_{0L} + H_{KL}d\xi_{0K}d\xi_{0L} \tag{1.272}$$

$$F_{KL} \equiv x_{,K}\, x_{,L} + X_{M,K} X_{N,L}\xi_{0M}\xi_{0N} + 2x_{,L}\, X_{M,K}\xi_{0M}dx_{0K}dx_{0L}$$

$$G_{KL} \equiv 2\left(x_{,K}\, X_L + X_L X_{M,K}\xi_{0M}\right) \tag{1.273}$$

$$H_{KL} \equiv X_K X_L$$

$$C_{KL} \equiv x_{k,K} x_{k,L}$$

$$C^*_{KL} \equiv x_{k,K} \chi_{kL}$$

$$\Gamma^*_{KLM} \equiv x_{k,M} \chi_{kL,M}$$

(1.274)

$$F_{KL} = C_{KL} + 2\xi_{0M}\Gamma^*_{KLM} + \xi_{0M}\xi_{0N}\Gamma^*_{PLM}\Gamma^*_{RNK}C^{-1}_{PR}$$

$$G_{KL} = 2\left(C^*_{KL} + \xi_{0M}C^*_{NL}\Gamma^*_{RMK} + C^{-1}_{NR}\right)$$

(1.275)

$$H_{KL} = C^*_{MK}C^*_{NL}C^{-1}_{MN}$$

Three alternative strain tensors are particularly useful to express the microcontinua constitutive equations: the deformation tensors $\underline{C}_{KL}$, the microdeformation tensor $\underline{C}_{KL}$ and the wryness tensor Γ_{KLM} (1.276). These three tensors are not independent but must respect three compatibility conditions given by (1.277), where ε_{KLM} is the Levi–Civita symbol or the permutation symbol. Conditions (1.277) follow from x_k and χ_{kK} derivability, as established by the Schwartz theorem (1.278).

$$\underline{C}_{KL} \equiv x_{k,K}\,\mathcal{X}_{LK}$$

$$\mathcal{C}_{KL} \equiv \chi_{kK}\chi_{kL}$$

(1.276)

$$\Gamma_{KLM} \equiv \mathcal{X}_{Kk}\chi_{kL,M}$$

$$\varepsilon_{KPQ}\left(\underline{C}_{PL,Q} + \underline{C}_{PR}\Gamma_{LRQ}\right) = 0$$

$$\varepsilon_{KPQ}\left(\Gamma_{LMP,Q} + \Gamma_{LRQ}\Gamma_{RMP}\right) = 0$$

(1.277)

$$\mathcal{C}_{KL,M} - \Gamma_{PKM}\mathcal{C}_{LP} - \Gamma_{PLM}\mathcal{C}_{KP} = 0$$

$$\frac{\partial^2 x_k}{\partial x_{0P}\partial x_{0Q}} = \frac{\partial^2 x_k}{\partial x_{0Q}\partial x_{0P}}\;; \qquad \frac{\partial^2 \chi_{kK}}{\partial x_{0P}\partial x_{0Q}} = \frac{\partial^2 \chi_{kK}}{\partial x_{0Q}\partial x_{0P}}$$

(1.278)

To calculate the speed $\underline{v}$ and the acceleration $\underline{a}$ of a point P of a microcontinuum body with coordinates x_0' in the reference configuration, the vector function $x'(x_0, \xi_0, t)$ must be partially differentiated twice with respect to time as in (1.279).

$$x' = x'\left(x_0, \xi_0, t\right)$$

$$\underline{v}\left(x_0, \xi_0, t\right) = \frac{\partial x'}{\partial t}$$

(1.279)

$$\underline{a}\left(x_0, \xi_0, t\right) = \frac{\partial v}{\partial t} = \frac{\partial^2 x'}{\partial t^2}$$

In (1.279), $\underline{v}$ and $\underline{a}$ are not vector fields; in fact, they do not define the two vectors associated with a geometric point of the space, but they define the speed and acceleration at the time t of an infinitesimal particle with mass, belonging to the reference configuration with coordinates x_0'. This approach of system representation is called Lagrangian. To obtain the speed vector field and the acceleration vector field, the vector function $x'(x_0, \xi_0, t)$ must be totally differentiated twice with respect to the time variable as in (1.280).

$$v\left(x',t\right) \equiv \frac{Dx'}{Dt} = \frac{dx'}{dt}$$

$$a\left(x',t\right) \equiv \frac{Dv}{Dt} = \frac{dv}{dt} = \frac{\partial v}{\partial t} + \mathbf{\nabla}vv \tag{1.280}$$

Definitions (1.280) give the speed field v and the acceleration field a of any point particle that crosses the space point localized by x' at the time t. This approach of system representation is called Eulerian, and it is applied by the microcontinuum field theory. Because point particles are not infinitesimal, they can rotate, and the speed field v and the acceleration field a are not sufficient to describe the kinematics of the system. To describe the microrotation field, the microgyration tensor υ_{kl} is defined and is given by (1.281).

$$\upsilon_{kl} \equiv \frac{D\chi_{kM}}{Dt} \chi_{Ml} \tag{1.281}$$

In a microcontinuum of grade one, a deformable vector ξ is associated with every point particle and it describes its deformation. Particularly, in micropolar continua, point particles are rigid: they can translate and rotate rigidly, and a generic vector ξ during the deformation can only change the direction but not the modulus; it behaves like a rigid beam, and its behavior is illustrated in Figure 1.46 and expressed by (1.282).

$$\left|\xi_0\right| = \left|\xi\right| \quad \forall t \tag{1.282}$$

In the case of micropolar continua, the motion of a vector ξ can be represented as the motion of a rigid segment rotated around an axis crossing its centroid, and its translation is given by the translation of the centroid. Furthermore, in the case of micropolar continua, the χ microdeformation tensor is orthogonal.

1.5.2 Dynamics and Balance of Energy

The previous section proposed a description of the movement of a microcontinuum according to a sufficient space–time continuity, but, in the domain of

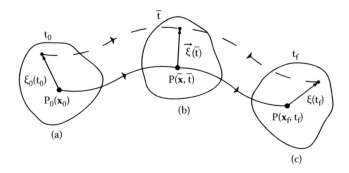

FIGURE 1.46
Deformation of a micropolar point particle. The point particle translates and rotates without dilatation and distortion, like the vector ξ that represents it. (a) Reference configuration; (b) generic configuration during the deformation; (c) final configuration.

a more general theory, cases of discontinuities on some surfaces should be considered. Therefore, with reference to Figure 1.47, a discontinuity surface Σ is observed with a normal unit vector $n(P)$ applied in the generic point P, the discontinuity of a generic field $G(P, t)$ on Σ is defined by the difference between the value of G calculated in points immediately after P (P^+), and the value of G calculated in points immediately before P (P^-) and is identified with the symbol $[|G(P, t)|]$ as shown in (1.283). An example of a discontinuity surface is given by the phase front of a wave perturbation that separates the perturbed area from the nonperturbed area at t time.

$$\left[\left|G(P,t)\right|\right] \equiv G(P^+,t) - G(P^-,t) \equiv G^+ - G^- \tag{1.283}$$

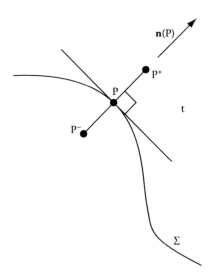

FIGURE 1.47
Discontinuity surface.

Some symbols and functions are defined successively in order to simplify the exposition of the balance equations: i.e., the directors χ_K and χ_k are vectorial properties, as defined in (1.284) expressed in function of i_k and i_{0K}, the respective unary vectors of the Cartesian reference systems (o, x_k) and (O, x_{0K}).

The shifters δ_{Kl} and δ_{kL} are tensors represented in (1.285): δ_{Kl} is a matrix formed by the row components of the three i_{0K} unit vectors with respect to the three i_k unit vectors. Likewise, δ_{kL} is the matrix of the row components of the three i_k unit vectors with respect to the three i_{0K} unit vectors. The shifters allow the introduction of a new formulation for the microdeformation tensor χ and for its inverse χ (1.286) that is more suitable to describe the point particles deformation than the vectors ξ.

$$\chi \equiv \chi_{kK}\left(x_0,t\right)i_k; \qquad \chi_k \equiv \chi_{Kk}\left(x,t\right)i_{0K} \qquad (1.284)$$

$$\delta_{Kl} \equiv i_{0K}\cdot i_l; \qquad \delta_{kL} \equiv i_k\cdot i_{0L} \qquad (1.285)$$

$$\chi_{Kl} \equiv \chi_{kK}\delta_{Kl}; \qquad \chi_{KL} \equiv \chi_{Kk}\delta_{kL} \qquad (1.286)$$

Thus, the field equations of balance for microcontinua can be constructed easily after these definitions: the conservation of mass, the conservation of microinertia, the balance of momentum, the balance of moments of momentum, and the balance of energy, observing, that the conservation of microinertia represents an innovation with respect to the classical theory. The deduction scheme of field equations is similar to the classical scheme with some additive complexities because there are new tensors and physical properties, i.e., contact actions in a microcontinuum between contiguous parts are represented by the second order, generally asymmetric, tensor of stress σ, and by second order, generally asymmetric, tensor of stress moment m. The former is represented in Figure 1.48, and its components are forces per unit of area; the latter is represented in Figure 1.49, and its components are couples per unit area. When classical continua respect the Cauchy hypothesis on the nullity of couples per unit area and couples per unit volume, their stress tensor becomes symmetric. Generally, in the case of microcontinua, external mechanical loads are forces per unit of mass, called f_k, and couples per unit of mass, called l_m, and, after some considerations, the third-order tensor m_{klm} replaces the second-order tensor m_{kl}, and the second-order tensor l_{km} replaces the vector l_k.

Further, the transport theorem is applied to deduce microcontinua field equations that take the form of (1.287); these allow replacing the total time derivative on a volume with the addition between a volume integral and a surface integral. In fact, an expression such as (1.288), with the transport theorem (1.287), can be transformed in (1.289); finally, the postulate of localization is applied to (1.289). This postulate affirms that, if the relation (1.289) is true for any volume portion, it must also be true for an infinitesimal volume

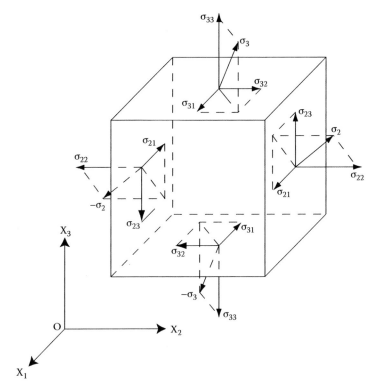

FIGURE 1.48
Representation of the stress tensor.

portion, and so two functions under the integral sign vanish as in (1.290). From a global equation like (1.289), a local equation like the first of (1.290) is obtained, and it is a field equation.

$$\frac{d}{dt}\int_{V-\zeta}f(x,t)d\tau = \int_{V-\zeta}h(x,t)d\tau + \int_{\zeta}\left[\!\left|g(x,t)\right|\!\right]\cdot da \qquad (1.287)$$

$$\frac{d}{dt}\int_{V-\zeta}f(x,t)d\tau = 0 \qquad (1.288)$$

$$\int_{V-\zeta}h(x,t)d\tau + \int_{\zeta}\left[\!\left|g(x,t)\right|\!\right]\cdot da = 0 \qquad (1.289)$$

$$h=0 \quad \forall x \in V-\zeta; \qquad \left[\!\left|g(x,t)\right|\!\right]=0 \quad \forall x \in \zeta \qquad (1.290)$$

With reference to Figure 1.50, the transport theorem establishes the identity (1.291) for any tensor Y of any order, where $Y_{lm...}$ is a generic component of Y,

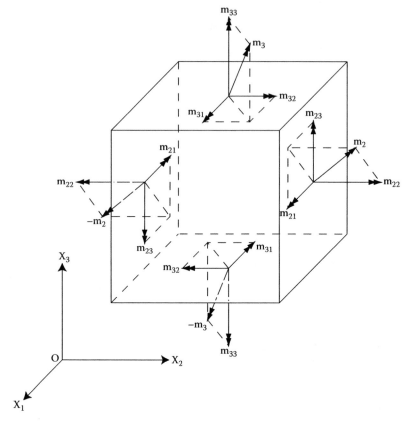

FIGURE 1.49
Representation of the stress moment tensor.

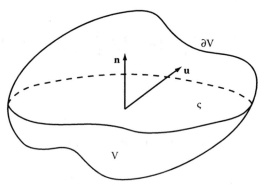

FIGURE 1.50
Control volume V with an internal surface ζ that has u speed.

v is the Eulerian speed of a generic point P that locates the volume infinitesimal element $d\tau$, ζ is a discontinuity surface with speed $\boldsymbol{u}$, da is an infinitesimal element of ζ surface, v-u is the relative speed of a point of ζ near a point P of the continuum, V is the volume of the continuum, and the symbol $[|\ldots|]$ denotes a discontinuity through ζ.

$$\frac{D}{Dt}\int_{V-\zeta}\Upsilon_{lm\ldots}d\tau = \int_{V-\zeta}\left(\frac{D\Upsilon_{lm\ldots}}{Dt}+\Upsilon_{lm\ldots}\nabla\cdot v\right)d\tau + \int_{\zeta}\left[\left|\Upsilon_{lm\ldots}\left(v-u\right)\right|\right]\cdot da \quad (1.291)$$

The proposed tools can be used now to deduce balance equations of microcontinua. The first equation deduced is the law of global balance of energy that affirms that in a microcontinuum the time ratio of the sum of the kinetic and internal energies is equal to the work done by all loads acting on the body per unit time (Eringen, 1999). For a micromorphic body, with reference to Figure 1.51, this law is expressed by (1.292), where v_l are the components of the speed vector, v_{kl} is the microgyration tensor, V is the volume of the body, ζ is a discontinuity surface, ε is the internal energy per unit mass, q_k is the flow heat coming to the body per unit time, h is the internal heat generated per unit time, and K is the kinetic energy per unit mass defined by (1.293).

$$\frac{D}{Dt}\int_{V-\zeta}\rho\left(\varepsilon+K\right)d\tau = \int_{\partial V-\zeta}\left(\sigma_{kl}v_l+m_{klm}v_{lm}+q_k\right)da_k$$
$$+\int_{V-\zeta}\rho\left(f_kv_k+l_{kl}v_{kl}+h\right)d\tau \quad (1.292)$$

$$K\equiv\frac{1}{2}\left(\frac{Dx}{Dt}+\frac{D\xi}{Dt}\right)^2 \quad (1.293)$$

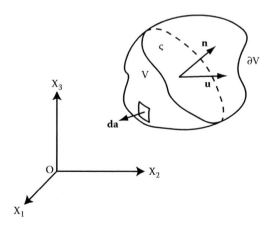

FIGURE 1.51
Micromorphic continuum with volume V and inside a surface ζ of discontinuity with speed $\boldsymbol{u}$.

In the equation (1.292), the left member is the time rate of the total internal and kinetic energies; on the other hand, the right member, under the surface integral, represents the stress energy, the energy of stress moments, and the heat energy per time unit, whereas, under the volume integral, it expresses, respectively, the energy of the body force, the energy of body moments, and the heat input per time unit. Some field tensors, such as m_{kln}, l_{km}, and v_{lm}, are not rigorously defined, but their kinematic nature is of sufficient knowledge for the intentions of this book. In the case of micropolar continua, the tensors m_{kln}, l_{kn} and v_{kl} reduce by one order and become, respectively, m_{kl}, l_k, and v_l, and the law (1.292) is simplified as in (1.294).

$$\frac{D}{Dt}\int_{V-\zeta}\rho\left(\varepsilon+K\right)d\tau = \int_{\partial V-\zeta}\left(\sigma_{kl}v_l+m_{kl}v_l+q_k\right)da_k$$
$$+\int_{V-\zeta}\rho\left(f_k v_k+l_k v_k+h\right)d\tau \tag{1.294}$$

We now draw attention to the second law of balance: it is the conservation of mass, expressed, in the global form, by the equation (1.295), where V is the volume, ζ is a discontinuity surface, and ρ is the density of microcontinuum. Because the equation (1.295) has the form (1.288), the transport theorem can be applied to obtain the equation (1.296) and the postulate of localization produces the equations (1.297), expressing the local law of mass conservation. Furthermore, we observe that the first equation of (1.297) is a field equation.

$$\frac{D}{Dt}\int_{V-\zeta}\rho d\tau = 0 \tag{1.295}$$

$$\int_{V-\zeta}\left(\frac{D\rho}{Dt}+\rho\mathbf{V}\cdot v\right)d\tau + \int_{\zeta}\left[\!\left[\rho\left(v-u\right)\right]\!\right]\cdot da = 0 \tag{1.296}$$

$$\frac{D\rho}{Dt}+\rho\mathbf{V}\cdot v = 0 \quad \forall x \in V-\zeta; \qquad \left[\!\left[\rho\left(v-u\right)\right]\!\right]\cdot n = 0 \quad \forall x \in \zeta \tag{1.297}$$

The third law of balance is the conservation of microinertia, introduced by A. C. Eringen in 1964 (Eringen, 1964) and representing the major innovation of microcontinuum field theories with respect to classical theory. In (1.298), microinertia symmetric tensors i_{0KL} and i_{kl} are defined, and the kinetic energy K, expressed by (1.293), can be also expressed by the relation (1.299); then, according to the law mass of conservation (1.300), the global law of conservation of microinertia (1.301) can be written in the form of the equation (1.302). Finally, applying the transport theorem and postulate of localization, and after some calculation is omitted, the local law of conservation of microinertia is obtained and is given by equations (1.303). The first relation of (1.303) represents nine field equations but, due to the symmetry of i_{kl}, only six equations are independent.

$$\rho_0 i_{0KL} \Delta V \equiv \int_{\Delta V} \rho_0' \Xi_K \Xi_L dV'; \qquad \rho i_{kl} \Delta v \equiv \int_{\Delta v} \rho' \xi_k \xi_l dv' \qquad (1.298)$$

$$K = \frac{1}{2} v^2 + \frac{1}{2} i_{kl} v_{mk} v_{ml} \qquad (1.299)$$

$$i_{kl} = i_{0KL} \chi_{kK} \chi_{lL}; \qquad i_{0KL} = i_{kl} \chi_{Kk} \chi_{Ll} \qquad (1.300)$$

$$\frac{D}{Dt} \int_{V-\zeta} \rho i_{0KL} d\tau = 0 \qquad (1.301)$$

$$\frac{D}{Dt} \int_{V-\zeta} \rho i_{kl} \chi_{Kk} \chi_{Ll} d\tau = 0 \qquad (1.302)$$

$$\frac{D i_{kl}}{Dt} - i_{kr} v_{lk} - i_{lr} v_{kr} = 0 \quad \forall x \in V - \zeta; \quad \left[\left|\rho i_{kl}(v-u)\right|\right] \cdot n = 0 \quad \forall x \in \zeta \quad (1.303)$$

The fourth law is the balance of momentum and is expressed in a global form by (1.304). Further, after the application of the postulate of localization, the local law of balance of momentum is obtained (1.305), where the first relation represents three field scalar equations.

$$\int_{V-\zeta} \sigma_{kl,k} + \rho \left(f_l - \frac{Dv_l}{Dt} \right) d\tau + \int_{\zeta} \left[\left|\sigma_{kl} - \rho v_l (v_k - u_k)\right|\right] n_k da = 0 \qquad (1.304)$$

$$\sigma_{kl,k} + \rho \left(f_l - \frac{Dv_l}{Dt} \right) = 0 \qquad \forall x \in V - \zeta;$$

$$\left[\left|\sigma_{kl} - \rho v_l (v_k - u_k)\right|\right] n_k = 0 \quad \forall x \in \zeta \qquad (1.305)$$

Finally, the fifth law is the balance of moments of momentum, expressed in a global form by (1.306), where s_{ml} is a symmetric tensor, and ς_{lm} is spin inertia tensor (1.307). The application of the postulate of localization produces the local law of balance of moments of momentum (1.308), where the first relation represents nine field scalar equations.

$$\int_{V-\zeta} m_{klm,k} + \sigma_{ml} - s_{ml} + \rho (l_{lm} - \varsigma_{lm}) d\tau + \int_{\zeta} \left[\left|m_{klm} - \rho i_{rm} v_{lr} (v_k - u_k)\right|\right] n_k da = 0 \quad (1.306)$$

$$\varsigma_{kl} \equiv i_{ml}\left(\frac{Dv_{km}}{Dt} + v_{kn}v_{nm}\right) \tag{1.307}$$

$$m_{klm,k} + \sigma_{ml} - s_{ml} + \rho\left(l_{lm} - \varsigma_{lm}\right) = 0 \qquad \forall x \in V - \zeta$$

$$\left[\left|m_{klm} - \rho i_{rm}v_{lr}\left(v_k - u_k\right)\right|\right]n_k = 0 \qquad \forall x \in \zeta \tag{1.308}$$

The local law of energy balance for micromorphics can be deduced from the global law of energy balance (1.292) after applying the transport theorem and the divergence theorem to obtain (1.309), where the three tensors a_{kl}, b_{klm}, and c_{kl} are defined in (1.310), and the postulate of localization produces the local law of energy balance (1.311), where the first relation of (1.311) is a field scalar equation.

$$\int_{V-\zeta}\left(-\rho\frac{D\varepsilon}{Dt} + \sigma_{kl}a_{kl} + s_{kl}c_{kl} + m_{klm}b_{lmk} + q_{k,k} + \rho h\right)d\tau +$$

$$\int_{\zeta}\left[\left|\sigma_{kl}v_l + m_{klm}v_{lm} + q_k - \left(\rho\varepsilon + \frac{1}{2}\rho v^2 + \frac{1}{2}\rho i_{rl}v_{mr}v_{ml}\right)\left(v_k - u_k\right)\right|\right]n_k da = 0 \tag{1.309}$$

$$a_{kl} \equiv v_{l,k} - v_{lk}; \qquad b_{klm} \equiv v_{kl,m}; \qquad c_{kl} \equiv \frac{1}{2}(v_{k,l} + v_{l,k}) \tag{1.310}$$

$$-\rho\frac{D\varepsilon}{Dt} + \sigma_{kl}a_{kl} + s_{kl}c_{kl} + m_{klm}b_{lmk} + q_{k,k} + \rho h = 0 \qquad \forall x \in V - \zeta \tag{1.311}$$

$$\left[\left|\sigma_{kl}v_l + m_{klm}v_{lm} + q_k - \left(\rho\varepsilon + \frac{1}{2}\rho v^2 + \frac{1}{2}\rho i_{rl}v_{mr}v_{ml}\right)\left(v_k - u_k\right)\right|\right]n_k = 0 \qquad \forall x \in \zeta$$

In the case of micropolar continua, some tensors reduce by one order, i.e., m_{klm}, l_{km}, v_{kl}, and ς_{lm} become, respectively, m_{kl}, l_k, v_l, and ς_l; moreover, only for micropolar the microinertia tensor i_{km} is indicated with j_{km}. With these mathematical properties the local balance equations for micropolar assume a simplified form: the local law of mass conservation is given by (1.312), that of microinertia conservation is given by (1.313), that of balance momentum is given by (1.314), that of moments of momentum balance is given by (1.315), and the local law of energy balance is given by (1.316).

$$\frac{D\rho}{Dt} + \rho\mathbf{V}\cdot v = 0 \quad \forall x \in V - \zeta; \qquad \left[\left|\rho(v-u)\right|\right]\cdot n = 0 \quad \forall x \in \zeta \tag{1.312}$$

$$\frac{Dj_{kl}}{Dt} + \left(\varepsilon_{kpr}j_{lp} + \varepsilon_{lpr}j_{kp}\right)v_r = 0 \quad \forall x \in V-\zeta; \quad \left[\!\left[\rho j_{kl}\left(v-u\right)\right]\!\right]\cdot n = 0 \quad \forall x \in \zeta \quad (1.313)$$

$$\sigma_{kl,k} + \rho\left(f_l - \frac{Dv_l}{Dt}\right) = 0 \quad \forall x \in V-\zeta; \quad \left[\!\left[\sigma_{kl} - \rho v_l\left(v_k - u_k\right)\right]\!\right]n_k = 0 \quad \forall x \in \zeta \quad (1.314)$$

$$m_{kl,k} + \varepsilon_{lmn}\sigma_{mn} + \rho\left(l_l - \varsigma_l\right) = 0 \quad \forall x \in V-\zeta;$$

$$\left[\!\left[m_{kl} - \rho j_{pl}v_p\left(v_k - u_k\right)\right]\!\right]n_k = 0 \quad \forall x \in \zeta \quad (1.315)$$

$$-\rho\frac{D\varepsilon}{Dt} + \sigma_{kl}\left(v_{l,k} + \varepsilon_{lkr}v_r\right) + q_{k,k} + \rho h = 0 \quad \forall x \in V-\zeta$$

$$\left[\!\left[\sigma_{kl}v_l + m_{kl}v_l + q_k - \left(\rho\varepsilon + \frac{1}{2}\rho v^2 + \frac{1}{2}\rho i_{rl}v_{mr}v_{ml}\right)\left(v_k - u_k\right)\right]\!\right]n_k = 0 \quad \forall x \in \zeta \quad (1.316)$$

These last ten equations are valid for all micropolar continua but are not sufficient to describe their physical behavior, as for all microcontinua classes due to the lack of constitutive equations, as shown in the following section.

1.5.3 Thermodynamics in Microcontinuum Fields

The second law of thermodynamics expressed by the Clausius inequality is given by (1.317), where ΔS is the entropy variation of the system during a generic transformation from the state A to the equilibrium state B, and θ is the absolute temperature of the source exchanging the heat dq.

$$\Delta S \geq \int_A^B \frac{dq}{\theta} \quad (1.317)$$

To extend the relation (1.317) to microcontinua, an absolute temperature and entropy function are associated with the microcontinuum evolution; thus, the global second law of thermodynamics for micromorphic continua can be expressed as (1.318), where η is the entropy density per unit mass. Further, the application of the transport theorem and divergence theorem furnishes the equation (1.319); finally, the postulate of localization produces the local second law of thermodynamics for micromorphics (1.320).

$$\frac{D}{Dt}\int_{V-\varsigma}\rho\eta \, d\tau - \int_{\partial V-\varsigma}\frac{q_k}{\theta}n_k \, da - \int_{V-\varsigma}\rho\frac{h}{\theta}d\tau \geq 0 \quad (1.318)$$

$$\int_{V-\varsigma} \rho \frac{D\eta}{Dt} - \left(\frac{q_k}{\theta}\right)_{,k} - \rho \frac{h}{\theta} d\tau - \int_{\varsigma} \left[\left|\rho\eta(v_k - u_k) - \frac{q_k}{\theta}\right|\right] n_k da \geq 0 \qquad (1.319)$$

$$\rho \frac{D\eta}{Dt} - \left(\frac{q_k}{\theta}\right)_{,k} - \rho \frac{h}{\theta} \geq 0 \qquad \forall x \in V - \zeta$$

$$\left[\left|\rho\eta(v_k - u_k) - \frac{q_k}{\theta}\right|\right] n_k \geq 0 \qquad \forall x \in \zeta$$

$$(1.320)$$

The generalized Clausius–Duhem inequality (1.322) is obtained by introducing the Helmotz free-energy function per unit mass ψ (1.321); this inequality has a very important role in the development of the constitutive equations.

$$\psi \equiv \varepsilon - \theta\eta \qquad (1.321)$$

$$-\rho\left(\frac{D\psi}{Dt} + \eta\frac{D\theta}{Dt}\right) + \sigma_{kl}a_{kl} + s_{kl}c_{kl} + m_{klm}b_{lmk} + \frac{q_k\theta_{,k}}{\theta} \geq 0 \qquad \forall x \in V - \zeta \quad (1.322)$$

The functions describing the evolution system, in accordance with the axiom of causality, must depend on the kinematics of the system, on time, and on the temperature of the microcontinuum; these functions, i.e., the dependent variables, are represented by (1.323) for a micromorphic continuum. Further, the state of a body is in thermodynamic equilibrium when the inequality (1.322) acquires the equal sign. To utilize this definition, the set (1.323) of dependent variables is decomposed into two subsets called *recoverable (static)* and *dynamic*. The former is indicated with the subscript R, the latter with the subscript D, and they are defined in (1.324), according to (1.325).

$$\{\psi, \quad \eta, \quad \sigma_{kl}, \quad m_{klm}, \quad s_{kl}, \quad q_k\} \qquad (1.323)$$

$$I_R \equiv \left\{\psi, \quad -\rho\eta_R, \quad \left(\sigma_{kl}\right)_R, \quad \left(m_{klm}\right)_R, \quad \left(s_{kl}\right)_R, \quad \bullet\right\};$$

$$I_D \equiv \left\{\bullet, \quad -\rho\eta_D, \quad \left(\sigma_{kl}\right)_D, \quad \left(m_{klm}\right)_D, \quad \left(s_{kl}\right)_D, \quad q_k\right\}$$

$$(1.324)$$

$$\eta = \eta_R + \eta_D; \quad \sigma_{kl} = \left(\sigma_{kl}\right)_R + \left(\sigma_{kl}\right)_D;$$

$$m_{klm} = \left(m_{klm}\right)_R + \left(m_{klm}\right)_D; \quad s_{kl} = \left(s_{kl}\right)_R + \left(s_{kl}\right)_D$$

$$(1.325)$$

The set I_R is in condition of thermodynamic equilibrium, so the inequality (1.322) and the thermodynamic equilibrium definition lead to (1.326) and, with (1.322), to (1.327), which expresses the dissipation of energy and is typical in irreversible thermodynamics. It can also be represented in a compact form using a scalar product operator (1.328), where Y is the thermodynamic force and is associated with the functions set in (1.329).

$$-\rho\left(\frac{D\psi}{Dt}+\eta_R\frac{D\theta}{Dt}\right)+\left(\sigma_{kl}\right)_R a_{kl}+\left(s_{kl}\right)_R c_{kl}+\left(m_{klm}\right)_R b_{lmk}=0 \qquad \forall x \in V-\zeta \quad (1.326)$$

$$\rho\hat{\eta}\equiv-\rho\eta_D\frac{D\theta}{Dt}+\left(\sigma_{kl}\right)_D a_{kl}+\left(s_{kl}\right)_D c_{kl}+\left(m_{klm}\right)_D b_{lmk}+\frac{q_k\theta_{,k}}{\theta}\geq0 \qquad \forall x \in V-\zeta \quad (1.327)$$

$$\rho\hat{\eta}=I_D\cdot Y \qquad (1.328)$$

$$Y\equiv\left\{\frac{D\theta}{Dt} \quad a_{kl} \quad b_{lmk} \quad c_{kl} \quad \frac{\theta_{,k}}{\theta}\right\} \qquad (1.329)$$

Then, the Clausius–Duhem inequality for micropolar continua is shown in (1.330), where $a_{kl}\ b_{kl}$ are defined in (1.331), and the dissipation of energy is expressed by (1.332) in an explicit and compact form, with I_D and Y (1.333).

$$-\rho\left(\frac{D\psi}{Dt}+\eta\frac{D\theta}{Dt}\right)+\sigma_{kl}a_{kl}+m_{kl}b_{lk}+\frac{q_k\theta_{,k}}{\theta}\geq0 \qquad \forall x \in V-\zeta; \quad (1.330)$$

$$\left[\left|\rho\eta(v_k-u_k)-\frac{q_k}{\theta}\right|\right]n_k\geq0 \qquad \forall x \in \zeta$$

$$a_{kl}\equiv v_{l,k}-\varepsilon_{lkm}v_m; \qquad b_{kl}\equiv v_{k,m} \qquad (1.331)$$

$$\rho\hat{\eta}\equiv-\rho\eta_D\frac{D\theta}{Dt}+\left(\sigma_{kl}\right)_D a_{kl}+\left(m_{kl}\right)_D b_{lk}+\frac{q_k\theta_{,k}}{\theta}\geq0 \qquad \forall x \in V-\zeta; \quad (1.332)$$

$$\rho\hat{\eta}\equiv I_D\cdot Y\geq0 \qquad \forall x \in V-\zeta$$

$$I_D\equiv\left\{-\rho\eta_D, \ \left(\sigma_{kl}\right)_D, \ \left(m_{klm}\right)_D, \ q_k\right\}; \qquad Y\overset{def}{=}\left\{\frac{D\theta}{Dt} \quad a_{kl} \quad b_{lk} \quad \frac{\theta_{,k}}{\theta}\right\} \quad (1.333)$$

The fundamental laws of micromorphic continua consist of a system of 20 partial differential equations, constituted by (1.297), (1.303), (1.305), (1.308), (1.311), and an inequality, (1.320), furnishing a tool to understand whether the mathematical solution can have a physical meaning.

Under the external loads f_k, l_{kl} and the internal heat generated per unit time h, there are 67 unknown functions: the mass density ρ, three components of the speed v_k, six independent components of the microinertia tensor i_{kl}, nine components of v_{kl} microgyration tensor, nine components of σ_{kl} stress tensor, twenty-seven components of the couple stress tensor m_{klm}, six components of the symmetric tensor s_{kl}, three components of the heat flow q_k, the internal energy function ε, the entropy function η, and the temperature function θ. Therefore, the system is highly indeterminate, and 47 independent additional equations are needed for the determination of the motions and the temperature of a micromorphic body. Further, the balance equations are valid for all micromorphic bodies irrespective of their physical constitutions, i.e., for fluids, oils, blood, liquid crystals, elastic solids, gases, plastics, polymers, or bones.

In fact, the characterization of physical properties is established by the empirical formulation of constitutive equations that are functional equations and that express relations between dependent and independent constitutive variables. To construct constitutive equations, there are three necessary conditions according to field theories: the first condition is the axiom of causality; the second is given by actions continuity, which affirms that the physical state of an infinitesimal part of a continuum depends on neighborhoods; the third condition is due to the covariant character of constitutive equations: their mathematical structure must always be the same with reference to all inertial rigid frames.

For local micromorphic elastic solids with no memory, the independent and dependent constitutive variables sets are expressed by (1.334). The ψ free energy *a priori* is represented by the function (1.335), and, substituting the function (1.335) in the relation (1.322), the inequality (1.336) can be written with definitions in (1.337). The relation (1.336) is linear in a_{kl}, b_{lmk}, c_{kl}, and θ derivatives, and it must remain in one sign for all independent variations of these quantities. This involves that coefficients of a_{kl}, b_{lmk}, c_{kl}, and θ derivatives vanish obtaining (1.338), which represents the constitutive equations.

$$y = \left\{ \underline{C}_{KL}, \quad \in_{KL}, \quad \Gamma_{KLM}, \quad \theta_{,K}, \quad \frac{D\theta}{Dt}, \quad \theta, \quad x_0 \right\};$$

$$(1.334)$$

$$Z = \left\{ \sigma_{kl}, \quad m_{klm}, \quad s_{kl}, \quad q_k \quad \varepsilon \right\}$$

$$\psi = \psi \left(\underline{C}_{KL}, \quad \in_{KL}, \quad \Gamma_{KLM}, \quad \theta_{,K}, \quad \frac{D\theta}{Dt}, \quad \theta, \quad x_0 \right) \qquad (1.335)$$

$$-\rho \left(\frac{\partial \psi}{\partial t} + \eta \right) \frac{D\theta}{Dt} + \left(\sigma_{kl} - \rho \frac{\partial \psi}{\partial \underline{C}_{KL}} x_{k,K} \chi_{Ll} \right) a_{kl}$$

$$+ \left(m_{klm} - \rho \frac{\partial \psi}{\partial \Gamma_{KLM}} x_{k,K} \chi_{Ll} \chi_{mM} \right) b_{lmk} \qquad (1.336)$$

$$+ \left(s_{kl} - 2\rho \frac{\partial \psi}{\partial \in_{KL}} \chi_{kK} \chi_{iL} \right) c_{kl} - \rho \frac{\partial \psi}{\partial \theta_{,k}} \dot{\theta}_{,K} - \rho \frac{\partial \psi}{\partial \dot{\theta}} \ddot{\theta} + \frac{q_k \theta_{,k}}{\theta} \geq 0 \qquad \forall x \in V - \zeta$$

$$\dot{\theta} \equiv \frac{D\theta}{Dt}; \qquad \ddot{\theta} \equiv \frac{D^2\theta}{Dt^2} \tag{1.337}$$

$$\sigma_{kl} = \rho \frac{\partial \psi}{\partial \underline{C}_{KL}} x_{k,K} \chi_{Ll}; \qquad m_{klm} = \rho \frac{\partial \psi}{\partial \Gamma_{KLM}} x_{k,K} \chi_{Ll} \chi_{mM}; \qquad s_{kl} = 2\rho \frac{\partial \psi}{\partial \in_{KL}} \chi_{kK} \chi_{iL};$$

$$\frac{\partial \psi}{\partial \theta_{,k}} = 0; \qquad \frac{\partial \psi}{\partial \dot{\theta}} = 0; \tag{1.338}$$

$$\eta_R = -\frac{\partial \psi}{\partial \theta}; \qquad -\rho \eta_D \dot{\theta} + \frac{q_k \theta_{,k}}{\theta} \ge 0 \quad \forall x \in V - \zeta$$

Constitutive equations of micropolar solids can be expressed in an alternative form (1.339), with the definitions of strain tensors of micropolar (1.340) and with the dissipation potential Φ, a scalar function shown in (1.341). Further, the compact form (1.342) can be obtained substituting micropolar relations in the local law of energy balance (1.316).

$$\psi = \psi\left(\underline{C}^*_{KL}, \ \Gamma_{KL}, \ \theta, \ x\right); \qquad \eta_R = -\frac{\partial \psi}{\partial \theta};$$

$$\sigma_{kl} = \rho \frac{\partial \psi}{\partial \underline{C}^*_{KL}} x_{k,K} \chi_{iL}; \qquad m_{kl} = \rho \frac{\partial \psi}{\partial \Gamma_{KL}} x_{k,K} \chi_{iL} \tag{1.339}$$

$$q_k = \frac{\partial \Phi}{\partial\left(\dfrac{\theta_{,k}}{\theta}\right)}; \qquad -\rho \eta_D = \frac{\partial \Phi}{\partial \dot{\theta}}$$

$$\underline{C}^*_{KL} \equiv x_{k,K} \chi_{LK}; \qquad \Gamma_{KL} \equiv \frac{1}{2} \varepsilon_{KMN} \chi_{kM,L} \chi_{kN} \tag{1.340}$$

$$\Phi = \Phi\left(\frac{D\theta}{Dt}, \ \frac{\nabla\theta}{\theta}, \ \underline{C}^*_{KL}, \ \Gamma_{KL}, \ \theta, \ x_0, \right) \tag{1.341}$$

$$\nabla_\gamma \Phi = I_D$$

$$-\rho\left(\dot{\theta}\eta_D + \theta\dot{\eta}_D + \theta\dot{\eta}_R\right) + \nabla \cdot q + \rho h = 0 \tag{1.342}$$

1.5.4 Electromagnetic Interactions

In the presence of electromagnetic fields, the balance equations must be reformulated together with the constitutive equations because external electromagnetic fields are modified by the properties of polarization and magnetization of the microcontinuum body; they stress the body and depend on its physical–chemical structure.

Local laws of balance for micromorphic continua without electromagnetic interactions are given by (1.297), (1.303), (1.305), (1.308), and (1.311), whereas, in presence of electromagnetic fields, some of these equations change. For the law of balance of momentum (1.305), only the jump condition changes, and it is replaced by (1.343), where σ^E_{kl} and G_l are relative to electromagnetic interactions; σ^E_{kl} is the electromagnetic tensor and is similar to the classical electromagnetic tensor, deducible from Maxwell equations; and G_l is the electromagnetic momentum (1.344). For the local law of energy balance given by equations (1.311), both equations change and are replaced by (1.345), where W^E is the electromagnetic energy and S_k is the Poynting vector.

$$\left[\left|\sigma_{kl}+\sigma^E_{kl}+u_k G_l-\rho v_l\left(v_k-u_k\right)\right|\right]n_k=0 \quad \forall x\in\zeta \tag{1.343}$$

$$G\equiv\frac{1}{c}E\wedge B \tag{1.344}$$

$$-\rho\frac{D\varepsilon}{Dt}+\sigma_{kl}a_{kl}+s_{kl}c_{kl}+m_{klm}b_{lmk}+q_{k,k}+\rho h+W^E=0 \quad \forall x\in V-\zeta$$

$$\left[\left|\left(\sigma_{kl}+\sigma^E_{kl}+u_k G_l\right)v_l+m_{klm}v_{lm}+q_k-S_k-\left(\rho\varepsilon+\frac{1}{2}\rho v\cdot v+\frac{1}{2}\rho i_{rl}v_{mr}v_{ml}\right.\right.\right. \tag{1.345}$$

$$\left.\left.\left.+\frac{1}{2}E\cdot E+\frac{1}{2}B\cdot B\right)\left(v_k-u_k\right)\right|\right]n_k=0 \quad \forall x\in\zeta$$

In the case of micropolar continua in the presence of electromagnetic fields, the balance of momentum for the same as micromorphic continua (1.305), and the local energy balance is given by equation (1.346).

$$-\rho\frac{D\varepsilon}{Dt}+\sigma_{kl}a_{kl}+m_{kl}b_{lk}+q_{k,k}+\rho h+W^E=0 \quad \forall x\in V-\zeta$$

$$\left[\left|\left(\sigma_{kl}+\sigma^E_{kl}+u_k G_l\right)v_l+m_{kl}v_l+q_k-S_k-\left(\rho\varepsilon+\frac{1}{2}\rho v\cdot v+\frac{1}{2}\rho i_{mn}v_m v_n\right.\right.\right. \tag{1.346}$$

$$\left.\left.\left.+\frac{1}{2}E\cdot E+\frac{1}{2}B\cdot B\right)\left(v_k-u_k\right)\right|\right]n_k=0 \quad \forall x\in\zeta$$

1.5.5 Micropolar Elasticity

A linearization is applicable when, during the system evolution from the reference configuration to the deformed configuration, the displacements, rotations, and temperature variations are small; therefore, in order to describe adequately the system dynamic behavior, the displacement vector function u, the microdisplacement vector function φ, and the temperature change function T are introduced and defined by (1.347).

$$u \equiv x - x_0; \qquad \varphi \equiv \xi - \xi_0; \qquad T \equiv \theta - T_0 \qquad |T| \ll T_0 \quad \wedge \quad T_0 > 0 \quad (1.347)$$

Functions u, φ, and T represent variations with respect to the reference configuration that has T_0 temperature, and point particles are located by the vector functions x_0 and ξ_0. The array W (1.348) is essential for the development of the linear theory, and its norm is indicated by δ, as shown in (1.349).

$$W \equiv \{u_{,k}, \quad \varphi, \quad \varphi_{,k}, \quad T_{,k}, \quad T\} \qquad (1.348)$$

$$\delta \equiv \|W\| = W \cdot W = u_{i,k}u_{j,k}\delta_{ij} + \varphi_i\varphi_l\delta_{li} + T_{,k}T_{,l}\delta_{kl} + T^2 \qquad (1.349)$$

Linear expressions of functions $x_{k,K}$, of microdeformation tensor χ_{kK}, temperature change T, speed vector v, microgyration vector υ, microinertia tensor j_{kl}, and density ρ are, respectively, given by (1.350); the linear expressions of deformation tensors (1.340) are given by (1.351). After fixing a common system of reference between the reference configuration and the deformed configuration, δ_{kK} and δ_{lL} become Kronecker deltas, and relations (1.351) can be simplified as in (1.352). Further, expressions of deformation tensors (1.352) are replaced by analogous deformation tensors (1.353).

$$x_{k,K} = \left(\delta_{kl} + u_{k,l}\right)\delta_{lK} + O\left(\delta^2\right); \qquad \chi_{kK} = \left(\delta_{kl} + \phi_{kl}\right)\delta_{lK} + O\left(\delta^2\right);$$

$$T_{,k} = \theta_{,k}; \qquad v_k = \frac{\partial u_k}{\partial t} + O\left(\delta^2\right); \qquad \mathrm{v}_k = \frac{\partial \phi_k}{\partial t} + O\left(\delta^2\right); \qquad (1.350)$$

$$j_{kl} = j_{0kl} + j_{0km}\phi_{lm} + j_{0lm}\phi_{km} + O\left(\delta^2\right); \qquad \frac{\rho}{\rho_0} = 1 - u_{k,k} + O\left(\delta^2\right)$$

$$\underset{\sim}{C}^*_{KL} = \left(u_{l,k} + \varepsilon_{klm}\phi_m\right)\delta_{kK}\delta_{lL} + \delta_{KL}; \qquad \Gamma_{KL} = \phi_{k,l}\delta_{kK}\delta_{lL} \qquad (1.351)$$

$$\underset{\sim}{C}^*_{KL} = u_{l,k} + \varepsilon_{klm}\phi_m + \delta_{KL}; \qquad \Gamma_{KL} = \phi_{k,l}; \qquad K = k \quad \wedge \quad L = l \qquad (1.352)$$

$$\varepsilon_{kl} \equiv u_{l,k} + \varepsilon_{klm}\phi_m; \qquad \gamma_{kl} \equiv \phi_{k,l} \qquad (1.353)$$

The linear approximation for the stress, couple stress, and temperature fields requires a quadratic approximation for the free energy ψ; thus, its expression is given by (1.354), where Σ_0, η_0, C_0, A_{kl}, B_{kl}, A_{klmn}, B_{klmn} and C_{klmn} are constitutive moduli of a solid elastic linear micropolar continua generally depending on ρ_0, T_0, j_{0KL}, and x_0 (1.355).

$$\rho_0 \psi = \Sigma_0 - \rho_0 \eta_0 T - \left(\frac{\rho C_0}{2T_0}\right) T^2 - A_{kl} T \varepsilon_{kl} - B_{kl} T \gamma_{kl}$$

$$+ \frac{1}{2} A_{klmn} \varepsilon_{kl} \varepsilon_{mn} + \frac{1}{2} B_{klmn} \gamma_{kl} \gamma_{mn} + C_{klmn} \varepsilon_{kl} \gamma_{mn} \tag{1.354}$$

$$A_{klmn} = A_{mnkl}; \qquad B_{klmn} = B_{mnkl}; \qquad K_{kl} = K_{lk} \tag{1.355}$$

The dissipation potential Φ for solid elastic linear micropolar continua is approximated by (1.356), where K_{kl} is the thermal conductivity tensor; then, constitutive equations of solid elastic linear micropolar bodies are obtained (1.357), substituting the expressions (1.354) and (1.356) in equations (1.339).

$$\Phi = \frac{1}{2T_0^2} K_{kl} T_{,K} T_{,l} \tag{1.356}$$

$$\eta_R = \eta_0 + \frac{C_0}{T_0} T + \frac{1}{\rho} A_{kl} \varepsilon_{kl} + \frac{1}{\rho} B_{kl} \gamma_{kl}; \qquad \sigma_{kl} = A_{kl} T + A_{klmn} \varepsilon_{mn} + C_{klmn} \gamma_{mn};$$

$$m_{kl} = -B_{lk} T + B_{lkmn} \gamma_{mn} + C_{klmn} \varepsilon_{mn}; \qquad q_k = \frac{1}{T_0} K_{kl} T_{,l}; \qquad \eta_D = 0 \tag{1.357}$$

Generally, the number of independent constitutive moduli for solid elastic linear micropolar continua is 196, for solid elastic linear isotropic micropolar continua, it decreases to nine, and the microinertia tensor becomes diagonal with only one independent component. These assertions are expressed by relations (1.358). The isotropic material moduli C_0, β_0, λ, μ, α, β, γ, and K, and the microinertia j, are functions of x for inhomogeneous materials, and they are constants for homogeneous materials.

$$A_{kl} = \beta_0 \delta_{kl}; \qquad B_{kl} = 0; \qquad \frac{K_{kl}}{T_0} = K \delta_{kl};$$

$$A_{klmn} = \lambda \delta_{kl} \delta_{mn} + (\mu + \kappa) \delta_{km} \delta_{ln} + \mu \delta_{kn} \delta_{lm};$$

$$B_{klmn} = \alpha \delta_{kl} \delta_{mn} + \beta \delta_{kn} \delta_{lm} + \gamma \delta_{km} \delta_{ln}; \qquad C_{klmn} = 0; \tag{1.358}$$

$$j_{kl} = j \delta_{kl}$$

Constitutive equations of solid linear elastic isotropic micropolar can be expressed as in (1.359), substituting relations (1.358) in (1.354) and in

(1.357), whereas relations (1.360) can be deduced from thermodynamic considerations.

$$\rho_0\psi = \Sigma_0 - \rho\left(\eta_0 T + \frac{\rho C_0}{2T_0}T^2\right) - \beta_0 T\varepsilon_{mm} + \frac{1}{2}\left[\lambda\varepsilon_{kk}\varepsilon_{ll} + (\mu+\kappa)\varepsilon_{kl}\varepsilon_{kl} + \mu\varepsilon_{kl}\varepsilon_{kl}\right]$$

$$+\frac{1}{2}\left[\alpha\gamma_{kk}\gamma_{ll} + \beta\gamma_{kl}\gamma_{kl} + \gamma\gamma_{kl}\gamma_{kl}\right];$$

$$\eta = \eta_R + \eta_D = \eta_0 + \frac{C_0}{T_0}T + \frac{\beta_0}{\rho}\varepsilon_{mm}; \tag{1.359}$$

$$\sigma_{kl} = -\beta_0 T\delta_{kl} + \lambda\delta_{kl}\varepsilon_{mm} + (\mu+\kappa)\varepsilon_{kl} + \mu\varepsilon_{lk};$$

$$m_{kl} = \alpha\delta_{kl}\gamma_{mm} + \beta\gamma_{kl} + \gamma\gamma_{kl};$$

$$q_k = KT_{,k}$$

$$3\lambda + 2\mu + \kappa \geq 0; \quad 2\mu + \kappa \geq 0; \quad \kappa \geq 0; \quad 3\alpha + \beta + \gamma \geq 0;$$

$$\gamma + \beta \geq 0; \quad \gamma - \beta \geq 0; \quad C_0 \geq 0; \quad K \geq 0; \tag{1.360}$$

$$\rho \geq 0; \quad j \geq 0$$

Linear approximations of kinematic quantities and constitutive equations allow a linearization of micropolar field equations. In linear theory, the equations of conservation of mass and microinertia are considered to be given properties of the body at the natural state. Thus, linear equations of balance of momentum, balance of moments of momentum, and energy balance are, respectively, represented by (1.361), (1.362), and (1.363).

$$\sigma_{kl,k} + \rho\left(f_l - \frac{\partial^2 u_l}{\partial t^2}\right) = 0 \quad \forall x \in V - \zeta \tag{1.361}$$

$$m_{kl,k} + \varepsilon_{lmn}\sigma_{mn} + \rho\left(l_l - j_{kl}\frac{\partial^2\phi_k}{\partial t^2}\right) = 0 \quad \forall x \in V - \zeta \tag{1.362}$$

$$\rho T_0\frac{\partial\eta}{\partial t} - q_{k,k} - \rho h = 0 \quad \forall x \in V - \zeta \tag{1.363}$$

In the case of solid linear elastic isotropic homogeneous microplar continua, material moduli are constant; therefore, a field equation can be written in terms of the displacement u, rotation Φ, and temperature change T. Thus, substituting relations (1.254), (1.255), and (1.256) in (1.361), (1.362), and (1.363), respectively, the equations (1.364), (1.365), and (1.366) represent the linear equations of balance of momentum, balance of moments of momentum, and the linear law of balance of energy.

$$(\lambda+2\mu+\kappa)\nabla\nabla\cdot u-(\mu+\kappa)\nabla\wedge\nabla\wedge u+$$

$$\kappa\nabla\wedge\phi-\beta_0\nabla T+\rho\left(f-\frac{\partial^2 u}{\partial t^2}\right)=0 \qquad\qquad \forall x\in V-\zeta \qquad (1.364)$$

$$(\alpha+\beta+\gamma)\nabla\nabla\cdot\phi-\gamma\nabla\wedge\nabla\wedge\phi+\kappa\nabla\wedge u$$

$$-2\kappa\phi+\rho\left(l-j\frac{\partial^2\phi}{\partial t^2}\right)=0 \qquad\qquad \forall x\in V-\zeta \qquad (1.365)$$

$$\rho C_0\dot{T}+\beta_0 T_0\nabla\cdot\frac{\partial u}{\partial t}-\nabla\cdot(K\nabla T)-\rho h=0 \qquad \forall x\in V-\zeta \qquad (1.366)$$

2

Modeling Actuators

2.1 A Preliminary Outlook

To provide an idea of a meso- and micro-actuator, we refer to the context of a meso-/micromachine, i.e., a system that uses small control energy to cause an observable or a controllable perturbation to the environment. Within this context, we are interested only in machines able to generate an observable mechanical perturbation to the environment, i.e., a machine able to generate a perturbation on environmental mechanical properties such as position, velocity, acceleration, force, pressure, and work. These types of machines can be defined as "mechanical machines." Within the group of mechanical machines, we are specifically interested in machines able to generate meso-/microperturbations to the environment, i.e., perturbations in the environmental mechanical properties that are recorded, respectively, in meso-/microunits, or in multiples of meso-/microunits, defined in terms of SI units (e.g., a 1 μm/mm change of the position of an object or a generation of a 1 μN/mN force on an object). Figure 2.1 illustrates how mechanical meso-/micromachines fit into the scheme of machines and meso-/micromachines.

From a functional standpoint, a mechanical machine is a system composed of an actuator, a transmission, and a user. The actuator generates the mechanical work; the transmission transforms this work and connects the actuator with the user; and the user acts directly on the environment (Figure 2.2).

A meso- and micro-actuator can have macrodimensions to generate a meso-/microenvironmental perturbation; however, for our purposes, we will consider actuators with at least one dimension on the order of millimeters/microns. So, we are referring to meso-/micromechanical actuators in the strictest sense (Figure 2.3).

In this chapter, the term meso- and micro-actuator will be used to describe a meso-/micromechanical actuator of less than 1 cm/mm or almost in the strictest sense (a very small meso-/micromechanical actuator larger than 1 cm/mm).

The functional characteristics of an actuator are those of its mechanical output. These can be described by the relationship between generated torque and angular speed (if the actuator produces rotary motion) or by the relationship between generated force and linear speed (if the actuator produces linear motion). For this discussion, we will assume a rotary motion. If the

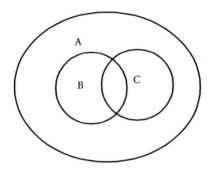

FIGURE 2.1
Definition of a mechanical meso-/micromachine, where *A* is the set of all machines, *B* is the set of mechanical machines, and *C* is the set of meso-/micromachines; the intersection of *B* and *C* is the set of mechanical meso-/micromachines.

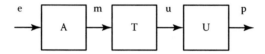

FIGURE 2.2
Functional definition of a mechanical machine, where *A* is the actuator, *T* is the transmission, *U* is the user, *e* is an energy, *m* is a mechanical function, *u* is a usable mechanical function, and *p* is an environmental perturbation.

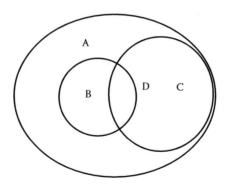

FIGURE 2.3
Definition of a meso-/micromechanical actuator in strict sense, where *A* is the set of all the actuators, *B* is the set of mechanical actuators, *C* is the set of meso- and micro-actuators, and *D* is the set of meso- and micro-actuators in the strictest sense; the set of meso-/micromechanical actuators in the strictest sense is the intersection of *B*, *C*, and *D*.

motor is operating under steady-state conditions (input power, load to overcome, and the environment are constant), the relationship between generated torque and angular speed can be graphed by a line on the torque-speed plane. This relationship is called the actuator characteristic (Figure 2.4).

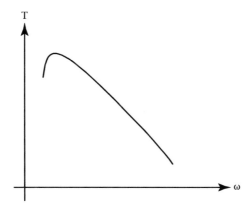

FIGURE 2.4
Mechanical characteristics of an actuator: T is the generated torque and ω is the generated angular speed.

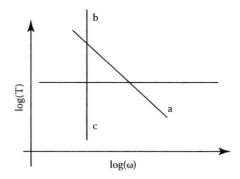

FIGURE 2.5
Ideal actuator: *a* is a torque generator, *b* is a speed generator, and *c* is a power generator.

On a bilogarithmic scale, it is possible to define three groups of ideal actuators (Figure 2.5): torque generators, described by horizontal straight lines (where T is constant); speed generators, described by vertical straight lines (where ω is constant); and, power generators, described by oblique straight curves (where the product between T and ω is constant).

Ideal characteristics are, in some respects, approximations of properties of real actuators, and usually, they can be manipulated with one or more control variables allowing adjustments or important transformations; therefore, a set of possible actuator characteristics can be described by an extended portion of the T–ω plane (Figure 2.6). Further, the actuator is also able to provide power for a limited amount of time, outside the standard field of the actuator characteristics, creating a joint field or an overloaded field (Figure 2.7).

We want to describe an actuator in terms of its mechanical output and, referring to Figure 2.2, this is represented by the user motion and load; furthermore, the description of the desired mechanical user input is similar to

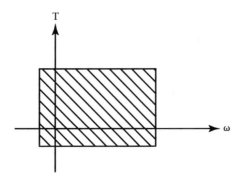

FIGURE 2.6
Field of the actuator characteristics.

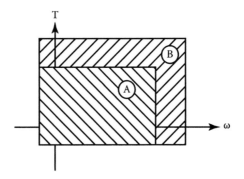

FIGURE 2.7
Fields of the actuator characteristics: A is the nominal field and B is the overloaded field.

the mechanical actuator output. In fact, the two related variables are the user torque (or force) and the user angular speed (or linear speed).

The same reference should be used to compare the desired user motion and load with the available actuator; hence, referring to Figure 2.2, we use the transmission function to transform the actuator output to the user input (2.1), where T_u and ω_u are, respectively, the user torque and speed, whereas f and g are the direct transmission functions, which can be dependent upon many variables and parameters synthesized into vector x.

$$T_u = f(T, x) \qquad \omega_u = g(\omega, x) \tag{2.1}$$

The inverse transmission function can be used to transform the desired user input into the desired actuator output (2.2), where T_{du} and ω_{du} are, respectively, the desired user torque and speed; T_d and ω_d are, respectively, the desired actuator torque and speed; while f^{inv} and g^{inv} are the inverse transmission functions.

$$T_d = f^{inv}(T_{du}, x) \qquad \omega_d = g^{inv}(\omega_{du}, x) \tag{2.2}$$

Now, we can compare the desired actuator characteristic (T_d vs. ω_d) with the available actuator characteristic (T vs. ω): if the desired actuator characteristic is within the available actuator field of characteristics, we can use this available actuator for our needs (Figures 2.8 and 2.9).

The optimal selection of an actuator should take into account the transmission, and they are usually selected simultaneously. Some suggestions for this can be found in Histand and Alciatore (1999), Gross et al. (2000), and Legnani, Adamini, and Tiboni (2002). The optimal selection of a micro-actuator is a very difficult task. Many parameters, such as environmental conditions, should be considered. Thus, big approximations are frequently acceptable.

There are two additional problems that complicate the correct selection of an actuator. First, the micro-actuators are not always well characterized from a mechanical viewpoint; therefore, a precise selection is sometimes impossible. Second, in the microtransmission, a scale factor can introduce undesired dynamic phenomena such as high friction or an elastic joint can be used to avoid complex connections; however, it exhibits nonlinear dynamic

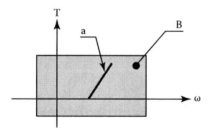

FIGURE 2.8
Comparing desired actuator with available actuator: *a* is the desired actuator characteristic and *B* is the field of the available actuator characteristics; *a* is within the field *B*, so the available actuator can be used for our needs.

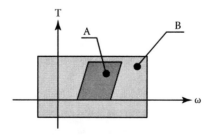

FIGURE 2.9
Comparing desired actuator with available actuator: *A* is the field of the desired actuator characteristics and *B* is the field of the available actuator characteristics. *A* is within field *B*, so the available actuator can be used for our needs.

behavior. For these reasons, some specialized optimization algorithms can be applied via a custom software program to achieve an optimal selection. These selection techniques should take into account many other properties of the considered actuators; one of the most interesting characteristics is the linearity, that refers to the input–output function of the actuator, where the input is a physical property of the input energy (current, tension, temperature, etc.) and the output is a physical property of the mechanical output energy (position, speed, acceleration, power, etc.). The maximum difference between a reference linear line and the actuator output is a measure of the linearity (Figure 2.10); in fact, a high linearity is synonymous with a simple relationship between input and output and it implies facility of commanding the actuator.

An actuator is used to generate a motion, that can differ, in general, from the commanded motion according to the positioning precision, or, specifically, to the accuracy, the repeatability, and the resolution (Figure 2.11). Accuracy is the distance between the average attained position and the target position; repeatability is the average distance between the different attained positions and the same target position, whereas resolution is the minimum incremental motion of the actuator.

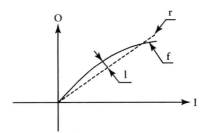

FIGURE 2.10
Definition of linearity: *O* is the actuator output, *I* is the actuator input, *r* is the reference line, *f* is the input–output function, and *l* is a measure of the linearity.

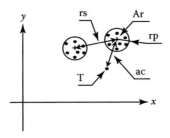

FIGURE 2.11
Definition of planar positioning precision: *x* and *y* are two spatial coordinates, *T* is the target position, *a* is a measure of the accuracy, *Ar* is the average reached position, *rs* is a measure of the resolution, and *rp* is a measure of the repeatability.

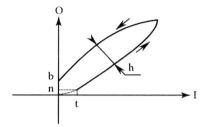

FIGURE 2.12
Definition of hysteresis, threshold, noise, and backlash: O is the actuator output, I is the actuator input, t is a measure of the threshold, h is a measure of the maximum hysteresis, b is a measure of the backlash, and n is a measure of the noise.

The dynamics of the actuator can be influenced by different undesired phenomena, including hysteresis, threshold fluctuations, noise, and backlash; any of these can reduce the performance of the system: hysteresis is the difference in the actuator output when it is reached from two opposite directions (Figure 2.12); the threshold is the smallest initial increment of the input able to generate an actuator output; the noise is a measure of the fluctuations in the output with zero input; and backlash is the lost motion after reversing direction.

When a meso- and micro-actuator is composed of two or more parts mutually connected by mechanical joints, additional mechanical properties of this connection should be taken into account; in particular, the connection can be affected by compliance in terms of mechanical backlash, level of stiffness, friction, or mechanical hysteresis, affecting the dynamics of the actuator. Mechanical backlash is the jarring reaction or striking back of a mechanical part in the actuator when the motion is not uniform or when a sudden force is applied; the level of stiffness is the ability to convert a force into a displacement without any deformations; friction is the resistance that any body meets when moving over another body; and mechanical hysteresis is the difference in the actuator displacement when it is reached from two opposite directions.

The particular properties of the actuator can be dependent on some parameters or can be related to one another, i.e., referring to the quasi-static behavior of the actuator (the force-displacement curve), the load is able to change this curve: the stiffness is dependent upon the applied load. Similarly, the sensitivity of the actuator, i.e., the ratio of change in the actuator output to an incremental change in its input, is usually a function of the temperature. Many general or specialized dynamic models were developed to describe these characteristics, and some of them will be proposed in this chapter. It is significant to note how the introduction of smart materials into the development of actuation technology has changed the way actuators are regarded, not only during the design stage, but also during the performance evaluation stage: the use of smart materials can reduce the number of bodies forming the actuator to one or two, minimizing the problem of connections; the number of design possibilities is significantly expanded by the introduction of smart materials, owing to their ability to generate different shapes and movements. For these reasons, the development of a micro-actuator needs to remain "open-minded" and the evaluation of its performance should take new factors into consideration. Therefore, some of these promising smart materials are listed in Table 2.1 and some of their characteristics are evaluated to understand their degree of usability.

Undoubtedly, many meso- and micro-actuators are present on the market and in research laboratories; therefore, a classification would be useful to frame this chapter into a schematic knowledge system. A first definition of meso- and micro-actuators can be based on their mechanical output, particularly on the generated motion. If the actuator is embedded within the other components of the machine, it is difficult to distinguish the actuator from the machine; so, the complete machine can be thought of as the actuator and this embedded actuator is, sometimes, also able to generate complex motions. If the actuator is only connected to the transmission of the machine, it is able to generate only simple motions; therefore, this second choice is usually selected to increase the modularity of the system. A scheme of the classification, based on generated motion, is shown in Table 2.2.

Some actuation technologies allow for the integration of many identical micro-actuators (or nanoactuators) combined to form a unique actuator (or micro-actuator): every integrated micro-actuator is able to exert a force and to generate a simple motion (a linear motion for this example); if the

TABLE 2.1

Properties of Some Actuators Made with Smart Materials

Properties	Electrostrictive	Electrorheological	Magnetostrictive	NiTi SMA	Piezoceramic
Cost	Moderate	Moderate	Moderate	Low	Moderate
Maturity	Moderate	Moderate	Moderate	Fair	Good
Networkable	Yes	Yes	Yes	Yes	Yes
Embedability	Good	Fair	Good	High	High

TABLE 2.2

Classification of Micro-Actuators Based on the Output Motion

Type of Embedding	Generated Motion
Actuator embedded in the machine	Complex three-dimensional Complex bi-dimensional motion Linear motion Rotational motion
Modular actuator	Linear motion Rotational motion
Special unmodular and unembedded actuator	Complex three-dimensional Complex bi-dimensional motion

TABLE 2.3

Classification of Micro-Actuators Based on Integration/Digitalization Level

Integration of the Design	Digitalization of Mechanical Output
Single component	Analogical output
Serial integration	Analogical force and possible digital stroke
Parallel integration	Possible digital force and analogical stroke
Serial and parallel integration	Possible digital force and possible digital stroke

micro-actuators are integrated in a parallel configuration, the force exerted by the new actuator is the sum of the forces exerted by every single micro-actuator and the stroke of the new actuator is the same stroke of each single micro-actuator (Mavroidis 2002); if the micro-actuators are integrated in a serial configuration, the force exerted by the new actuator is the same force exerted by each single micro-actuator and the stroke of the new actuator is the sum of the strokes of each single micro-actuator. If each integrated micro-actuator is able to exert only a null force or its maximum exertable force and if it is able to generate only a null motion or the total stroke motion, then the resulting actuator is able to exert a digitalized force or a digitalized stroke, depending on the integration configuration. Therefore, it is possible to define an actuator based on the digitalization level of the mechanical output or on the integration level of the design (see Table 2.3).

Finally, a classification of micro-actuators can be based on the type of input energy (Table 2.4): the most important are electrical, fluidic, thermal, chemical, optical, and acoustic. This chapter will adopt a classification of meso- and micro-actuators based on the form of the input energy and will deal with only some of the actuators listed in Table 2.4.

2.2 Electrostatic Actuators

The electrostatic actuators are able to accumulate electrostatic energy and, subsequently, to convert it into mechanical work. To perform these operations a schematic system is assumed, and it is composed by two conducting bodies at different electric potential levels for a sufficient period to accumulate the desired quantity of electrostatic energy; at this point, everybody sensible to the generated electrostatic field is subjected to a force and its eventual motion is directed to an energetic transformation tending to rebalance the whole system from an energy viewpoint. Before deepening the characteristics of such an energetic transformation, a distinction between different typologies of actuators is proposed on the basis of their principal functional and morphological properties. We can discriminate between actuators for prismatic or revolute movements. From a morphological viewpoint, the first

TABLE 2.4

Classification of Meso- and Micro-Actuators Based
on the Input Energy

Input Energy	Physical Class	Actuator
Electrical	Electric and magnetic field	Electrostatic
		Electromagnetic
	Molecular forces	Piezoelectric
		Piezoceramic
		Piezopolymeric
		Magnetostrictive
		Electrostrictive
		Magnetorheological
		Electrorheological
Fluidic	Pneumatic	High pressure
		Low pressure
	Hydraulic	Hydraulic
Thermal	Thermal expansion	Bimetallic
		Thermal
		Polymer gels
	Shape memory effect	Shape memory alloys
		Shape memory polymers
Chemical	Electrolytic	Electrochemical
	Explosive	Pyrotechnical
Optical	Photomechanical	Photomechanical
		Polymer gels
Acoustic	Induced vibration	Vibrating

ones exhibit an axis of symmetry that lies on a plane of symmetry, and the prismatic motion occurs along this axis, whereas the others show an axis of symmetry, identified by a sheaf of symmetry planes, and it represents the revolute axis of the actuator. These devices can be realized with various shapes, but, for technological reasons, simple shapes are usually preferred and this constraint does not excessively reduce the kinematic possibilities, because the most demanded ones are prismatic and revolute: particularly, the configuration made up of parallel plates is commonly used to realize linear motions, whereas a configuration made up of circular coplanar and concentric conductors is commonly used to realize rotary motions. Furthermore, from a functional viewpoint, a distinction between continuative and oscillating movement can be useful: the first is characterized by the absence of motion inversion, whereas the latter repeats itself with a constant frequency. For a special class of device, the movement produced by the actuator can go out of the plane where the actuator is realized: this interesting property is obtained through a proper diaphragm that, after compression, bends and rises from the actuating plane. The inversion of the motion can be gained through a passive external force, usually an elastic spring or a constant force, or through an active external force, i.e., an opposite actuator.

Therefore, a wide range of device typologies can be deduced, including the most common models but neglecting some prototypes for special purposes. The combination of elementary actuators can generate complex devices that can be identified with a unique actuator or with a machine for special purpose: generally, the difference in the definition is associated with the major or minor functional flexibility of the system or with the presence of a special purpose in addition to the movement. Examining the working principle of these devices, a major characteristic is the ability to accumulate electrostatic energy through a capacitor effect that produces a potential difference V between the faces of two conductors accumulating an electric charge $+Q$ on one of them and an opposite charge $-Q$ on the other one; therefore, the capacitance C that allows this function is defined by the ratio between Q and V, and it depends only on the geometrical shapes and on the material constituting the system. The expression of the electric potential energy U_e stored in the capacitor can be obtained as the work W required to move a charge dq from a conductor to the other in the presence of a potential $v(q)$, integrating until the final charge Q is reached (2.3); therefore, using the just enunciated definition of capacitance, the expression (2.4) can be gained to describe the electrostatically stored energy U_E for every pair of conductors independent of their geometrical configuration.

$$U_E = W = \int_0^Q v(q)dq \tag{2.3}$$

$$U_E = \int_0^Q \frac{q}{C} dq = \frac{Q^2}{2C} = \frac{1}{2}QV = \frac{1}{2}CV^2 \tag{2.4}$$

Then, a pair of planar parallel conductors is examined to simplify the problem without losing too much of the level of applicability; furthermore, the ratio between the distance d between the parallel plates and the principal dimensions of their area A is small. One of these conductors has a charge $+Q$; therefore, we can find the produced electric field (Figure 2.13) by examining an infinitesimal volume dV in it with having a charge dq and applying the Gauss–Green theorem.

If the surface is homogeneously and uniformly charged, neglecting the border effects, the expression (2.5) allows the quantification of the electric field intensity E_1 near the faces of the conductor.

$$\oint E \cdot dA = (-E_1)(-A) + E_1 A = 2E_1 A = \frac{Q}{\varepsilon} \tag{2.5}$$

Examining both oppositely charged plates (Figure 2.14) under the hypotheses just enunciated and adding the superposition principle, the value of

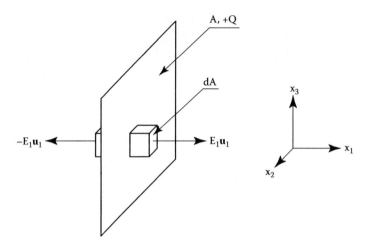

FIGURE 2.13
Electric field produced by a planar conductor homogeneously charged.

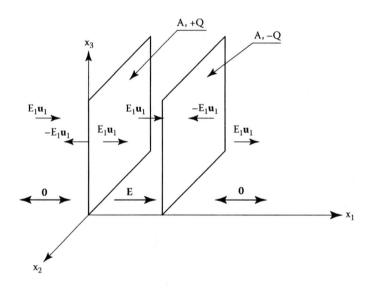

FIGURE 2.14
Electric field produced by two planar parallel conductors.

the constant electric field **E** between the conductors can be computed (2.6), whereas the outer field is null.

$$\begin{cases} E = 2E_1 u_1 = \dfrac{Q}{A\varepsilon} u_1, & 0 < x_1 < d \\ E = E_1 u_1 - E_1 u_1 = 0, & x_1 < 0 \vee x_1 > d \end{cases} \tag{2.6}$$

Therefore, through the expression (2.7), the potential difference from one end of the capacitor to the other can be calculated; then, with the definition of capacity C as the ratio between stored charge Q and voltage V, C can be simply obtained (2.8) and, applying (2.4), the potential energy stored in the capacitor can be computed (2.9).

$$V \equiv -\int_d^0 E dx_1 = \frac{Q}{A\varepsilon} d, \quad 0 < x < d \tag{2.7}$$

$$C \equiv \frac{Q}{V} = \frac{A\varepsilon}{d} \tag{2.8}$$

$$U_E = \frac{1}{2}\frac{A\varepsilon}{d}V^2 \tag{2.9}$$

The negatively charged conductor is supposed to be rigidly connected to the frame with a fixed mechanical constraint, whereas the positively charged conductor can be moved perpendicularly to the faces of the capacitor (Figure 2.15); therefore, because the distance d between the two faces is variable, it is renamed as x_1, and F_1 (2.10) expresses the force exerted by the electric field to produce the motion of the positive conductor corresponding to an equivalent mechanical reaction on the fixed negative conductor.

$$F_1 = -\frac{\partial U_E}{\partial x_1} = \frac{1}{2}V^2 \frac{A\varepsilon}{(x_1)^2} \tag{2.10}$$

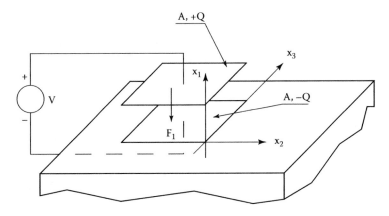

FIGURE 2.15
Electrostatic linear actuator composed by planar faces.

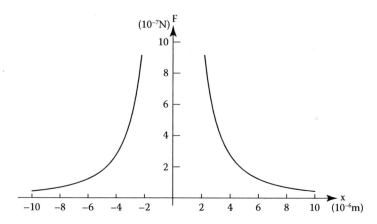

FIGURE 2.16
Force versus position curve for a planar-plates electrostatic actuator.

Equation (2.10) can be used to identify a relation intervening between the exerted attraction force and the relative position of the mobile face with respect to the fixed face; furthermore, an interesting property is constituted by the independence of the sign of the electric force F_1 from the sign of the electric tension V: it should be interpreted as a univocally attractive force between the two plates (Figure 2.16). Incrementing the geometrical parameter identified by the area A or the material parameter identified by the electric constant ε, the curve F_1 versus V depicted in Figure 2.16 is subjected to a distancing from the Cartesian axes. An analogous result can be obtained with an increment of the modulus of the electric tension V, allowing control of the characteristic curve of the actuator to set the desired position on the mobile plate with a proper compensation of the effects produced by external forces.

Another interesting structure can be constituted with two oppositely charged plates, both mechanically fixed to the frame to avoid relative motions; furthermore, a solid dielectric mobile core is interposed between the plates (Figure 2.17).

This system tends toward configurations of energetic equilibrium attracting the dielectric material into the structure with a force tangential to the plates; as a matter of fact, denoting by U_E the electrostatic energy stored between two parallel capacitors containing air and dielectric material (2.11), the attraction force of the resulting capacitor can be expressed by (2.12). In this model the attraction force F_2 is independent of the actuation variable x_2 and its modulus increases with increments of the structural parameters ε_d and h, or with a reduction of the gap d between the plates or incrementing the control variable V.

$$U_E = \frac{1}{2}V^2\left[\varepsilon_d \frac{x_2 h}{d} + \varepsilon_{air} \frac{(l - x_2)h}{d}\right] \qquad (2.11)$$

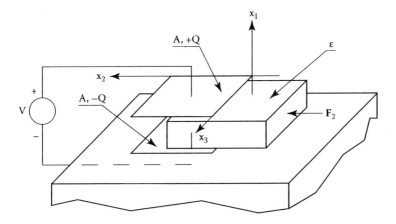

FIGURE 2.17
Electrostatic linear actuator constituted by planar plates with a dielectric core.

$$F_2 = -\frac{\partial U_E}{\partial x_2} = \frac{1}{2}V^2\frac{(\varepsilon_d - \varepsilon_{air})}{d}h \tag{2.12}$$

The electrostatic techniques are usually used to produce actuators of small dimensions that can also be parts of a complex system. We can consider, for example, a series of elementary plane-plates actuators (as proposed in Figure 2.15) made up of a pile of modular actuators fed only at the extremities with a tension V (Figure 2.18); therefore, the resulting capacity of the system can be expressed by (2.13), and it is able to produce an actuation force identified by (2.14).

$$C = \left(\sum_{i=1}^{n}\frac{1}{C_i}\right)^{-1} = \left(\sum_{i=1}^{n}\frac{x_{1i}}{\varepsilon \cdot A_i}\right)^{-1} = \frac{\varepsilon \cdot A_i}{n \cdot x_{1i}} = \frac{\varepsilon \cdot A_i}{x_1} \tag{2.13}$$

$$F = F_i = \frac{1}{2}\frac{\varepsilon \cdot A_i}{x_{1i}^2}V_i^2 = \frac{1}{2}\frac{\varepsilon \cdot A_i}{x_{1i}^2}\frac{V^2}{n^2} = \frac{1}{2}\frac{\varepsilon \cdot A_i}{x_1^2}V^2 \tag{2.14}$$

Equation (2.14) shows how the series of electrostatic actuators is able to multiply the stroke executable by a single elementary actuator; in fact, from an energetic viewpoint, the work performed by n actuators is n times the work performed by an elementary actuator and the multiplication of the work is revealed mechanically on the stroke and electrically on the tension (Table 2.5).

Then, the expression (2.14) shows also that a series of n electrostatic actuators is equivalent to a single actuator with a stroke equal to n times the stroke of an elementary actuator. An increment of mechanical work is due to an

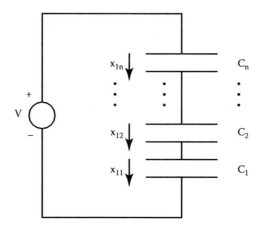

FIGURE 2.18
Pile of electrostatic linear actuators with plane plates.

TABLE 2.5

Effect of a Series of Electrostatic Actuators

Mechanical	Electrical
$F_1 = F_{1i}$	$V = n \cdot V_i$
$x_1 = n \cdot x_{1i}$	$q = q_i$
$F_1 \cdot x_1 = n \cdot F_{1i} \cdot x_{1i}$	$V \cdot q = n \cdot V_i \cdot q_i$

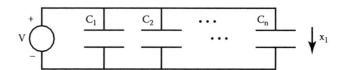

FIGURE 2.19
Parallel configuration of linear electrostatic actuators with planar plates.

increment of fed electric energy; therefore, the equivalent feeding tension V should be equal to n times the tension V_i of an elementary actuator.

An alternative setup can be identified with a parallel configuration of the plane plates actuators just proposed in Figure 2.15, as shown schematically in Figure 2.19, where the moving plates are electrically connected and mechanically united.

The mechanical work of a single elementary actuator can also be multiplied in this system, obtaining an increment of the exerted force, whereas the corresponding increment of electric energy is associated with a higher level of electric charge (Table 2.6).

TABLE 2.6

Effect of a Parallel Setup

Mechanical	Electrical
$F_1 = n \cdot F_{1i}$	$V = V_i$
$x_1 = x_{1i}$	$q = n \cdot q_i$
$F_1 \cdot x_1 = n \cdot F_{1i} \cdot x_{1i}$	$V \cdot q = n \cdot V_i \cdot q_i$

The resulting force F_1 is equal to n times the elementary force F_{1i}, as listed in Table 2.6 and in (2.15), whereas the resulting capacity C is equal to the sum of the elementary capacities C_i (2.16).

$$F = n \cdot F_i = n \frac{1}{2} \frac{\varepsilon \cdot A_i}{x_{1i}^2} V_i^2 = \frac{1}{2} \frac{\varepsilon \cdot (n \cdot A_i)}{x_{1i}^2} V_i^2 = \frac{1}{2} \frac{\varepsilon \cdot A}{x_1^2} V^2 \qquad (2.15)$$

$$C = \sum_{i=1}^{n} C_i = n \frac{\varepsilon \cdot A_i}{x_{1i}} = \frac{\varepsilon \cdot (n \cdot A_i)}{x_{1i}} = \frac{\varepsilon \cdot A}{x_1} \qquad (2.16)$$

The increment of work is produced by a higher afflux of the charge q at the faces of the equivalent capacitor without an increment of feeding electric power; therefore, we expect a slowing of the overall actuator with respect to an elementary actuator, unless there is an overfeeding of tension or a control of the feeding electric current with a proper current generator.

Complex systems of elementary actuators can also be constituted by electrostatic transversal actuators, as represented in Figure 2.16; in this connection, a pile of electrostatic transversal actuators is assumed to be provided with dielectric mobile cores mechanically jointed each to other to form a single rigid body (Figure 2.20).

Also, this configuration of elementary actuators is able to multiply the mechanical work accomplished by a single actuator, incrementing the exerted force F_2 (2.17) corresponding to an increment of electrical tension as shown in Table 2.7.

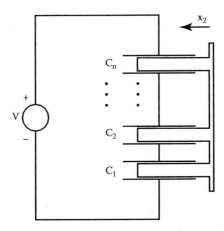

FIGURE 2.20
Series of transversal linear electrostatic actuators with planar plates and dielectric core.

TABLE 2.7

Effect of a Pile of Transversal Actuators

Mechanical	Electrical
$F_2 = n \cdot F_{2i}$	$V = n \cdot V_i$
$x_2 = x_{2i}$	$q = q_i$
$F_2 \cdot x_2 = n \cdot F_{2i} \cdot x_{2i}$	$V \cdot q = n \cdot V_i \cdot q_i$

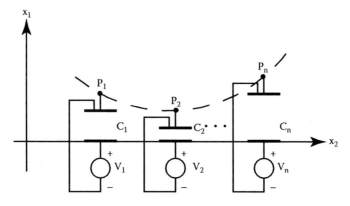

FIGURE 2.21
Parallel geometrical configuration of autonomously fed actuators.

$$F = n \cdot F_i = n \frac{1}{2} V_i^2 \frac{\varepsilon_d - \varepsilon_{air}}{d_i} h = n \frac{1}{2} \frac{V^2}{n^2} \frac{\varepsilon_d - \varepsilon_{air}}{d_i} h = \frac{1}{2} V^2 \frac{\varepsilon_d - \varepsilon_{air}}{d} h \qquad (2.17)$$

Changing the approach, the combination of elementary actuators can be devoted, instead of multiplying the mechanical work, to actively controlling the high number of mechanical degrees of freedom to obtain complex functional characteristics with simple modular elements. Therefore, n electrostatic actuators with planar plates (Figure 2.15) are geometrically arranged in a parallel configuration and are autonomously fed (Figure 2.21).

Referring to Figure 2.21, the movements x_{1i} of the mobile plates are associated with the control tensions V_i and to the external loads F_i with the relation (2.18).

$$x_{1i} = V_i \sqrt{\frac{A \cdot \varepsilon}{F_i}} \qquad (2.18)$$

Thus, denoting the midpoint of the mobile part of the capacitor with P_i, its coordinates are identified by x_{1i} along the axis x_1 and by the product between the length of the segment l and the index i associated with P_i (2.19).

$$p_i\left(V_i, F_i\right) \equiv \begin{bmatrix} x_{1i}\left(V_i, F_i\right) \\ x_{2i}\left(V_i, F_i\right) \end{bmatrix} = \begin{bmatrix} V_i\sqrt{\dfrac{A \cdot \varepsilon}{F_i}} \\ l \cdot i \end{bmatrix} \tag{2.19}$$

Interpolating opportunely the coordinates p_i of the points P_i, a relation can be obtained to represent a one-value function $x_2(x_1)$ or a multivalue function $x_1(x_2)$ (2.20).

$$x_1\left(x_2, V(x_2), F(x_2)\right) = V(x_2)\sqrt{\dfrac{A \cdot \varepsilon}{F(x_2)}} \tag{2.20}$$

The function represented in (2.20) can be physically generated with a close approximation by using a high number of actuators compared with their dimensions, i.e., a high density. Incrementing the density of actuators, the possibility of mutual interactions between adjacent devices becomes more and more important; therefore, complex models can become useful to describe their behavior in order to properly control it. Furthermore, when the actuator density is high, a reduced number of possible feeding tensions $V(x_2)$ can be used to reduce the effect of electrostatic disturbances; particularly, selecting only two levels of tension, we can obtain a digitalization of the functional characteristic (2.20) for the device that can be achieved associating the usual logical values equal to zero and one.

2.3 Electromagnetic Actuators

The realization of a mechanical force on a conductor fed with an electric current can be applied to wide areas of instrumentation, of automation, and of power production at macroscopic or microscopic levels; therefore, these devices are particularly suitable to deepen scale problems. The variety of these devices and of their applications suggests how to plan properly the description of this class of actuators. From a geometrical viewpoint, the most common shapes can be identified with simple configurations, as planar, cylindrical, spherical, toroidal, and conical or, more generally, some complex geometries can be used for special purposes. The required trajectory for these actuators is represented, normally, by simple prismatic or revolute movements, but, for special purposes, complex movements can be realized with one or more degrees of freedom. The stroke is usually limited in actuators able to produce trajectories that can be represented with open paths, whereas it can be unlimited when the producible trajectories are represented with closed paths: these simple observations can be invalidated by mobile actuators, that can realize virtually unlimited strokes also for open trajectories; on the other hand, design problems can suggest the realization of limited strokes for some actuators that are conceptually associated with unlimited strokes.

From a structural viewpoint, these devices can be realized using conductors fed with electric current, derived from an external supply or induced by an electromagnetic field; furthermore, the use of permanent magnets is common in different models.

Only some electromagnetic devices are shown in the following to make clear their working principle: a linear actuator represented by a mobile conductor immersed in an electric field, another linear actuator realized with an electromagnet, and three classes of rotary actuators, respectively, DC, AC, and synchronous. Particularly, the rotary actuators are made up of a mobile component, the rotor, and an immobile component, the stator.

2.3.1   Linear Actuator with a Mobile Conductor

The mobile conductor is geometrically represented by an element with cylindrical symmetry and with a length associated with the preferential axial dimension; therefore, the properties of the body can be concentrated on the neutral axis of the cylinder, obtaining a 1-D structure identified with the segment l and its material points are locally provided with geometrical properties, as the local section; and with electrical properties, as the charge density. This conductor is immersed in a constant magnetic field B normal to the movement plane and it is immobile at the beginning of the observation period; furthermore, to realize an electric circuit, it is constrained to translate on two conductive slide-guides normal to the conductor and to the magnetic field B and rigidly connected with the immobile observer. These two slide-guides are set at a proper electric potential difference with a generator of constant voltage V combined in series with a resistor identified with the equivalent resistance R that represents synthetically the resistance of the whole circuit (Figure 2.22).

The potential difference V shown in Figure 2.22 produces an electric current i that flows into the electric circuit; therefore, the mobile conductor is subjected to a force F that causes its translating movement with an increasing speed u (2.21).

$$F = i \cdot l \times B \tag{2.21}$$

When the conductor begins its movement under the action of the force F, the field B induces an electromotive force V_C opposed to V in accordance with the relation (2.22), incrementing its value with an increment of the speed u, till the value V is matched as shown in (2.23).

$$V_C = B \times u \cdot l \tag{2.22}$$

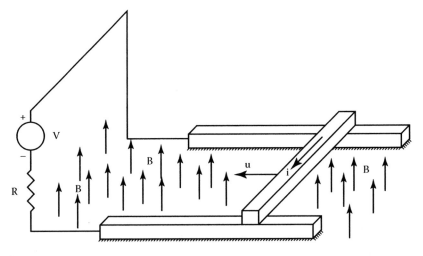

FIGURE 2.22
Translating conductors immersed in an electromagnetic field **B** normal to the movement plane.

$$v = \frac{V}{L \cdot B} \tag{2.23}$$

From an electric viewpoint, the whole system can be represented by a simplified electric circuit as appears in Figure 2.23, generating the electric power $P_{E,tot}$ represented in (2.24), where the first term P_E imparts a prismatic movement to the conductor, although the latter term P_J is dissipated by the Joule effect.

$$P_{E,tot} = V \cdot i = i \cdot l \times B \cdot u + R \cdot i^2 = P_E + P_J \tag{2.24}$$

The proposed geometrical configuration is particularly simple because every vectorial quantity that is used in the electrometrical model is parallel or perpendicular to the others; thus, we can use only a scalar quantity to iden-

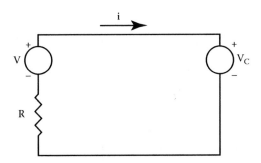

FIGURE 2.23
Equivalent electric circuit of a translating conductor immersed in an electromagnetic field **B**.

tify the net electric power P_E usable for the conversion into mechanical power P_M. Therefore, neglecting the friction phenomena, the expression (2.25) can be a reference to identify an ideal system of electromechanical conversion.

$$P_E = V \cdot i = Blu \cdot i$$
$$P_M = F \cdot u = Bli \cdot u$$

(2.25)

The expression (2.25), which represents the equivalence between electric power P_E and mechanical power P_M exchanged into the ideal actuator, opens the door to the concept of ideal electromechanical transformer (Lynch and Truxal 1962), allowing an interesting circuital representation (Figure 2.24), where the mechanical component is constituted by an ideal force generator that depends on the electric current i flowing the bar, although the electric component is constituted by an ideal tension depending on the mechanical speed u of the bar.

To improve the accuracy of the lumped parameters model, an electric resistance R and the coefficient r of mechanical friction can be introduced to represent the dissipative factors, whereas the inertial properties can be identified with a mass M and the respective inductive properties can be associated with an inductance L, resulting in an increment of one order of differentiability for the mathematical model expressed by (2.26) and represented by the circuital scheme in Figure 2.25.

$$\begin{cases} V - Ri - L\dfrac{di}{dt} = Blu \\[2mm] F - ru - M\dfrac{du}{dt} = -Bli \end{cases}$$

(2.26)

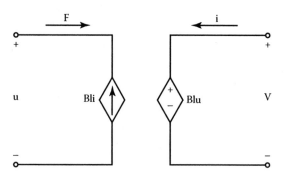

FIGURE 2.24
Circuital model of an ideal electromechanical transformer.

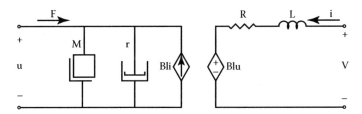

FIGURE 2.25
Circuital model of an electromechanical transformer.

2.3.2 Electromagnetic Transducer

A magnetic circuit is constituted by an electromagnet provided with a component rigidly constrained to the frame and electrically fed, and by a mobile component hinged to the previous one; this configuration identifies, rigorously, a rotary actuator, but, for the slightness of the stroke, it can be approximated to a prismatic actuator. Between the fixed armature and the mobile core there is a gap, with a depth *l* and an area *A*, that mechanically allows the mobility of the moving component and magnetically represents a source of reluctance. The armature is fed by an electric circuit winded on it with *N* coils flowed by a current *i* and wholly provided with a resistance *R* (Figure 2.26).

The electromechanical conversion can be analyzed under a virtual displacement *dl* of the mobile component and, then, equalling the energy that enters in the system to the sum of the first term, representing the variation of stored energy, and of the latter term, describing the lost energy. The entering mechanical energy is equal to the product between the motor force *F* and the virtual displacement *dl*, and because it is è work exerted by the actuator, it assumes conventionally a negative sign (exiting energy). If, during the displacement the feeding current *i* is regulated to maintain a constant magnetic flux Φ into the magnetic circuit, an induced electromotive force is not present

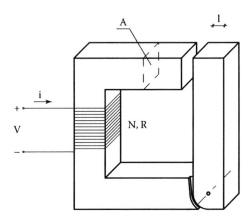

FIGURE 2.26
Electromagnetic transducer.

in the system because the variation of magnetic flux is null. The electrical energy entering the system is equal to the product of the entering electric power and the time period *dt* associated with the displacement; the supply circuit is provided with a resistance *R*, thus expressing the electric power as the product of the resistance *R* and the square of the feeding current. Neglecting the inertial effects, the variation of stored energy has only a magnetic nature and it is due to the reduction of gap; particularly, neglecting the border effects, it is equal to the output between the volume of gap and the density of magnetic energy produced by the magnetic field *B*. Finally, energetic losses can be associated only with the Joule effect in the feeding circuit; therefore, a simplified form can be identified by the expression (2.27).

$$-Fdl + Ri^2 dt = -\frac{1}{2}\frac{B^2}{\mu_0} Adl + Ri^2 dt \qquad (2.27)$$

The expression (2.27) allows computation of the force produced by the actuator, resulting in equation (2.28), which is also valid when the magnetic flux Φ or the magnetic field *B* is not constant; furthermore, the expression (2.28) represents the maximal developable force, because, practically, the induction is limited by the saturation.

$$F = \frac{1}{2}\frac{B^2}{\mu_0} A \qquad (2.28)$$

The magnetic permeability in the metal is higher than in the gap; therefore, the magnetomotive force is employed to overcome the reluctance of the gap. To determine the relation between produced mechanical force *F* and feeding current *i*, the magnetic intensity *B* can be indicated as a ratio between the magnetic flux Φ and the area of the gap *A*; furthermore, the magnetic flux Φ can be expressed as the ratio of the magnetomotive force *Ni* and the reluctance of the gap, obtaining the expression (2.29) for the magnetic field.

$$B = \frac{Ni}{(l/\mu_0 A) A} = \frac{\mu_0 Ni}{l} \qquad (2.29)$$

2.3.3 DC Actuators

Referring only to rotary actuators, different classes of DC devices can be discriminated depending on whether the statoric field is produced with electromagnetic wires or it is realized with permanent magnets. When electromagnets are used, different types of DC devices can be identified for different field excitations. When permanent magnets are used, the shape of the rotor, cylindrical or discoid, is a discriminant. Thus, a wide range of

TABLE 2.8

DC Actuators

Electromagnetic wires	Independent excitation	
	Series excitation	
	Shunt excitation	
	Compound excitation	
Permanent magnets	Cylindrical rotor	Slotless rotor
		Moving coil
		Slotted rotor
	Discoid rotor	With wrapped wires
		With printed windings

constructive classes can be noted, and their principal members can be listed as in Table 2.8.

DC meso-actuators are commonly used with or without brushes whereas DC micro-actuators can be produced with permanent magnets on the rotor, avoiding brushes because of thermal problems related with brush friction. To propose a model of the actuator, we can select a cylindrical geometry for rotational movements with an unlimited stroke (Figure 2.28). A single conductive coil on the cylindrical rotor is able to rotate around its axis while it is embedded in a magnetic field B, and it is fed with an electric current i as in Figure 2.27; therefore, due to the action of the magnetic field on the electric current, the coil is affected by the mechanical couple in (2.30), where τ is the mechanically generated couple, A is the area vector perpendicular to the coil, and B is the magnetic field intensity.

$$\tau = i \cdot A \times B \qquad (2.30)$$

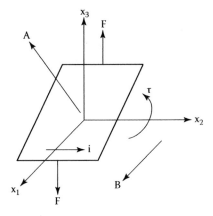

FIGURE 2.27
Coil with an area A, an electric current i, and embedded in a magnetic field B.

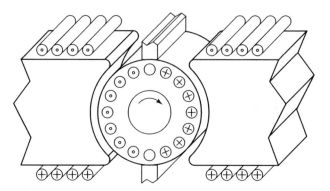

FIGURE 2.28
Cross-section scheme of a DC micro-actuator.

Then we can assume that the rotor is subject to the force computed (2.31), where θ is the angular coordinate that identifies the position of the rotor in respect to the micro-actuator's stator. It results in a set of n coils equally spaced on the rotor, affected by a set of p magnetic fields perpendicular to the rotor axis, electrically fed with the same current i in all the coils, and with the electric potential of the rotor applied to the coils by brushes, which are used to change the direction of the current every half-turn of the rotor.

$$\tau(\theta) = i \cdot |A| \cdot |B| \cdot \sum_{j=0}^{n-1} \sum_{k=0}^{p-1} \left| \cos\left(\theta + \pi \cdot \frac{j \cdot p + k \cdot n}{p \cdot n}\right) \right| \qquad (2.31)$$

When the number of coils or the number of magnetic fields is sufficiently high, the expression for mechanical torque can be further simplified, as in (2.32); therefore, under the assumption that the magnetic field is generated by the stator's current, the magnetic torque can be represented as the product of a constant term and two electric variables: the rotor's and the stator's currents (2.33).

$$\tau = i \cdot |A||B| \cdot n \cdot \frac{2p}{\pi} \qquad (2.32)$$

$$\tau = k_s \cdot i \cdot i_s \qquad (2.33)$$

To obtain the value of the angular speed of the rotor, we temporarily disregard dissipations and other "parasitic phenomena" due to their effect occuring only before and after the transformation from electric to mechanical energy. For this reason, all the input electrical energy from the rotor is converted into mechanical energy and the angular speed of the rotor can be expressed as in (2.34), where e is the electrical potential of the rotor.

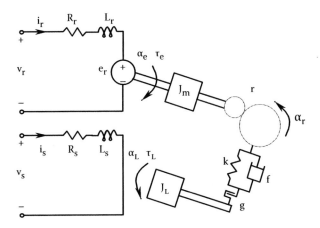

FIGURE 2.29
Circuit model of a DC micro-actuator. The subdivision takes place in three parts: (*s*) is the stator's electric circuit, (*r*) is the rotor's electric circuit, and (*m*) is the mechanical circuit of the rotor. (*R*) is the mechanical circuit of the rotor after the speed reduction stage is integrated in the micro-actuator.

$$\omega_m = e / \left(k_s i_s \right) \qquad (2.34)$$

With the aid of the transformation equations (2.33–2.34), the micro-actuator behavior can be described from a standpoint of circuit's approach as in Figure 2.29.

The symbols mentioned in Figure 2.29 are summarized as follows:

(*s*) Stator's circuit

v_s is the stator's electric potential

i_s is the stator's current

R_s and L_s are the resistance and the inductance of the stator

(*r*) Rotor's circuit

v_r is the rotor's electric potential

i_r is the rotor's current

R_r and L_r are the resistance and the inductance of the rotor

e_r is the rotor's potential that, linked with the rotor's current, is transformed in mechanical power whose value is related with the stator's current and with the theoretical motion speed as in (2.35)

(*m*) Mechanical stage

τ_e is the theoretic motor torque, which depends upon the rotor's and stator's currents (2.36)

α_e is the theoretic position of the rotor

J_m is the motor's inertia

J_L is the load's inertia

f is the reduced friction coefficient of the system

k is the reduced stiffness coefficient of the system

g is the reduced backlash coefficient of the system

α_R is the theoretic position after the reduction stage (2.37)

α_L is the position of the load

α_L is the applied torque of the load

$$e_r = k_s i_s \dot{\alpha}_e \qquad (2.35)$$

where k_s is the electric-mechanic transformation constant.

$$\tau_e = k_s i_s i_r \qquad (2.36)$$

$$\alpha_e = \alpha_r \cdot r \qquad (2.37)$$

where $r\,(>1)$ is the mechanical reduction ratio. Then, a lumped parameter model of the micro-actuator can be obtained with a simple circuit analysis (2.38).

$$\begin{cases} v_s = R_s i_s + L_s \dfrac{di_s}{dt} \\[2mm] v_r = R_r i_r + L_r \dfrac{di_r}{dt} + e_r \\[2mm] e_r = k_s i_s \dot{\alpha}_e \\[2mm] \alpha_R = \alpha_e / r \\[2mm] \tau_e = k_s i_s i_r \\[2mm] \tau_R \equiv k\left(\alpha_R - \alpha_L\right) + f\left(\dot{\alpha}_R - \dot{\alpha}_L\right) \\[2mm] \left(\tau_e - \dfrac{\tau_R}{r}\right) = J_m \ddot{\alpha}_e \\[2mm] \tau_R - \tau_L = J_L \ddot{\alpha}_L \end{cases} \qquad (2.38)$$

The result of the model in (2.38) is an approximation of the nonlinear model for the behavior of the electromechanical micro-actuator after a proper setting of the considered parameters. If any of the parameters of the model are not able to reach the desired predictive precision, more complex models could be considered in order to take into account neglected phenomena. The introduction of fictitious mechanical compliances, electrical parasitic phenomena could be followed by a calibration procedure ruled directly by real experimental results or by a finite element model. Should a model not be implemented in a controlling algorithm, a finite element model can be directly used to provide a good predictive result. Some useful finite element software is available on the market to achieve a proper model; thus, creating new finite element software is not generally necessary.

The electric input variables are related to the mechanical variables in the previous model without auxiliary magnetic parameters. In the following, a different model is proposed to stress the presence of induced magnetic phenomena for some design configuration of DC actuators. Particularly, the presence of an excitation current is made explicit with its ability to produce an induced electromotive force through a magnetization phenomenon. Then, inertial properties, backlashes, and compliances are neglected to focus attention on conversion effects, obtaining simple linear circuital models in stationary regimes. These models can be used also as a stand-alone simulator or as a module of a complex system. From a nomenclatural viewpoint, the electrical variables associated with the stator are also called field variables, because the stator is devoted to the generation of the electromagnetic field in which the rotor is immersed. The electrical variables referred to the rotor are also called armature variables, because of the geometrical distribution of the active electromagnetic material on the rotor's surface. In addition to these variables, the excitations identify two interesting characteristics: the magnetization that produces the magnetic flux Φ and the core losses that takes into account the hysteretic dissipations and the eddy currents. The excitation can be independent or it can be generated by the field sources that can also feed the armature; another useful distinction is between shunt excitation, where a low electric current flows in many coils with reduced dimensions, and between series excitation, where a higher current flows in a reduced number of coils. Therefore, a schematic representation of different types of excitation (Figure 2.30) can be obtained denoting the armature, excitation, and load characteristics, respectively, with the subscript A, F, and L.

In the case of a shunt actuator, a simple circuital scheme can be proposed for stationary conditions (Figure 2.31), and it can be algebraically represented with linear relations (2.39). Particularly, the supply tension V of the actuator produces a load current i_L that is partially used to excite the system with i_F and partially used to feed the rotor with i_A. Furthermore, the excitation current i_F, can be set within a certain range through a proper rheostat R_C. The excitation current i_F produces a field flux Φ and, consequently, an induced electromotive tension E that is related to i_F through an experimentally obtainable magnetization curve (Figure 2.32) or through a linearized expression. This curve or expression allows the determination of the functional characteristic of the actuator in terms of a relation between the produced torque T and speed Ω (2.40), as represented in Figure 2.33. Furthermore, the expression (2.40) approximates to a curve with constant speed, because the tension fall at the armature, equal to the product between armature current i_A and armature resistance R_A, is usually negligible comparing with the

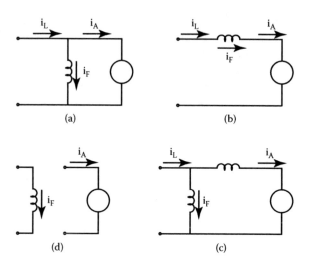

FIGURE 2.30
Examples of excitation connections, particularly (a), (b), (c), and (d) represents, respectively, the shunt, series, compound, and independent configurations.

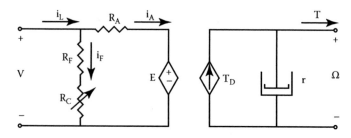

FIGURE 2.31
Circuital diagram of a DC actuator with shunt excitation.

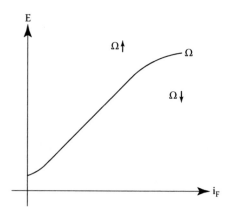

FIGURE 2.32
Experimental magnetization curve.

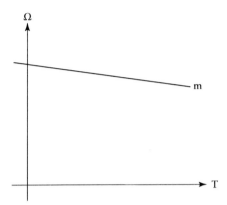

FIGURE 2.33
Functional characteristic of a DC actuator with shunt excitation.

tension V, whereas the friction loss due to r is necessarily less than the active component of the actuator.

$$\begin{cases} E = K\Phi\Omega; \quad T_d = K\Phi i_A \\ V = E + i_A R_A \\ T = T_d - \Omega \cdot r \end{cases} \tag{2.39}$$

$$T = \frac{K\Phi}{R_A} V - \left[\frac{(K\Phi)^2}{R_A} + r \right] \Omega \tag{2.40}$$

In the actuator with series excitation, the excitation current i_F assumes a value that can be compared with the armature current i_A; consequently, the functional behavior of this actuator is substantially different from the behavior of a shunt actuator. In fact, referring to the circuital model in Figure 2.34 and assuming, for the sake of simplicity, a linear relation between magnetization

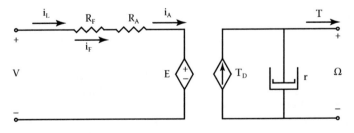

FIGURE 2.34
Circuital model of a DC actuator with series excitation.

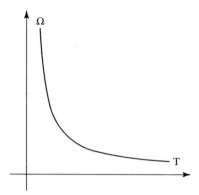

FIGURE 2.35
Functional characteristic of a DC actuator with series excitation.

and armature current (2.41), an algebraic representation can be obtained as in (2.42). This expression allows us to stress the prevalence of an inverse relationship between the produced torque T and the square of the speed Ω (2.43) in the functional characteristic, as shown in Figure 2.35, because the resistive term due to R_A and R_F assumes often a negligible value in (2.43), whereas the friction component is inferior to the active component if the actuator does not act as a brake.

$$\Omega = k \cdot i_A \qquad (2.41)$$

$$\begin{cases} E = K\Phi\Omega; \quad T_d = K\Phi i_A \\ V = E + i_A\left(R_A + R_F\right) \\ T = T_d - \Omega \cdot r \end{cases} \qquad (2.42)$$

$$T = V^2 Kk\left[Kk\Omega + \left(R_A + R_F\right)\right]^{-2} - \Omega \cdot r \qquad (2.43)$$

2.3.4 Induction Actuators

We observe a common configuration consisting of a cylindrical rotary actuator with unlimited strokes with a three-phase actuation system, where the transformation from electric to mechanical energy is due to the inductance of the mobile core and of the stator (Figure 2.36). The rotor has a cylindrical shape and is able to rotate around its axis, whereas the stator has the same axis as the rotor and is separated from it by an air gap; both are composed of ferromagnetic material and incorporate lengthwise holes carrying conductive wires that are close to the air gap.

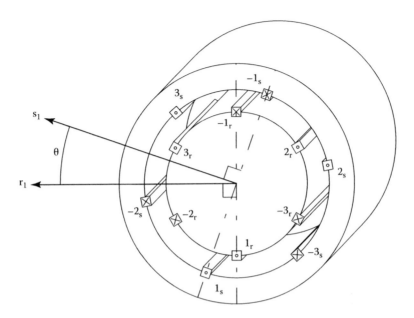

FIGURE 2.36
Definition of the position of the rotor's windings: i_r and j_s represent, respectively, the ith rotor's winding and the jth stator's winding; r_i and s_j represent, respectively, the axis of the ith rotor's winding and the axis of the jth stator's winding, and θ is the angular position of the first rotor's winding in reference to the position of the first stator's winding.

A reference system and an angular variable, which identifies the position of the rotor (Figure 2.36), should be defined to develop a mathematical model of the dynamic behavior for the considered actuator; in particular, the variable θ_i denotes the position of the ith rotor's winding (r_i) in respect to the first stator's winding (s_1) (2.44), where θ is indicated in Figure 2.36, as the angle between s_1 and r_1, and f is the number of phases.

$$\theta_i = \theta + \frac{2\pi}{f}(i-1) \tag{2.44}$$

The proposed dynamic model can be generalized as a multipole micromachine for which the ideal actuator speed can be calculated using (2.45), where p is the (even) number of poles, θ is indicated in Figure 2.36, and θ_a is the ideal position of the actuator. Therefore, the dynamic electromagnetic behavior of the system can be described by the matrix equation (2.46), where the vector e of electric potential is defined in (2.47), the vector i of electric currents is defined in (2.47), the vector φ of magnetic fluxes is defined in (2.47), the matrix R of electric resistances is defined in (2.48), and the matrix L of inductances is defined in (2.49).

$$\dot{\theta}_a = 2\dot{\theta}/p \tag{2.45}$$

$$\begin{cases} e = R \cdot i + \dot{\varphi} \\ \varphi = L \cdot i \end{cases} \tag{2.46}$$

$$e = \begin{bmatrix} e_{s1} & e_{s2} & e_{s3} & e_{r1} & e_{r2} & e_{r3} \end{bmatrix}^T,$$

$$i = \begin{bmatrix} i_{s1} & i_{s2} & i_{s3} & i_{r1} & i_{r2} & i_{r3} \end{bmatrix}^T, \tag{2.47}$$

$$\varphi = \begin{bmatrix} \varphi_{s1} & \varphi_{s2} & \varphi_{s3} & \varphi_{r1} & \varphi_{r2} & \varphi_{r3} \end{bmatrix}^T$$

where the index associated with each element of the three vectors refer to the stator/rotor winding shown in Figure 2.36.

$$R = \begin{bmatrix} I(3) \cdot R_s & 0(3) \\ 0(3) & I(3) \cdot R_r \end{bmatrix} \tag{2.48}$$

where R_s and R_r are, respectively, the stator and rotor winding resistances, and $I(3)$ and $0(3)$ are, respectively, the 3-D identity and zero matrixes.

$$L = \begin{bmatrix} L_{1,1} & L_{1,2} \\ L_{2,1} & L_{2,2} \end{bmatrix} \tag{2.49}$$

where the submatrixes $L_{1,1}$ and $L_{2,2}$, respectively, of statoric and rotoric inductances are defined in (2.50), whereas the submatrixes $L_{1,2}$ and $L_{2,1}$ of mutual inductances are defined in (2.51).

$$L_{1,1} = \begin{bmatrix} L_s & M_s & M_s \\ M_s & L_s & M_s \\ M_s & M_s & L_s \end{bmatrix}, \quad L_{2,2} = \begin{bmatrix} L_r & M_r & M_r \\ M_r & L_r & M_r \\ M_r & M_r & L_r \end{bmatrix} \tag{2.50}$$

where L_s and L_r are the self-inductance of, respectively, each stator's and each rotor's winding, and M_s and M_r are the mutual inductance of, respectively, two stator's or two rotor's windings.

$$L_{1,2} = L_{2,1}^T = \begin{bmatrix} l_{i,j} \end{bmatrix}_{i=1\ldots3, j=1\ldots3}, \quad l_{i,j} = M_{sr} \cos\left[\theta + \frac{2\pi}{3}(j-i) \right] \tag{2.51}$$

where M_{sr} is the mutual inductance between a stator's and a rotor's winding and θ is indicated in Figure 2.36. Coupled with the appropriate electric dynamics and with consideration of speed reduction stage, the mechanical equilibrium equations (2.52) would develop in the complete micromechanical model of the system.

$$\begin{cases} \left(\tau_e - \dfrac{\tau_R}{r} \right) = J_m \ddot{\theta} \\ \tau_R - \tau_L = J_L \ddot{\theta}_L \\ \tau_R \equiv k\left(\theta_R - \theta_L \right) + f\left(\dot{\theta}_R - \dot{\theta}_L \right) \end{cases} \qquad (2.52)$$

where τ_e is the electric torque generated by the electromagnetic interaction (2.56), τ_R is torque transferred by the compliance of the mechanical stage, τ_L is the torque of the load; J_m and J_L are, respectively, the inertia of the motor and the inertia of the load; r (> 1) is the mechanical reduction ratio; k and f are, respectively, the stiffness and the friction coefficients of the reduction stage; θ, θ_R, and θ_L are, respectively, the theoretical position of the rotor, the theoretical position after the reduction stage (2.53), and the position of the load.

$$\theta = \theta_R \cdot r \qquad (2.53)$$

The computation of the electric torque τ_e is then related to the calculation of the generated power (2.54).

$$p = e^T i$$

$$= \left(R \cdot i + \dot{\varphi} \right)^T i \qquad (2.54)$$

$$= \left(i^T R^T i \right)_1 + \left(i^T \dot{L}^T i \right)_2 + \left(i^T L^T i \right)_3$$

where e and i are, respectively, the vector of the electric potentials and the vector of the electric currents (2.47), R is the matrix of the electric resistance (2.48), φ is the vector of magnetic fluxes (2.47), L is the matrix of inductances (2.49). The three resulting terms of the power equation (2.54) are, respectively, the electric power converted into heat, the electric power converted into mechanical power, and the variation of electromagnetic power of the system. The second term, the conversion of electric power into mechanical power, is used to obtain the electric torque τ_e; only the evaluation of the time derivative of the transposed matrix of the inductances is required (2.55).

$$\dot{L}^T = \begin{bmatrix} 0 & \dot{L}_{21} \\ \dot{L}_{12} & 0 \end{bmatrix}, \quad \dot{L}_{12} = \left[\dot{l}_{i,j} \right]_{i=1\ldots3,\,j=1\ldots3}, \quad \dot{l}_{i,j} = -\dot{\theta} \cdot M_{sr} \sin\left[\theta + \frac{2\pi}{3}(j-i) \right] \qquad (2.55)$$

The electric torque is easily found with (2.56), because the time derivative of the transposed matrix of the inductances is linearly dependent on the time derivative of the angular position of the rotor

$$\tau_e = \frac{i^T \dot{L}^T i}{\dot{\theta}}$$
(2.56)

In every rotary electromagnetic actuator, the torque can be conceived as the effect of an interaction between a current and a magnetic field; when these entities are perpendicular to each other, the interaction is maximal and, therefore, the energetic conversion is optimal. In a synchronous machine, the phase-alternating currents at the stator produce a rotating magnetic field and the direct current field of the rotor rotates at synchronous speed; in an induction machine, the phase-alternating currents at the stator produce a rotating magnetic field and alternating currents are induced on the rotor. The statoric field rotates at the synchronous speed Ω_s determined by the feeding frequency and by the number of poles, whereas the rotor rotates at a lower speed Ω: the manifested speed difference is identified by the shift s and it is expressed in percentage form (2.57).

$$s = \left(\Omega_s - \Omega\right)/\Omega$$
(2.57)

According to these basic observations and the representations of the transformations between electric and mechanical energy (2.39–2.43), a circuital model can be proposed to describe the behavior of an induction actuator (Figure 2.37).

The circuital model represented in Figure 2.37 comprises three circuits, constituting an electric transformer and an electromechanical transformer: circuits (a) and (b) are, respectively, the electric primary and secondary of the electric transformer and, analogously, the circuits (b) and (c) are, respectively, the electric primary and the mechanical secondary of the electromechanical

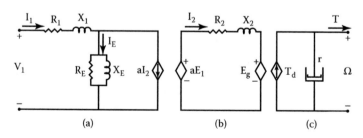

FIGURE 2.37
Circuital model of an induction actuator.

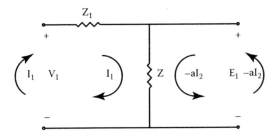

FIGURE 2.38
Electric primary as a two-port circuit.

transformer. The feeding tension and current V_1 and I_1 are present in the first electric circuit (a) with their electric resistance R_1 and reactance X_1; the feeding current separates in an excitation current I_E and in a principal current, that represents the primary of an ideal electric transformer. Furthermore, the primary and the secondary of this ideal transformer are linked by the coil ratio a. The coil (b) identifies the rotor that receives an induced electromotive force and, after the losses R_2 and X_2, produces mechanical work through an electromechanical transformer between (b) and (c). Finally, the mechanical circuit (c) is provided with the usual mechanical quantities, as just seen in the model represented in Figure 2.31. To analyze the circuit (a), the resistances and the reactances are incorporated, on the same line, in a unique impedance; so, R_1 and X_1 constitute Z_1, whereas R_E and X_E constitute Z; then the circuit (a) can be represented as a two-port circuit (Figure 2.38), and through a ring analysis, the impedance matrix is recognized to relate the input–output tensions with the input–output currents (2.58).

$$\begin{bmatrix} V_1 \\ E_1 \end{bmatrix} = \begin{bmatrix} Z_1 + Z & Z \\ Z & Z \end{bmatrix} \begin{bmatrix} I_1 \\ -a \cdot I_2 \end{bmatrix} \tag{2.58}$$

The same approach cannot be used for the circuit (b), because the equivalent two-port circuit does not admit an impedance matrix; however, the tension induced by the stator is partially dispersed by the impedance of the circuit and partially reached to feed the ideal electromechanical transformer. Therefore, the ring current I_2 can be computed with (2.59).

$$I_2 = \left(a \cdot E_1 - E_g \right) / Z_2 \tag{2.59}$$

Incorporating (2.59) into (2.58) and properly elaborating the obtained expressions, an algebraic representation of the variables I_1 and E_1 can be obtained as a function of the electromechanical variables through E_g and through some known parameters of the system (2.60).

$$I_1 = \frac{-\left(aZE_g - a^2ZV_1 - Z_2V_1\right)}{a^2ZZ_1 + Z_1Z_2 + ZZ_2}$$

$$E_1 = \frac{Z}{a^2Z + Z_2}\left[\frac{-aZE_g + a^2ZV_1 + Z_2V_1}{a^2ZZ_1 + Z_1Z_2 + ZZ_2}Z_2 + aE_g\right]$$

(2.60)

The rotary electromagnetic field is shifted with respect to the static ones and the output mechanical power is equal to the input electric power for the ideal electromechanical transformer; thus, the expression (2.61) can be used to analyze the electromechanical energy conversion.

$$E_g = I_2Z_2\left(1-s\right)/s; \qquad T_d = E_gI_2/\Omega$$

(2.61)

The current I_2 can be expressed in terms of the input tension V_1 and of the parameters of the system (2.62), inserting E_g, as shown in (2.61), and E_1, referring to (2.60), into (2.29); therefore, using the expression of the ideal torque identified by (2.61), a relation between the input–output electrical/mechanical variables can be obtained through the parameters of the system (2.62).

$$I_2 = \frac{aZs}{Z_1Z_2 + ZZ_2 + a^2Z_1Zs}V_1$$

$$T_d = \frac{a^2Z^2Z_2s\left(1-s\right)}{\left(Z_1Z_2 + ZZ_2 + a^2Z_1Zs\right)^2}\frac{V_1^2}{\Omega}$$

(2.62)

Finally, subtracting the friction losses from the ideal mechanical power, the produced mechanical torque can be obtained as in (2.63).

$$T = \frac{a^2Z^2Z_2s\left(1-s\right)}{\left(Z_1Z_2 + ZZ_2 + a^2Z_1Zs\right)^2}\frac{V_1^2}{\Omega} - r\Omega$$

(2.63)

For meso-actuators, some simplifications can be implemented to obtain a reduced circuital model: a first step consists in the assumption of a unary coil ratio and in a consequential redefinition of the values of R_2 and of X_2. This hypothesis allows the omission of the electric transformer, whereas a further simplification is associated with the transfer of mechanical losses into the electrical losses Z_E only for low rotation frequency. Finally, the electric losses are properly reduced to obtain the target model shown in Figure 2.39.

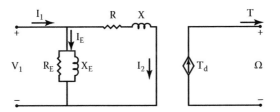

FIGURE 2.39
Simplified circuital model for an induction meso-actuator.

The parameters shown in Figure 2.39 are related to the circuital model in Figure 2.38 by the expression (2.64) that stresses also some electrical simplifications.

$$T_d = \frac{I_2^2 R_2}{s\Omega_s}$$

$$R = R_1 + \frac{R_2}{s}; \quad X = 2X_1$$

(2.64)

2.3.5 Synchronous Actuators

The denomination of "synchronous machine" is due to the synchronization between the magnetic fields of rotor and stator. As in section 2.3.4, to simplify the analysis, we observe only a cylindrical rotary actuator with unlimited stroke with a three-phase two-pole actuation system. In Figure 2.39, a rotary synchronous actuator is shown: the mobile part (rotor) can rotate around its axis, the static part (stator) has the same axis of the rotor, and it is separated from it by a gap. The stator is composed of ferromagnetic material and is furnished with linear conductive wires, whereas the composition of the rotor will be given in the following.

As previously, a reference system and an angular variable that identifies the position of the rotor (Figure 2.40) are chosen to develop a mathematical model of the dynamic behavior for the proposed micro-actuator; in particular, the position of the rotor in respect to the ith stator's winding (s_i) is indicated by the variable θ_i (2.65).

$$\theta_i = \theta + \frac{2\pi}{f}(i-1)$$

(2.65)

where θ indicated in Figure 2.40 is the angle between s_1 and d, whereas f is the number of phases.

In mathematical terms, the dynamic electromagnetic behavior of the system can be described by the matrix equation (2.66), which takes the same form as the equation (2.47).

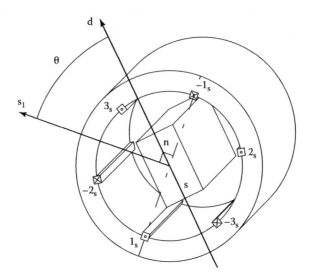

FIGURE 2.40
Definition of the rotor's position: i_s represents the ith stator's winding; s_i represents the axis of the ith stator's winding; d and q are, respectively, the direct and the quadrature magnetic axis, while θ represents the angular position of the direct magnetic axis with reference to the position of the first stator's winding.

$$\begin{cases} e = R \cdot i + \dot{\boldsymbol{\varphi}} \\ \boldsymbol{\varphi} = L \cdot i \end{cases}$$

$$e = \begin{bmatrix} e_{s1} & e_{s2} & e_{s3} & e_e & 0 & 0 \end{bmatrix}^T,$$

$$i = \begin{bmatrix} i_{s1} & i_{s2} & i_{s3} & i_e & i_D & i_Q \end{bmatrix}^T, \qquad (2.66)$$

$$\boldsymbol{\varphi} = \begin{bmatrix} \varphi_{s1} & \varphi_{s2} & \varphi_{s3} & \varphi_e & \varphi_D & \varphi_Q \end{bmatrix}^T$$

where the vector e, i, and φ represent, respectively, the electric potential, the electric current, and the magnetic flux, the matrix R of the electric resistance and the matrix L of the inductance are defined in (2.67); the indices s1, s2, and s3 associated with each element refer to the stator's winding shown in Figure 2.40. The index e is associated with the excitation winding on the rotor (Figure 2.41), whereas the indices D and Q are, respectively, associated with the equivalent damping circuits of the direct and quadrature axis (Figure 2.41).

$$R = Diag(R_s, R_s, R_s, R_e, R_D, R_Q), \qquad L = \begin{bmatrix} L_{1,1} & L_{1,2} \\ L_{2,1} & L_{2,2} \end{bmatrix} \qquad (2.67)$$

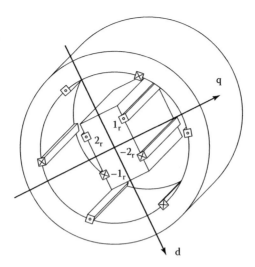

FIGURE 2.41
Equivalent damping windings on the rotor: the electromagnetic characteristics of the two rotors' windings are projected on the quadrature and direct axes to form two equivalent virtual windings.

where $Diag(\ )$ is a diagonal matrix, while R_s, R_e, R_D, and R_Q are, respectively, the resistance of the stator's winding, the excitation's winding, the direct axis winding, and the quadrature axis winding, and the submatrixes $L_{1,1}$, $L_{2,2}$, $L_{1,2}$, and $L_{2,1}$ are defined in (2.68–2.70).

$$L_{1,1} = \left[l_{i,j} \right]_{i=1\ldots3,\, j=1\ldots3}, \quad l_{i,j} = \begin{cases} l_0 + L_{00} + L_2 \cos\left[2\theta - \dfrac{4\pi}{3}(i-1) \right], & i = j \\[2mm] -\dfrac{L_{00}}{2} + L_2 \cos\left[2\theta - \dfrac{4\pi}{3}(5-i-j) \right], & i \neq j \end{cases} \tag{2.68}$$

where l_0 is the statoric dispersion self-inductance, L_{00} is the net statoric self-inductance, and L_2 is the amplitude of the second Fourier term of each self-inductance of the stator.

$$L_{2,2} = \begin{bmatrix} L_e & L_{eD} & 0 \\ L_{eD} & L_D & 0 \\ 0 & 0 & L_Q \end{bmatrix} \tag{2.69}$$

where L_e, L_D, and L_Q are, respectively, the excitation, the direct axis, and the quadrature axis self-inductance, whereas L_{eD} is the mutual inductance between excitation and direct axis windings.

$$L_{1,2} = L_{2,1}^T = \left[l_{i,j} \right]_{i=1\ldots3,\,j=1\ldots3}, \quad l_{i,j} = \begin{cases} L_{mf}\cos(\theta_i), & j=1 \\ L_{mD}\cos(\theta_i), & j=2 \\ -L_{mQ}\sin(\theta_i) & j=3 \end{cases} \quad (2.70)$$

where L_{mf}, L_{mD}, and L_{mQ} are the maximum values of the mutual inductance between the rotor's and stator's windings. The electric couple exerted by the actuator can be calculated as in (2.54) and (2.56), considering electric input power converted into mechanical power divided by the angular speed. The resulting real mechanical torque can be obtained with an equilibrium equation, like (2.52), which takes into consideration the friction, stiffness, and backlash of the mechanical components.

2.4 Piezoelectric Actuators

Niezrecki et al. (2001) proposed a review of the state of the art of piezoelectric actuation. This section will take this scheme and provide some explanations of most common actuation systems.

Piezoelectric actuators are composed of PZT elements that can fall in three categories, depending on the used piezoelectric relation: axial, transversal, and flexural components (Figure 2.42). Axial and transversal components

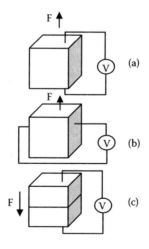

FIGURE 2.42
PZT element, where F is the exerted force and V is the electric potential difference between two faces. (a), (b), (c) are, respectively, an axial, a transversal, and a flexural actuator.

are characterized by greater stiffness, reduced stroke, and higher exertable forces, while flexural components can achieve larger strokes, but exhibit lower stiffness.

Although we have shown in Figure 2.42 piezoelectric elements with a parallelepiped shape, piezoelectric materials are produced in a wide range of forms using different production techniques—from simple forms, such as rectangular patches or thin disks, to customized very complex shapes. Because of the reduced displacements, piezoelectric materials are not usually used directly to generate a motion, but are connected to the user by a transmission element. Therefore, the piezoelectric actuators are not just "simple actuators," they are complete machines with an actuation system (the PZT element) and a transmission allowing the transformation of mechanically generated power in a desired form. In fact, the primary design parameters of a piezoelectric actuator (referring to the entire actuating machine, not only the elementary PZT part) include:

- The functional parameters—displacement, force, frequency
- The design parameters—size, weight, and electrical input power

Underlining only the functional aspects, the generated mechanical power is essentially a trade-off between these three parameters; the actuator architecture is devoted to increment one or two of these characteristics at the cost of the other ones. Piezoelectric actuators are characterized by salient exerted forces and high frequencies, but also significantly reduced strokes; therefore, the architectural designs aim to improve the stroke, reducing force or frequency. A first distinction can then be made between

- force-leveraged actuators and
- frequency-leveraged actuators

The leverage effect can be gained with an integrated architecture or with external mechanisms; so, another distinction can be made between

- internally leveraged actuators and
- externally leveraged actuators

The most common internally force-leveraged actuators include:

1. Stack actuators
2. Bender actuators
3. Unimorph actuators
4. Building-block actuators

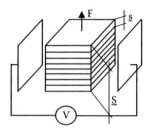

FIGURE 2.43
Stack actuator: V is the electrical potential difference applied to each piezoelectric element; F is the total exertable force. The stroke s of a single piezoelectric element is proportional to $\underline{s}$, while the total stroke S is proportional to $\underline{S}$.

The most common externally force-leveraged actuators can be subdivided as

1. Lever arm actuators
2. Hydraulic amplified actuators
3. Flextensional actuators
4. Special kinematics actuators

The frequency-leveraged actuators can be, in general, led back to inchworm architecture.

Stack actuators consist of multiple layers of piezoceramics (Figure 2.43): each layer is subjected to the same electrical potential difference (electrical parallel configuration), so that the total stroke results the sum of the stroke of each elementary layer, whereas the total exertable force is the force exerted by a single elementary layer. The leverage effect on the stroke is linearly proportional to the ratio between the elementary piezoelectric length and the actuator length.

The most common stack architectures can have a stroke of several microns, exerting some kilonewtons forces, with about ten microseconds time responses.

Bender actuators consist of two or more layers of PZT materials subjected to electric potential differences, which induce opposite strains on the layers (Figure 2.44). The opposite strains cause a flexion of the bender, due to the induced internal moment in the

FIGURE 2.44
Bender actuator, where S is the total stroke.

structure. This architecture is able to generate an amplification of the stroke as a quadratic function of the length of the actuator, resulting in a stroke of more than one millimeter. Different configurations of bender actuators are available, such as end-supported, cantilever, and many other configurations, with different design tricks to improve stability or homogeneity of the movement.

Unimorph actuators are a special class of bender actuators composed of a PZT layer and a nonactive host; two common unimorph architectures are Rainbow, developed by Heartling (1994), and Thunder, developed at Nasa Langley Research Center (Wise 1998). These are characterized by a pre-stressed configuration and are stackable; thus, they are able to gain important strokes (some millimeters).

Building-block actuators consist of various configurations characterized by the ability to combine the elementary blocks in series or parallel configurations to form an arrayed actuation system with improved stroke by series arrays and improved force by parallel arrays. There are various state-of-the-art elementary blocks available such as C-blocks, recurve actuators, and telescopic actuators.

The first class of externally leveraged actuators to be examined is the lever arm actuator class. Lever arm actuators are machines composed of an elementary actuator and a transmission able to amplify the stroke and to reduce the generated force; the transmission utilized is a leverage mechanism or a multistage leverage system, that is generally composed of two simple elements to reduce design complexity: a thin and flexible member (the fulcrum) and a thicker, more rigid, long element (the leverage arm). Another interesting externally leveraged architecture is the hydraulic amplification: in this configuration, a piezoelectric actuator moves a piston, which pumps a fluid into a tube moving another piston of a reduced section. The result is a very high stroke amplification (approximately 100 times); however, this amplification involves some problems due to the presence of fluids, microfluidic phenomena, and high frequency mechanical waves that are transmitted to the fluids. The third class of externally amplified actuators to be examined is flextensional actuators, characterized by the presence of a flexible component with a proper shape, able to amplify displacement. It differs from the lever arm actuators approach, because of its closed-loop configuration, resulting in a higher stiffness but reduced amplification power. This class of actuators can be used in a building-block architecture, to increment the stroke amplification: a typical example is a stack of Moonie actuators. The research on actuation architecture suggests always new design solutions emerging in literature and on the market; therefore, it would be improper to generalize our classifications based on only the three above-described classes of externally leveraged actuators: lever arm, hydraulic, and flextensional.

The final class of externally leveraged actuators uses the frequency leverage effect. These actuators are basically reducible to inchworm systems (Figure 2.45): in general, they are composed of three or more actuators, alternatively contracting, to simulate an inchworm movement. The resulting system is a very precise actuator, with very high stroke (more than 10 mm), but with a reduced natural frequency.

The behavior of PZT actuators can be affected by undesired physical phenomena such as hysteresis; in fact, hysteresis can account for as much as 30% of the full stroke of the actuator (Figure 2.46). An additional problem is the

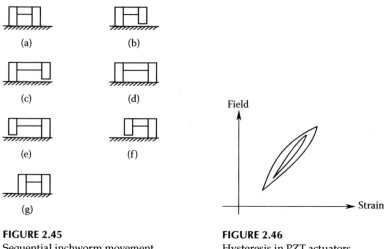

FIGURE 2.45
Sequential inchworm movement.

FIGURE 2.46
Hysteresis in PZT actuators.

occurrence of spurious additional resonance frequencies under the natural frequencies. These additional frequencies introduce undesired vibrations, reducing positioning precision, and the overall performance of the actuator. Furthermore, the depoling effect, which results in an undesired depolarization of artificially polarized materials, occurs when a too high temperature of the PZT is reached, a too large potential is imposed to the actuation system, or a too high mechanical stress is applied. To avoid undesired phenomena, the actuator should be maintained within a proper range of temperatures, mechanical stresses, and electrical potential. A design able to counteract such undesired effects could be studied and a control system implemented. The piezoelectric effect could be implemented to sense mechanical deformations; therefore, a controlled electromechanical system can be developed with a very compact design and this is one of the many reasons piezoelectric actuators have become so successful. Their wide range of applications can be spread to include different sectors (Bansevicius and Tolocka 2002), such as suppression of oscillations, microrobotic, micropumps, microgrippers, micromanipulators, and microdosage devices as well as many different other branches.

2.4.1 Elementary Actuators

As just shown, the core element of piezoelectric actuators is constituted by an elementary component with piezoelectric nature able to transform electric energy into mechanical energy and, particularly, able to generate a movement. For the sake of simplicity, a geometrically simple element, such as a piezoelectric parallelepiped, is observed; a face of this elementary component is rigidly constrained to a fixed frame to avoid movements with respect to the observer from a fixed reference system. A polarizing electric tension is applied from the fixed face and to its opposite to realize a polarization

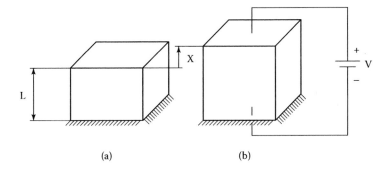

FIGURE 2.47
Piezoelectric element with parallelepiped shape, respectively, without (a) and with (b) a polarizing tension V.

perpendicular to the fixed frame. Therefore, due to the effect of piezoelectricity, the structure of the parallelepiped assumes a different geometrical configuration producing a deformation; particularly, we are interested in the movement of the face in front of the fixed one: this moving face is subjected to a movement x in the direction of the polarization (Figure 2.47) as indicated in (2.71), where s represents the strain and d is the piezoelectric coefficient.

$$x = s \cdot L \cong d \cdot V \tag{2.71}$$

An analogous result can be gained (Bansevicius and Tolocka 2002) also with other common configurations (shown in Figure 2.48), obtaining corresponding the movements described in (2.72).

$$\begin{cases} (a): & x = d_{31} \dfrac{L}{T} V; & L > 2W, L > 3T \\ (b): & x = d_{33} V; & D > 5T \\ (c): & x = d_{31} \dfrac{D}{T} V; & D > 5T \\ (d): & x = d_{33} V; & L > 3D \\ (e): & x = d_{15} V; & W > 5T, L > 5T \end{cases} \tag{2.72}$$

The expressions (2.71) and (2.72) are subjected to an error due to hysteresis and it can be estimated between 10% and 20% of the forecasted value for the entity of the movement x, also in presence of a polarizing tension V imposed with a virtually absolute precision. To remove this error, a feedback drive allows us to go beyond micrometric precision to obtain a nanometric positioning accuracy. Another undesired phenomenon emerging in opened-loop actuators is the creep, or drift, due to the realignment of piezoelectric domains after a time interval t in function of the creep factor γ, that assumes values of the order of 0.01–0.02, as shown in the expression (2.73).

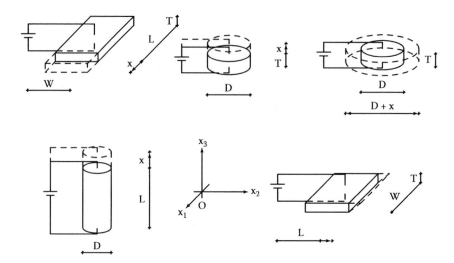

FIGURE 2.48
Example of configurations for elementary piezoelectric components where (a), (b), (c), (d), and (e) denote, respectively, transverse length, thickness extension, radial, and longitudinal *e* thickness shear modes.

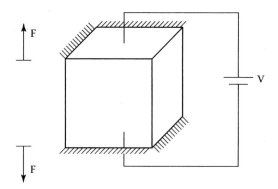

FIGURE 2.49
Blocked force configuration of a piezoelectric actuator.

$$x(t) = x\big|_{t=0.1}\left(1+\gamma \lg\left(10\cdot t\right)\right) \tag{2.73}$$

We described one of the functional characteristics of a piezoelectric actuator, i.e., the ability to realize a movement; it is also able to exert an external force, because it can convert an electric energy into a mechanical energy. To verify this property, a piezoelectric element with parallelepipedal shape is constrained at two opposite faces to a fixed reference frame and these faces are set to a potential difference V (Figure 2.49); therefore, the actuator can realize the maximal force F, called also blocked force, depending on the stiffness k of the actuator and on the maximal realizable stroke s_{max} (2.74).

$$F = k \cdot s_{max} \tag{2.74}$$

The stiffness of the actuator is an important parameter to describe its dynamic behavior; particularly, with an effective mass m_{eff} equal to a third of the total mass of the actuator, its natural frequency f_r can be expressed by (2.75), that is an indicator of the passband.

$$f_r = \sqrt{\frac{k}{m_{eff}}} \Big/ 2\pi \tag{2.75}$$

Then, applying an oscillating field able to produce a vibration of the actuator at a frequency f with an amplitude of the movement equal to x halves around the uncharged position, the produced dynamic force is equal to (2.76).

$$F = \left(2\pi f\right)^2 \frac{x}{2} m_{eff} \tag{2.76}$$

The actuator response is out of phase with respect to input electric signal depending on the distance between the imposed frequency f and the natural frequency of the system f_r, according to (2.77), that represents the phase difference.

$$\phi = 2 \cdot \arctan\left(f / f_r\right) \tag{2.77}$$

From an electric viewpoint, the response of a piezoelectric element is not instantaneous, because, at low frequencies, it can be approximated by a capacitor, where the displacement x is proportional to the accumulated charge q (2.78), by means of a proportionality coefficient k_q.

$$z = k_q \cdot q \tag{2.78}$$

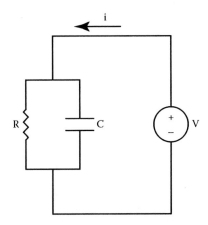

The expression (2.79) correlates the movement x with the feeding tension V by means of a simple first-order model that allows a representation of the piezoelectric element as a shunt circuit between a capacitor with a capacity C and a resistor with resistance R (Figure 2.50), and observing that the time derivative of the charge q is the electric current that flows through the piezoelectric element.

FIGURE 2.50
Circuital model of a piezoelectric actuator.

$$V = R\left(i - C\frac{dV}{dt}\right)$$

$$= \frac{R}{k_q}\frac{dx}{dt} - RC\frac{dV}{dt} \tag{2.79}$$

The corresponding electric temporal constant τ is equal to the ratio between the capacity C and the resistance R of the circuit, then, denoting with k the product between the capacity C and the charge factor k_q, the transfer function between input tension V_{IN} and output movement x_{OUT} can be represented by (2.80).

$$\begin{cases} \dfrac{x_{OUT}}{V_{IN}} = k\dfrac{\tau s + 1}{\tau s} \\[2ex] \tau \equiv RC; \quad k \equiv k_q C \end{cases} \tag{2.80}$$

The lumped parameter expression (2.80) is valid for low frequencies or in a limited range of frequencies; with an increase of the frequency band and, especially, approaching the natural frequency, the proposed model loses its validity. Therefore, in these conditions, more precise models can be taken from literature, an interesting model is proposed by Georgiu and Ben Mrad (2004) that presents a reasonable computational simplicity combined with a forecasting ability in a wide frequency range. They propose to extend the concept of relative permittivity ε^*_r, to the complex domain; therefore, expressing it as a ratio between the complex permittivity of the material ε^* and the vacuum permittivity ε_0, we observe its dependence from two parameters k' and k'', associated with the working frequency.

$$\varepsilon^*_r \equiv \frac{\varepsilon^*}{\varepsilon_0} \equiv k' - k''j \tag{2.81}$$

C_0 denotes the capacity of the capacitor in the vacuum; thus, the fundamental properties of the piezoelectric element shown as an RC shunt circuit drives to the expression (2.82).

$$\frac{i}{V} = C_0 k'\omega i + C_0 k''\omega \tag{2.82}$$

Observing the expressions (2.82) and (2.79), Georgiu and Ben Mrad (2004) grasp the direct proportionality between C and k' and the inverse proportionality between R and k'' proposing a physical meaning for their model.

Part of the electric energy is not converted into mechanical energy, due to frictions developing during movements and transformations of piezoelectric domains; therefore, a factor of conversion K can be defined (Kasap 1997) to evaluate the efficiency of the transformation from electric power W_E and mechanical power W_M (2.83).

$$W_M = K^2 W_E \qquad\qquad (2.83)$$

2.4.2 Load Effects

Piezoelectric actuators are used to produce displacements for different applications, whereas they can produce only forces if there are constraints blocking their strains. If an electric potential difference is applied to the ends of a piezoelectric actuator loaded with a resisting mechanical force, the resulting displacement is shorter than in the unloaded condition and the difference is proportional to the mechanical load. The maximal force that can be exerted by the actuator is measured in the configuration of blocking-stroke, while the maximal displacement realizable by the same actuator is measured in the absence of external load, i.e., in the blocking-force configuration. These quantities depend linearly on the used electric potential difference; thus, their ratio remains constant with fixed geometrical unchanged dimensions of the actuator and with fixed material characteristics. This ratio between a force and a displacement describes the characteristic stiffness of the actuator and allows a preliminary linear model between the displacement and the external mechanical load by means of this constant that can be easily evaluated as the ratio between the force generated in blocking-stroke configuration and the displacement in blocking-force configuration. The hypothesis of linearity can be valid when the intensity of the electric tension V is low and the mechanical load is not excessive.

With this approach, different models can be proposed to describe the behavior of a wide range of piezoelectric actuators subjected to a potential difference and loaded with an external force, when the relation used to calculate blocking-force and blocking-stroke are known. As shown in Figure 2.51, the stiffness coefficient of the actuator is geometrically associated with the slope of the line with null tension; thus, changing the mechanical load, the characteristic line is followed, whereas changing the tension, parallel lines are followed, as shown in Figure 2.52, describing an example of typical behavior of a PZT actuator.

The actuator can also be loaded with a variable force that linearly depends on the produced displacement and represented by a linear spring acting in the direction of the actuator elongation. Thus, the displacements depend on, besides the characteristics of the actuator and the feeding tension, also the stiffness of the resistant load. As shown in Figure 2.53, the final elongation and the equilibrium force are determined by the intersection of two lines that describes the behavior of the load and of the actuator.

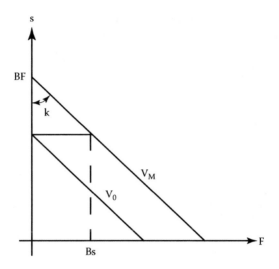

FIGURE 2.51
Relation between produced force F and produced displacement s; k is the stiffness of the actuator, BF is the maximal stroke with blocking-force, Bs is the maximal force with blocking-stroke, V_0 is the null electric tension, whereas V_M is the maximal feeding tension.

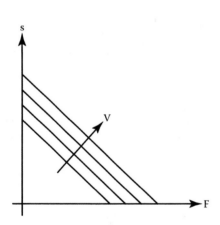

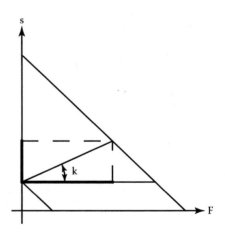

FIGURE 2.52
Effect of a change in the electric tension on the relation between displacement s and force F of the actuator.

FIGURE 2.53
Relation between displacement s and force F associated with an external linear elastic stiffness k.

The external force does not influence the entity of the blocking-stroke or of the blocking-force and, therefore, the stiffness of the piezoelectric element can be computed independently. The behavior of the system can be associated with a two-spring series and the working point is individuated by the intersection between the line with a slope equal to the stiffness of the actuator and the line with a slope equal to the stiffness of the external load. Then the intercepts of the two lines coincide with the condition of initial length of the piezoelectric element without external load, whereas the abscissa of the working point defines the final displacement of the piezoelectric and of the spring and the corresponding ordinate furnishes the equilibrium force acting on the PZT.

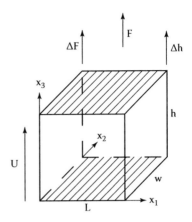

FIGURE 2.54
Axial actuator where the surface electrodes are highlighted.

The simplified case of a parallel-epipedal piezoelectric is shown in the following with a low external mechanical load and with a low constant electric feeding tension U. In these conditions the potential difference is generated in the direction of the natural polarization of the material and of the applied mechanical load, as described in Figure 2.54. The actuator, geometrically $h \cdot w \cdot l$, is composed by piezoelectric material with a piezoelectric constant d_{33} that expresses the elastic deformation for unity of electric field, and with a constant s_{33} that expresses the elastic deformation for stress unity.

The displacement blocking-force Δh is computed as the product between the piezoelectric constant d_{33} and the electric tension at the electrodes (2.84), although the blocking-stroke force ΔF depends linearly on the geometrical dimension, on the piezoelectric constants, and also on the electric tension (2.84).

$$\Delta h = d_{33}U; \quad \Delta F = \frac{d_{33}lw}{s_{33}^{E}h}U \tag{2.84}$$

The characteristic stiffness k_E of the actuator is a function only of the geometrical dimensions and of the piezoelectric constant s_{33} without dependence on the mechanical load or the electric tension (2.85).

$$k_E = \frac{\Delta F}{\Delta h} = \frac{lw}{s_{33}^{E}h} \tag{2.85}$$

The final displacement is calculated as the algebraic sum of the elongation due to the application of the electric tension and the shortening due to mechanical load, acting independently for the hypotheses of low loads and tensions (2.86).

$$h = \Delta h - \frac{F}{k_E} \tag{2.86}$$

When the applied load is due to a elastic force with stiffness k_L, the final elongation is computed with a stiffness series, as shown in (2.87).

$$h_L = \frac{\Delta F}{1 + \dfrac{k_L}{k_E}}$$

(2.87)

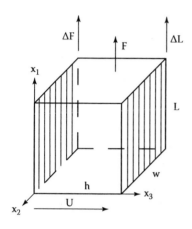

FIGURE 2.55
Transversal actuator.

Another interesting configuration is the transversal actuator, where the behavior of a parallelepipedal piezoelectric is observed when it is subjected to a low external mechanical load F and it is fed with a low electric tension U. The potential difference is generated in the same direction as the natural polarization of the material and perpendicularly to the mechanical load, as represented in Figure 2.55. The actuator, geometrically $h \cdot w \cdot l$, is composed by piezoelectric material with a piezoelectric constant d_{31} that expresses the elastic deformation for unity of electric field, and with a constant s_{11} that expresses the elastic deformation for stress unity.

The displacement blocking-force Δh depends on the piezoelectric constant d_{31}, on the geometrical dimensions, and its proportional to the electric tension at the electrodes (2.84), whereas the blocking-stroke force ΔF depends linearly on the geometrical dimension, on the piezoelectric constants, and also on the electric tension (2.88).

$$\Delta l = d_{31} \frac{l}{h} U; \qquad \Delta F = \frac{d_{31} w}{s_{11}^E} U$$

(2.88)

The characteristic stiffness k_E of the actuator is a function only of the geometrical dimensions and of the piezoelectric constant s_{11} without dependence on the mechanical load or the electric tension (2.89).

$$k_E = \frac{\Delta F}{\Delta l} \frac{hw}{s_{11}^E l}$$

(2.89)

The final displacement is evaluated as the algebraic sum of the elongation due to the application of the electric tension and the shortening due to mechanical load, acting independently for the hypotheses of low loads and tensions (2.90).

$$l = \Delta l - \frac{F}{k_E} \tag{2.90}$$

When the applied force is due to a variable load with characteristic stiffness k_L, the final elongation can be computed with a series of springs as in (2.91).

$$l_L = \frac{\Delta F}{1 + \dfrac{k_L}{k_E}} \tag{2.91}$$

A common configuration is the bending bracket composed by two piezo-electric layers and subjected by a constant load F and by a constant potential difference U in the same direction for both layers, as shown in Figure 2.56: this actuator is also called parallel bimorph. If the mechanical load and the electric tension are low, the mechanical and electrical contributions can be approximated with a linear relation. The length of a layer is L, while its thickness and its width are, respectively, h and w. The expressions proposed in the followings depend on two coefficients, A and B, associated with the adopted PZT material. The first coefficient, A, correlates the generated force with the furnished tension with blocked stroke, while the second coefficient, B, correlates the rise z with the generated force with blocked force (2.92).

$$\Delta F = A \frac{hw}{L} U; \qquad \Delta z = B \frac{L^2}{h^2} U \tag{2.92}$$

The characteristic stiffness k_E of the actuator can be computed only with the geometrical dimensions and with some empirical coefficients without dependence on the mechanical load and on the electric tension (2.93).

$$k_E = \frac{\Delta F}{\Delta z} = \frac{B}{A} \frac{h^3}{L^3} w \tag{2.93}$$

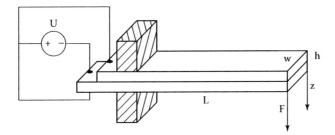

FIGURE 2.56
Parallel actuator bender.

Therefore, the final stroke can be expressed as an algebraic sum between the positive decrease due to the application of a positive tension and the negative increase due to a load acting upwards (2.94).

$$z = \Delta z - \frac{\Delta F}{k_E} \tag{2.94}$$

2.4.3 Amplification Systems

A common purpose of the research in the field of actuators is to produce an architecture able to generate important strains and stresses, with high frequency of response with a good energetic efficiency. Piezoelectric actuators can be suitable to one or more of these demands with proper configuration systems. In fact, stroke amplifiers can increment the stroke for linear actuators without an excessive loss in other areas of performance, such as exerted force, response frequency, size, and feeding electric power. The sole limit of the piezoelectric actuators, the stroke, can also be faced with the union of more similar actuators as shown in Figure 2.57 with a pile of elementary actuators.

This configuration allows an increment of linear deformations with a low demanded tension and the increment of the stroke is directly proportional to the length of the pile, whereas the exerted force is proportional to its section. A pile of elementary actuators can be used as a single actuator or as a piezoelectric element into a more complex amplification system, i.e., in a series with a three-bar system, as shown in Figure 2.58. A common amplification system can also be composed with different layers on a bender (Figure 2.58)

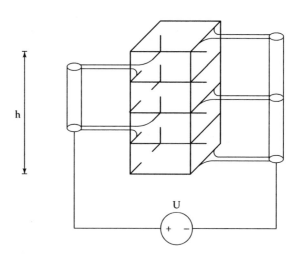

FIGURE 2.57
Pile of elementary actuators.

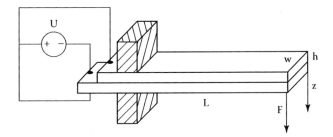

FIGURE 2.58
Parallel bending actuator.

that, when electrically fed, bends amplifying the displacement depending on its length.

These bending actuators allow a stroke amplification that is directly proportional to the square of the length, incrementing the rise of the system. Bending actuators can suggest a stiff configuration constituted by an arch composed of an inert layer and a piezoelectric layer (Figure 2.59).

The geometrical shape and the type of frame constraints allow a reasonable stroke amplification and the global configuration of the system suggests a further superposition of different amplifiers to multiply the displacement again. Among externally leveraged actuators, we can also find hydraulic amplifiers and special kinematic configurations, like the three-hinge arch, where a piezoelectric element produce a displacement of a lateral hinge inducing the elevation of the central hinge (2.60).

We do not mention frequency leveraged actuators; in fact, due to proper input signals and to high response, suitable architectures can be produced to obtain virtually unlimited linear/rotary displacement, i.e., ultrasonic motors and inchworm structures that allow infinite linear displacements through successive extensions and contractions of the piezoelectric elements.

In the kinematics of actuation mechanisms, schemes of different devices are proposed to gain very similar results with simplicity of use and without

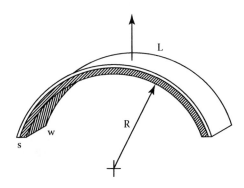

FIGURE 2.59
Arch amplifier.

an excessive idealization of structure, i.e., avoiding to neglect the mechanical resistance of flexure hinges. Two external amplification devices are observed: the first is constituted by a three-hinge arch where a mobile hinge is actuated by a piezoelectric element; and the second is composed by two connected arch-shaped layers: an inert one and a piezoelectric one. The proposed approach in the two configurations is devoted to obtain the relation between deformation of the piezoelectric element and force loaded on it.

The first amplification system, the three-bar amplifier, is shown in Figure 2.60 and it is composed by two inert bars and by an active piezoelectric bar; these three bars are connected by three flexure hinges, forming a three-hinge arch. The actuator is loaded with an external vertical force directed downwards and applied in the central hinge, while the piezoelectric element is subjected to a constant potential difference that produces an elongation. The kinematic scheme of this amplifier is represented by a three-hinge arch, as shown in Figure 2.61: a relative hinge and two hinges connected to the frame. A frame-hinge can yield and its displacement δ can be associated with the horizontal reaction F. The force acting horizontally on the frame corresponds to the load effectively resting on the parallelepipedal piezoelectric element. The external load P is applied on the central hinge between the two bars with a length equal to L, a is the disalignment between relative hinge and frame-hinges, and α is the angle between bars and frame.

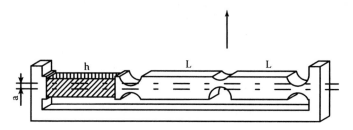

FIGURE 2.60
Three bars amplifier.

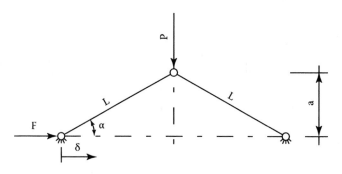

FIGURE 2.61
Three hinges arch.

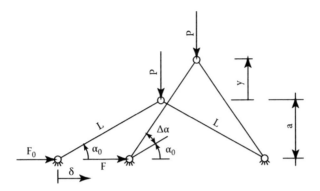

FIGURE 2.62
Effect of the displacement of the mobile hinge.

The geometrical relation between the displacement δ and the force F on the piezoelectric element can be obtained, observing the amplifier in the initial configuration and highlighting the varying quantities due to the application of the mechanical load and of the electric tension. This procedure allows a preliminary evaluation of the final results for an eventual iterative computation. A shown in Figure 2.62, the displacement of the mobile hinge is denoted by δ, whereas α_0 is the initial angle between the bar and the frame and $\Delta\alpha$ is the variation of the same angle. As a first attempt, the hinges are considered as ideal.

The initial load F_0 applied on the frame-hinge and the initial angle α_0 depend on the disalignment of the hinges, as shown in (2.95), whereas the angle variation $\Delta\alpha$ is associated with the displacement δ of the frame-hinge with the expression (2.96).

$$F_0 = \frac{P}{2} ctg(\alpha_0); \quad \alpha_0 = \arcsin\left(\frac{a}{L}\right) \tag{2.95}$$

$$\Delta\alpha = \arccos\left(\frac{2L\cos(\alpha_0) - \delta}{2L}\right) - L \tag{2.96}$$

The rotary equilibrium equation of the left bar with respect to the central hinge produces a relationship between the force F on the piezoelectric and the varying geometry of the actuator (2.97).

$$F = \frac{P}{2}\left(\cos(\alpha_0) - \frac{\delta}{2L}\right)\left[\sin\left(\arccos\left(\cos(\alpha_0) - \frac{\delta}{2L}\right)\right)\right]^{-1} \tag{2.97}$$

This expression (2.97) describes the axial load on the actuator with a characteristic stiffness k_E (2.98) that can relate the actuator load P with the

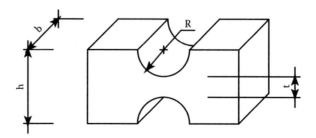

FIGURE 2.63
Flexure hinge.

displacement δ, under the effect of varying geometry and of the dynamic properties (identified with k_E).

$$\delta = \Delta h - \frac{F}{k_E} \tag{2.98}$$

An improvement of the model can be obtained with flexure hinges realized with a local weakening of the structure due to a local section reduction of the inert bars, i.e., circular reduction are shown in Figure 2.63 with an indication of the their principal geometrical parameters.

The behavior of these types of hinges is described by expressions that correlate the bending moment acting on them with the relative rotation produced by the compliance C (2.99), completely derived in appendix C.2.

$$C = \frac{3}{2EbR^2} \cdot \frac{1}{\beta^{\frac{5}{2}} \cdot (\beta+2)^{\frac{5}{2}}} \cdot \left\{ \frac{\sqrt{\beta \cdot (\beta+2)} \cdot \sqrt{1-(1+\beta-\gamma)^2}}{2 \cdot \gamma} \cdot \left[\beta(2+\beta-\gamma) \right. \right.$$

$$\left. \cdot \frac{(\beta+1)}{\gamma} + 3 \cdot \beta + 3 + 2 \cdot \beta^2 \right] + 3 \cdot (\beta+1) \cdot \arctan\left[\sqrt{\frac{\beta+2}{\beta}} \cdot \frac{\gamma-\beta}{\sqrt{1-(1+\beta-\gamma)^2}} \right] \right\} \tag{2.99}$$

In the equation (2.99), E is the elastic modulus of the inert material; b, R, and h are the geometrical dimensions shown in Figure 2.63; β and γ correspond, respectively, to the ratio between t and the diameter of the striction and to the ratio between h and the same diameter. Furthermore, the expression (2.99) can be simplified as (2.100) for every hinge where the thickness t and the height h are much smaller than R.

$$\frac{\Delta\alpha}{T} = \frac{9\pi\sqrt{R}}{2Eb\sqrt{t^5}} \tag{2.100}$$

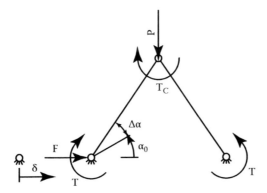

FIGURE 2.64
Arch with moments produced by torsional springs.

This simplified expression can be used to evaluate the internal resistance of the hinges introducing fictitious torsional springs with the characteristic described in (2.100); the torsional spring can be described in the dynamic model as lumped moments applied in the hinges, with an intensity that depends on the torsional angle due to the strain of the piezoelectric element, as shown in Figure 2.64.

The three lumped moments describe the behavior of the flexure hinges simulating the behavior of torsional springs with a characteristic stiffness k_t (2.101) obtained from equation (2.100).

$$k_t = \frac{2Eb\sqrt{t^5}}{9\pi\sqrt{R}} \tag{2.101}$$

Geometrical observations suggest that a rotation $\Delta\alpha$ of the frame-hinges is associated with a double rotation of the central relative hinge; therefore, the applied torsional moment also is double in the central hinge with respect to the frame-hinges. The rotary equilibrium equation of the left bar with respect to the central hinge expresses the relation that correlates the force on the piezoelectric with the variable geometry (2.102).

$$F = \frac{3k_t\Delta\alpha + \dfrac{P}{2}2L\cos(\alpha_0 + \Delta\alpha)}{L\sin(\alpha_0 + \Delta\alpha)} \tag{2.102}$$

This expression of the mechanical load acting horizontally on the axial actuator with a characteristic stiffness k_E can be related to the piezoelectric displacement by equation (2.103).

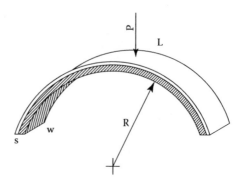

FIGURE 2.65
Thunder actuator.

$$\delta = \Delta h - \frac{F}{k_E} \qquad (2.103)$$

This procedure describes a relation between the piezoelectric elongation that coincides with the displacement of the mobile hinge, and the load P, considering a variable geometry. Finally, the expression (2.104) is useful to obtain the vertical displacement δ of the central hinge with ideal or with flexure hinges.

$$\delta = L \sin\left(\alpha_0 + \Delta\alpha\right) - a \qquad (2.104)$$

Another interesting amplification system is the arch amplifier, commonly known as a thunder actuator and shown in Figure 2.65; it is essentially constituted by an inert layer and by a piezoelectric layer and it is loaded with a vertical force directed downwards and applied in the central point of the arch, while the piezoelectric layer is subjected to a constant potential difference, causing a deformation.

A scheme represented by a curved beam is associated with the actuator, as shown in Figure 2.66, where a circumference arch is constrained to the frame at its ends. Then, a relation can be individuated correlating the elongation ΔL of the beam induced by the piezoelectric layer with the reaction F on the frame and tangent to arch in its ends. The reaction force on the frame corresponds to the effective load on the piezoelectric arch. The external load P is applied in the central point of the arch with length L and radius r; α is its semiamplitude.

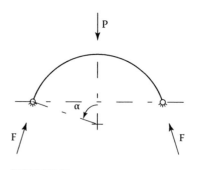

FIGURE 2.66
Curved beam.

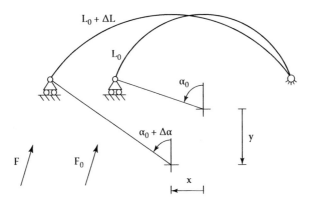

FIGURE 2.67
Isostatic structure.

The constraints are preliminary approximated with an isostatic structure produced by two ideal joints: a hinge and a slider, then a hyperstatic structure with two ideal hinges on the ends of the arch. The isostatic model shown in Figure 2.67 highlights the type of deformation on the structure and the variable entity with an elongation of the beam from the initial configuration.

The tangential component F_0 of the initial reaction and its associated initial angle α_0 can be correlated to the initial length and radius of the arch (2.105).

$$F_0 = \frac{P}{2}\sin(\alpha_0); \qquad \alpha_0 = \frac{L_0}{2r_0} \tag{2.105}$$

After an elongation ΔL of the circular arch, a system of geometrical equations can be written to relate this variation to the consequent displacement of the midline of the arch and to the variation of its radius (2.106). The first equation expresses the length of the final arch in terms of the final radius and angle; the second describes the horizontal displacement of the center of the arch; and the third represents the vertical displacement of the same center.

$$\begin{cases} L_0 + \Delta L = 2(\alpha_0 + \Delta\alpha)(r_0 + \Delta r) \\ x = (r_0 + \Delta r)\sin(\alpha_0 + \Delta\alpha) - r_0\sin(\alpha_0) \\ y = r_0\cos(\alpha_0) - (r_0 + \Delta r)\cos(\alpha_0 + \Delta\alpha) \end{cases} \tag{2.106}$$

The tangential component of the reaction on the frame is equal to the force produced by the piezoelectric layer with a variable geometry (2.107).

$$F = \frac{P}{2}\sin(\alpha_0 + \Delta\alpha_0) \tag{2.107}$$

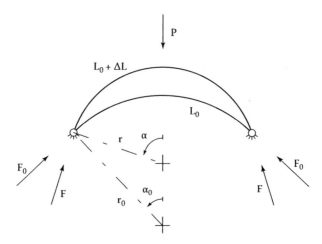

FIGURE 2.68
Hyperstatic structure.

This is the expression of the axial load acting on the piezoelectric layer with a characteristic stiffness k_E with its associated relation (2.108).

$$\Delta L = \Delta h - \frac{F}{k_E}$$

(2.108)

As just mentioned, an accurate model should be associated with a hyperstatic model, as shown in Figure 2.68 highlighting the type of deformation on the structure and the variable entity with an elongation of the beam from the initial configuration.

The tangential component F_0 of the initial reaction applied on the hinge and the initial angle α_0 can be expressed as function of the initial arch length and radius. After an elongation ΔL of the arch, a system of geometrical equations can be written to relate this variation to the consequent displacement of the center of the arch and to the variation of its radius. The first equation correlates the final length of the arch to the final radius and angle (2.109), whereas the second produces a relation between final/initial radius and angle because the two arches have the same subtended chord (2.110).

$$L_0 + \Delta L = 2 \cdot \alpha \cdot r$$

(2.109)

$$r_0 \sin(\alpha_0) = r \sin(\alpha)$$

(2.110)

The other dynamic relations should be gained with a suitable approach, because the structure is hyperstatic; therefore, a principle of virtual work is

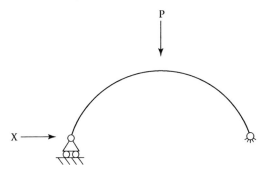

FIGURE 2.69
Fictious isostatic structure with the hyperstatic force X converted into an external unknown force.

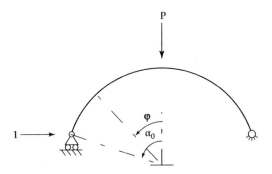

FIGURE 2.70
Fictious isostatic structure with conventionally unit hyperstatic force.

proposed as a solution method. The structure is virtually rendered isostatic highlighting the unknown hyperstatic force X and the equilibrium is evaluated on the fictitious isostatic structure as shown in Figure 2.69.

Then, the internal reactions into the fictious structure can be expressed as in (2.111), where the hyperstatic force is conventionally unitary and no external forces are applied to the system as shown in Figure 2.70.

$$\begin{cases} T' = -1 \cdot \left[r_0 \cos(\varphi) - r_0 \cos(\alpha_0) \right] \\ N' = -1 \cdot \cos(\alpha_0) \\ S' = 1 \cdot \sin(\alpha_0) \end{cases} \qquad (2.111)$$

The real load is put on the isostatic structure to evaluate the produced internal reactions (2.112) as a function of the arch angle (Figure 2.71).

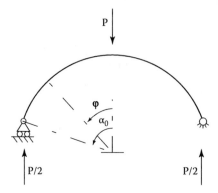

FIGURE 2.71
Fictious isostatic structure with real loads.

$$\begin{cases} T^0 = \dfrac{P}{2} \cdot \left[r_0 \sin(\alpha_0) - r_0 \sin(\varphi) \right] \\[2mm] N^0 = -\dfrac{P}{2} \cdot \sin(\alpha_0) \\[2mm] S^0 = \dfrac{P}{2} \cdot \cos(\alpha_0) \end{cases} \tag{2.112}$$

These relations can be used, with the help of the superposition principle, to compute the internal work of the system (2.113), where A, A^*, and J are, respectively, the area, the reduced area, and the inertia moment of the section of the inert layer. Then, the internal work should be equal to the external work of the load and of the hyperstatic that can be expressed by (2.114), where y denotes the displacement of the application point for the load P that coincides with the midpoint of the arch.

$$W_i = 2 \int_0^\varepsilon T' \frac{T^0 + X \cdot T'}{E \cdot J} r \cdot d\varphi + 2 \int_0^\varepsilon N' \frac{N^0 + X \cdot N'}{E \cdot A} r \cdot d\varphi + 2 \int_0^\varepsilon S' \frac{S^0 + X \cdot S'}{G \cdot A^*} r \cdot d\varphi \tag{2.113}$$

$$W_e = P \cdot y \tag{2.114}$$

$$y = \left(r - r\cos(\alpha) \right) - \left(r_0 - r_0 \cos(\alpha_0) \right) \tag{2.115}$$

The axial forces that compress the piezoelectric material are due to the superposition of the tangential components of the vertical reaction and of the hyperstatic (2.116).

$$F = \frac{P}{2}\sin(\alpha) - X\cos(\alpha) \tag{2.116}$$

The load effect on the piezoelectric conducts to the usual expression (2.117), where k_E is the characteristic stiffness of the active layer.

$$\Delta L = \Delta h - \frac{F}{k_E} \tag{2.117}$$

The proposed approaches for the two described amplification systems, a three-bar and a thunder actuator, used a single piezoelectric element with an applied potential difference, resulting in low elementary deformations, but different elementary actuators, as just mentioned, can compose a pile, as just shown for axial actuators subjected to the same tension U (Figure 2.57).

This configuration increments the elongation produced by the applied tension allowing a wider field of application than in the single elementary actuator.

The behavior of the overall system is similar to the one of the single axial actuator with only two differences: the height of the element and the feeding tension are multiplied by the number of elements into the pile. Therefore, the blocking-force elongation can be described as the product between the piezoelectric constant d_{33} and the n times amplified tension at the electrodes (2.118).

$$\Delta h = n \cdot d_{33} \cdot U \tag{2.118}$$

The generated force in blocking-stroke configuration depends on the geometrical and piezoelectric parameters and, linearly, by the electric tension (2.119), where the result is unaffected by the contemporaneous n times amplification of the height h and of the tension U.

$$\Delta F = \frac{d_{33}lw}{s_{33}^E h} U \tag{2.119}$$

The characteristic stiffness of the actuator depends only on the geometrical dimensions and on the piezoelectric constant s_{33} as shown in (2.120).

$$k_E = \frac{\Delta F}{\Delta h} = \frac{l \cdot w}{n \cdot s_{33}^E \cdot h} \tag{2.120}$$

Therefore, the final elongation is computed as the algebraic sum between elongation due to n times amplified tension and the shortening due to the mechanical load; this superposition is allowed by the hypotheses low loads and tension (2.121) and the final displacement of the pile is amplified by the number of elementary actuators.

$$h = \Delta h - \frac{F}{k_E} \qquad (2.121)$$

Furthermore, another interesting displacement amplification can be obtained with a series of complete actuators with their own amplification system feeding all the actuators with the same tension; therefore, the displacement of a single actuator is amplified n times, where n is the number of the actuators composing the pile. This simple method is especially used for modular actuators that are just designed to be mounted in a series.

2.4.4   Bender Actuators

Piezoelectric bimorph benders are a particular class of piezoelectric devices, which are characterized by the ability to produce flexural deformation much larger than the length or thickness deformation of a single piezoelectric layer. We describe here an analytical model for a piezoelectric bimorph. The model combines an equivalent single-layer theory for the mechanical displacements (without relative sliding between electrical layers) with layerwise-type approximation for the electric potential. First-order Timoshenko shear deformation theory kinematics and quadratic electric potentials are assumed in developing the analytical solution. Mechanical displacement and electric potential Fourier-series amplitudes are treated as fundamental variables, and full electromechanical coupling is maintained. According to Timoshenko theory, a shear correction factor is introduced with a value proposed by Timoshenko (1921) and by Cowper (1966).

We are interested in piezoelectric bimorphs because they can be easily used as an actuator to generate mechanical movements or, specifically, vibrations; for this reason, the paper is devoted to the analysis of the kinematic behavior of benders in terms of natural frequencies, under different conditions. A piezoelectric bimorph bender is a structure composed by two superimposed active piezoelectric beam-shaped layers. Under the effect of an electric field, the polarization of piezoelectric layers is able to generate a bending of the beam, so the proposed structure can be classified as a mechanical actuator, or a device able to produce mechanical power. These sorts of devices are diffusely studied in literature and applied in different industrial sectors. We are interested in kinematic and dynamic characteristics of the system (Hutchinson 1981), especially in natural frequencies (Cowper 1966; Kawashima 1996).

In the undeformed and initial condition, the considered bar has its neutral axis coincident with the x_1 axis of a Cartesian reference system (0) composed by the origin O and three orthogonal axes x_1, x_2, and x_3; furthermore, under appropriate hypotheses, only a 2-D problem is considered, so the bar is studied only in the plane identified by the axes x_1 and x_3. The considered bar is composed of two piezoelectric layers, mechanically and electrically connected, the bar has a length l and each layer has a thickness h. Under mechanical or electric actions, the bar can be deformed so that a cross section

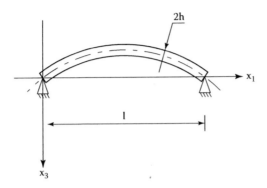

FIGURE 2.72
Simply supported piezoelectric bimorph.

of the bar is subjected to a w translation along the x_3 axis and to a Ψ rotation on the x_2 axis (Figure 2.72).

An element of the beam is considered and, under plane stress, small strain hypotheses and with a symmetric stress tensor, the well-known mechanical equilibrium equations can be derived (2.122). Because the electric beam does not hold an electric charge, the Gauss theorem and the divergence theorem give the electric equilibrium equation (2.123). For an orthotropic piezoelectric material, the constitutive equations in condensed notation are expressed in (2.124).

$$\frac{\partial \sigma_{11}}{\partial x_1} + \frac{\partial \sigma_{31}}{\partial x_3} + f_1^b = \rho \frac{d^2 u_1}{dt^2}$$

$$\frac{\partial \sigma_{13}}{\partial x_1} + \frac{\partial \sigma_{33}}{\partial x_3} + f_3^b = \rho \frac{d^2 u_3}{dt^2} \tag{2.122}$$

where σ_{ij} are the components of the stress tensor, f^b_i are the components the body force, ρ is mass density of the bar and u_i are the components of the displacement of a bar element.

$$\frac{\partial D_1}{\partial x_1} + \frac{\partial D_3}{\partial x_3} = 0 \tag{2.123}$$

where D is the electric displacement.

$$\sigma_{ij} = C_{ij} s_j - (-1)^r e_{ji} E_j, \qquad D_i = (-1)^r e_{ij} s_j + \varepsilon_{ij} E_j \tag{2.124}$$

where C is the elastic stiffness matrix, s is the strain tensor organized in a vector, e is the piezoelectric matrix, ε is the dielectric matrix, E is the electric field, while the poling direction coefficient r assumes the value 1, when the poling direction is coincident with x_3 direction, while it is equal to 2, when the poling direction is opposite to x_3 direction.

The strain tensor is related to displacements (2.125) and these are approximated with Timoshenko theory (2.126), on the electric side, the field intensity is related to electric potential (2.127) and electric potential is approximated with (2.128), as proposed by Zhou, Chen, and Ding (2005).

$$S_{ii} = \frac{\partial u_i}{\partial x_i}; \quad S_{ij} = \frac{\partial u_i}{\partial x_j} + \frac{\partial u_j}{\partial x_i} \quad i \neq j \tag{2.125}$$

where S is the strain tensor.

$$u_3(x_1, t) = w(x_1, t) \quad u_1(x_1, x_3, t) = -x_3 \psi(x_1, t) \tag{2.126}$$

where ψ is the rotation of a cross section of the bar.

$$E_i = -\frac{\partial \phi}{\partial x_i} \tag{2.127}$$

where ϕ is the electric potential.

$$\phi(x_1, x_3, t) = g(x_3) V(x_1, t) + f(x_3) \Phi(x_1, t) \tag{2.128}$$

where V is the amplitude of the applied electric potential, g is the thickness distribution of V, Φ is the electric potential amplitude on the layer midline, and f is the thickness distribution of Φ.

As observed by Smits, Dalke, and Cooney (1991), the thickness distributions g and f could be expressed as in (2.129, 2.130).

$$g(x_3) = (-1)^r |x_3|/h \tag{2.129}$$

where h is the thickness of a piezoelectric layer.

$$f(x_3) = (-1)^r \left[1 - \left(1 - 2|x_3|/h \right)^2 \right] \tag{2.130}$$

We can consider a parallel layer arrangement, with r always equal to 2, under zero normal stress in x_3 direction, integrating the equilibrium equations (2.122, 2.123) over the cross section and substituting the expressions (2.124, 2.130), so we can find the motion equations of the bimorph. The boundary conditions can be expressed by (2.131). Then, expanding w, ψ, and Φ with

Fourier series (2.132) and substituting in the motion equation, avoiding the effect of electromechanical loads, we can find the Fourier equation (2.133).

$$w\big|_{x_1=0,l} = 0 \qquad \sigma_{11}\big|_{x_1=0,l} = 0 \qquad \phi\big|_{x_1=0,l} = 0 \qquad (2.131)$$

where w is the translation of a cross section and l is the length of a piezoelectric layer.

$$w = \sum_n W_n \sin\left(n\pi x/l\right)e^{i\omega t} \qquad \psi = \sum_n \Psi_n \cos\left(n\pi x/l\right)e^{i\omega t}$$

$$\Phi(x,t) = \sum_n \Phi_n \sin\left(n\pi x/l\right)e^{i\omega t} \qquad (2.132)$$

where ω_n are the natural frequencies of the system.

$$A \cdot \left[W_n \quad \Psi_n \quad \Phi_n\right]^T = 0 \qquad (2.133)$$

where the A matrix is defined and completely derived by Zhou, Chen, and Ding (2005). When the determinant of A is set to zero, the free vibration frequencies can be calculated.

2.5 SMA Actuators

2.5.1 SMA Properties of NiTi Alloys

The most common alloys with SMA effect usually contain the NiTi binary intermetallic compound with different additional elements used to modify the behavioral properties of the system, in accordance with the desired performances for the material. A nickel excess, i.e., up to 1%, is frequently added to reduce the transformation temperature and to increment the ultimate tensile stress of the austenitic phase. Other common additives are: iron and chromium to reduce the transformation temperature, copper to reduce the hysteretic cycle and to decrease the stress due to deformation in the martensitic phase. On the other hand, the presence of some contaminating agents should be limited, i.e., oxygen and carbon can modify the transformation interval, worsening the mechanical properties of the material.

The classical stoichiometry of the nickel titanium SMA alloys provides for a 55% Ni and a 45% Ti; this alloy is commonly known as Nitinol (Nickel Titanium Naval Ordinance Laboratory) and possesses interesting SMA properties and superelasticity; furthermore, it presents good electrical and mechanical properties together with a considerable resistance to corrosion and fatigue. In addition, the characteristics of this compound allow an electrical activation through the Joule effect, i.e., when the electric current flows through an NiTi component, an internal heat is generated due to its electric resistance, resulting, after a time lapse, in a phase transformation. In different alloys used for their SMA effect, the transformation temperature is higher than the ambient temperature; furthermore, the advantages of NiTi alloys appear evident when used for small components: wires or thin films.

The shape memory effect consists in restoring an initial configuration after a proper heating over the transformation temperature; this macroscopic effect is the result of a crystalline phase change: the thermoelastic martensitic transition. The maximal recoverable deformation can be up to 8% for these alloys realizing a movement and, if external mechanical loads are applied, realizing a mechanical work during the transformation process. The transition temperature can also be set to activate the transformation with specific physical temperatures, i.e., the corporeal temperature or the water ebullition temperature.

The superelasticity is used for applications demanding extraordinary flexibility or torsion elasticity to absorb deformation energy during the application of external actions releasing it after the removal of mechanical loads. In fact, the Nitinol martensite can be induced by stress at temperatures higher than A_f (temperature of austenite-finish, as shown below), because at this temperature the austenitic phase is stable, then the material recovers rapidly its original configuration when the stress is nullified. Furthermore, the stress–strain diagram of SMA alloys is influenced by the temperature that can especially change the slope of the curves, because over A_f it is subjected to the superelastic effect. The elasticity of NiTi is approximately 10 times the elasticity of iron; furthermore, it resists torsional loads well and can furnish a constant force in a wide range of deformations; these peculiar characteristics are suitable for medical and orthodontic applications, such as surgical instruments, bone suture, and orthodontic arches.

Solid state transformations can be diffusive or displasive; in diffusive transformations, the new phase is formed through an atomic movement for a relatively long distance, because the new phase has a chemically different composition if compared to the initial matrix, and the advancement of the transformation depends on the time and on the temperature; although the displasive transformations do not demand these long movements, the atoms are reorganized in a new crystalline structure without a change of the chemical composition from the initial phase. The absence of atomic migrations in displasive transformations induces an advancement that does not depend on time. Furthermore, the quantity of the new phase depends only on the gained temperature and not on the isothermal permanence; therefore, they are athermic.

The martensitic transformations can be recognized as displasive and are obtained from cooling of the initial austenitic phase that releases heat with a martensite formation (i.e., it is a first-order transformation); furthermore, a hysteretic cycle is associated with the transformation and there is a temperature interval where martensite and austenite coexist.

The martensite preserves the chemical composition and the atomic order of the initial phase, and crystallographically, its transformation from austenite can be subdivided in two stages: Bain Strain and Lattice-Invariant Shear. The first stage is a deformation of the crystalline cell and includes every movement to produce the new structure. In Figure 2.73 an austenitic structure is sketched (a) and its progression towards a completely martensitic structure is illustrated from (b) to the completion. When the interface achieves an atomic plane, a modest movement (c) is demanded of every atom and the final result of all these small Bain Strain coordinated displacements is a new martensitic structure.

The second stage of the martensitic transformation, the Lattice-Invariant Shear, serves as an adjustment because the generated structure has a shape, and also a volume, that changes if compared with the austenitic phase, as can be observed comparing the austenitic structure in Figure 2.73 with the martensitic structure in Figure 2.74. The shape memory effect is associated with a shape variation and, to rearrange the new structure, the new phase can be altered; the deforming mechanisms are slip and twinning, as shown in Figure 2.74, the first is permanent whereas the latter is reversible.

A complete shape memory effect should be associated with a reversible change of shape; thus, the twinning process should be preferred for NiTi alloys. The geminate borders, i.e., the atomic planes between cells oriented in different manners (Figure 2.74b), have a low energy and are rather mobile.

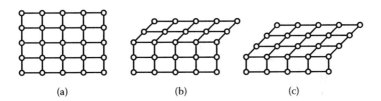

(a)　　　　　(b)　　　　　(c)

FIGURE 2.73
Deformation due to the displacement of the martensitic interface.

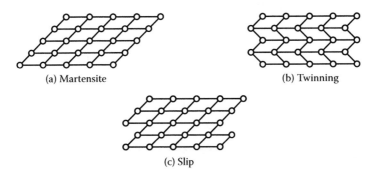

(a) Martensite　　　　　(b) Twinning

(c) Slip

FIGURE 2.74
Slip and twinning.

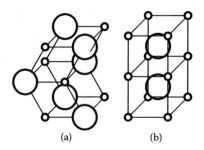

FIGURE 2.75
Elementary martensite and austenite cells.

Furthermore, we do not distinguish the effect of different elements in the crystallographic lattice, although the different positions of atoms in the lattice should radically change the macroscopic behavior of the alloy, i.e., the body-centered cubic (BCC) structure of austenite (Figure 2.75b) is common in shape memory alloys.

From a macroscopic viewpoint, the physical properties of austenite and martensite are usually different; thus, some of these properties can be monitored to follow the advancement of the transformation.

The characteristic temperature shown in Figure 2.76 individuates the start and the finish of martensitic transformations, more precisely, M_s, M_f, A_s, and A_f are, respectively, the martensite start temperature, the martensite finish temperature, the austenite start temperature, and the austenite finish temperature. The hysteretic behavior is associated with a difference of the transformation temperature between heating and cooling and the intensity of the hysteresis depends on the alloy composition: a typical value can be 20°–40°C. Microscopically, this phenomenon can be associated with a friction during the movement of geminated borders.

The yield point changes significantly during the phase transformation; in fact, the martensite is deformed through the motion of a mobile geminate border and it exhibits less resistance than the austenite that is deformed by

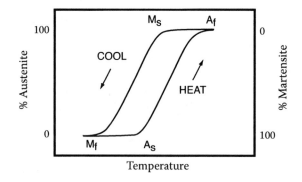

FIGURE. 2.76
Hysteresis in transformation temperature cycle.

a motion of dislocations. Only a quota of the martensitic deformation can be rearranged through a motion of geminates, over a critical value the structure can be deformed elastically eventually yielding again and irreversibly by a sliding of dislocations. The ratio between resistance to reversible and irreversible deformation, i.e., between geminate movement and dislocation sliding, is related to the ratio between martensite and austenite yielding points. Furthermore, SMA alloys should be designed to maximize the recoverable deformation after the heating.

The martensitic phase can be gained with different paths in the graphic shown in Figure 2.74b, where we can observe two types of martensite produced with different shear stresses, while a single way is represented in the same diagram to go back to austenite: this simple geometrical concept is basilar to describe the shape memory effect (SME).

A cooling of the austenitic structure (Figure 2.77) can be used to obtain martensite in two varieties (rhombi in different orientations); the deformation provokes the movements of geminates, driving to a new martensitic structure that complies with the external load (rhombi with a single orientation). A heating over A_f, independently on the martensite variety, produce only austenite (squares), recovering the original shape.

An interesting observation cycle can be performed on an SMA specimen; if it is cooled from a temperature higher than A_f to a temperature lower than M_f, it is not subjected to changes of shape; then, if the specimen is deformed with a mechanical load and, after a successive heating, it starts to recover its original shape when the temperature surpasses A_s and the transformation is completed after reaching temperature A_f (Figure 2.78). This phenomenon is commonly defined as one-way shape memory effect (OWSME) and the SMA alloys can produce a 7–8% recovery.

A specific training on the SMA alloys is sometimes used to obtain a two-way shape memory effect (TWSME) that can be described as the ability to change the shape of an SMA specimen with a heating and also with a cooling (Figure 2.79). The material can recover a high temperature shape and also

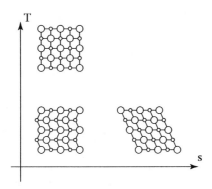

FIGURE 2.77
The martensitic microstructure can be modified and heated to the transformation temperature and produces austenite.

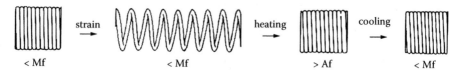

FIGURE 2.78
One way shape memory effect.

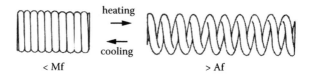

FIGURE 2.79
Two-way memory effect.

a low temperature shape without using any mechanical activation force, just as the TWSME can exhibit after a proper thermo-mechanical training treatment, but the resulting actuation system cannot produce a high mechanical work because it is not able to face the same loads of OWSME components.

Another interesting structure to be mentioned for the SMA alloy is the stress-biased martensite, produced with mechanical excitations on martensite under the M_f temperature or on austenite not much over A_f. The mechanical excitation generates an alteration of the martensitic microstructure and the corresponding new shape can be recovered at low temperature, and the accumulated stresses drive the material to this low temperature shape when it is cooled.

The TWSME can be obtained with different thermo-mechanical treatments: with an over-deformation in martensitic conditions, with an SME cycle, with a pseudo-elastic cycle, with a combined SME/pseudo-elastic cycle, or with a forced temperature cycle on deformed martensite. The over-deformation in martensitic conditions consists in a cooling of the martensitic microstructure under M_f, with a successive severe deformation over the yielding point. The SME cycle consists in a cooling of the martensitic microstructure under M_f, a deformation, and a successive heating of the austenitic microstructure over A_f to recover the high temperature shape; the cycle should be repeated 5–10 times until the TWSME appears.

The pseudo-elastic training cycle is realized with a cooling under the temperature M_d (described below), a mechanical deformation, the release of mechanical load with a consequent shape recovery; the last two stages should be repeated 5–10 times until the TWSME appears. A combined SME/pseudo-elastic cycle can be realized with the successive steps: cooling under M_d, mechanically loading the specimen with the production of a quota of martensitic microstructure, cooling the loaded specimen, mechanically releasing and heating over A_f. The cycle should be repeated until the TWSME appears. Finally, a forced temperature cycle on deformed martensite is induced: cooling the martensitic microstructure under M_f, mechanically loading the specimen with the production of a quota of martensitic microstructure, heating over A_f the mechanically loaded specimen, performing a thermal cycle over A_f and under M_f on the loaded specimen.

As just mentioned, an interesting property of NiTi alloys is the superelasticity. Generally, the martensite start to be produced at the M_S temperature without the need of mechanical loads. A new phase can be generated in the same material at higher temperatures than M_S due to the presence of an applied stress; therefore, the realized martensite is called stress-induced martensite (SIM), because the mechanical action replaces the thermodynamic driving force due to cooling. The demanded stress to produce SIM increments proportionally to temperature up to M_d and, over this critical temperature, the necessary stress is higher than the yield point; thus, M_d is the higher temperature where martensite can be found; furthermore, the superelasticity can be exhibited when the material is deformed at a temperature higher than A_S, but still lower than M_d. In this temperature range, the martensite phase can be stabilized applying a stress, but it becomes unstable again with a removal of the stress. From a microscopic viewpoint, the mechanical load on the specimen generates a SIM variety oriented along the application direction and produces the maximal deformation due to a twinning of the martensitic structure. During the SIM transformation, the demanded stress is constant, because all the formed martensite has the same structure; then, at the maximal deformation of the specimen, the percentage of SIM is high; finally, the deformation is recovered with a removal of the mechanical load.

From a macroscopic viewpoint the behavior of the martensite phase under M_f is associated with an induced deformation, approximately 4%, that is recovered with a heating between the temperatures A_S and A_f, after the removal of the mechanical load. At a temperature higher than M_S, but lower than M_d, SIM can be produced and the typical superelastic loop can be observed. Finally, when the temperature is higher than M_d, the martensite phase does not appear and the physical behavior of the system exhibits only the presence of austenite.

The knowledge of the transformation temperature that is influenced by the chemical composition of the alloy is a prominent parameter for the characterization and for the choice of the material because the functional characteristics of the SMA actuators are strictly related to the SMA transformations. In fact, the transformation temperature is a critical value to identify the starting/finishing point of the deformation and, consequently, of the realization of the actuating movement. A temperature of transformation can be determined with different procedures: a constant load test, the differential scanning calorimeter (DSC) and the active austenite finish (Active A_f).

The first method consists in applying a constant load to an SMA specimen and recording the deformation and the shape recovering, corresponding to a cooling and a heating during the transformation interval. Generally, these trials are realized with loads that can be imposed during the typical working conditions demanded by the application of this material. Finally, this test is usually suggested for applications utilizing the shape memory effect and not the superelasticity of the NiTi alloy.

The transformation temperature depends on the loading parameter; therefore, an interesting measurement can be obtained in absence of load to find the natural characteristics of the system; the DSC approach can be used in this situation to deduce an experimental diagram measuring the absorbed/released heat of an SMA specimen during a purely thermal transformation cycle. The principal disadvantage of the DSC analysis can be related to its poor results on materials that were subjected to mechanical treatment, i.e., to optimize the superelasticity effect, or that were subjected to thermal treatments at high temperatures. On the other hand, the DSC approach is recommended to choose an untreated NiTi alloy, because it can furnish its original characteristics exempting the effects of eventual thermal treatments or hot/cold workings.

The last interesting test is the Active A_f, or Water Bath, or Alcohol Bath Test; it is realized simply bending an SMA specimen, i.e., a wire, while its temperature is under M_s, and measuring its shape recovering during the heating. The heating is performed with an immersion in a liquid bath, then, controlling the temperature of the liquid and gradually varying it, an experimental relation can be measured between the realized deformation and the set temperature. This procedure, even if it is not sophisticated, can furnish interesting results with modest equipment and it is often used to determine the A_f temperature of superelastic materials.

The correct choice of a NiTi alloy depends on the demanded macroscopic characteristics to obtain the desired shape memory effect or superelasticity; for this reason, in Table 2.9 are listed some standard NiTi alloys with their composition and their common applications: the S, N, and C alloys are typically used for their superelastic properties, the M and H alloys exhibit an

TABLE 2.9

Some Standard NiTi Alloys

Alloy Code	A_S	A_f	% Weight	Applications
S	–5 to 15	10 to 20	55.8 Ni	Orthopaedic, orthodontic, and surgical instruments; cell phone antennas
N	–20 to –5	0 to 20	56.0 Ni	Wires, springs
C	–20 to –5	0 to 10	55.8 Ni, 0.25 Cr	Thin wires
B	15 to 45	20 to 40	55.6 Ni	Corporeal temperature activated, filters
M	45 to 95	45 to 95	55.1 to 55.5 Ni	Actuators, toys, springs
H	> 95	95 to 115	< 55.0 Ni	Actuators

interesting shape memory effect, while the B alloys present both the super-elastic and the shape memory effect.

The NiTi materials are subjected to hot workings, i.e., forging or hot roll-ing, to cold workings and to thermal treatments to obtain different results: improving the deformability of the martensitic phase, incrementing the mechanical resistance of the austenitic phase, enhancing the superelasticity, partially or totally conferring and recovering a shape, producing a TWSME, giving additive surface properties.

The titanium preserves a high reactivity in these alloys; therefore, an eventual fusion is usually performed into vacuum or into an inert atmosphere adopting electron beams or plasma furnaces; while hot working are executed between 450°C and 600°C in a normal atmosphere before the final thermal treatment.

The treatments are able to significantly modify the behavior of an alloy; thus, a classification based only on the thermal treatments or the hot or cold workings can be useful to characterize the macroscopic properties of an SMA alloy or of an SMA component. The cold-worked, i.e., cold-rolled, SMA alloys that are not subjected to a thermal treatment do not possess shape memory or superelastic properties; the straight annealed materials are properly thermal treated and wire shaped to realize SMA wires; the flat annealed materials are subjected to the same thermal treatments as the straight annealed materials and are lamina-shaped; some materials can also be distributed after a metal forming to assign a special shape; finally, other thermo-mechanical processes can include prestressing, and further thermal/vacuum treatments.

The realization of a component with a specific shape is performed introduc-ing the raw material into a desired mold and executing the proper thermal treatment to obtain shape memory or superelastic properties. The process parameters are usually set empirically within a theoretically predetermined range, i.e., a temperature over 400°C for some minutes, followed by a rapid cooling into a water bath or an air atmosphere, can set a desired shape for different NiTi alloys when the dimensions of the molded component are small. Too high temperatures or too long a period of exposure to heat sources can produce aggressive thermal reactions that can reduce the resistance to mechanical deformations; furthermore, a too slow heating process can retard the thermal equilibrium between mold and raw material, compromising the mechanical characteristics of the final component.

Important precautions should be taken during two-way shape memory induction techniques; in fact, they are really sensitive to process parameter and can realize limited performances with respect to the one-way shape memory components: the maximal recoverable stress is usually only 2%, the intensity of the tolerable mechanical stress is reduced after the thermal cool-ing, the shape memory effect can be cancelled by an overheating, the fatigue effects and the long-time stability of the thermo-mechanical properties are not easily forecastable; therefore, OWSME should be preferred to TWSME, when it is possible.

The NiTi materials are often subjected to surface oxidation due to chemi-cal reactions between titanium and oxygen that produce titanium dioxide

(TiO$_2$) compound; this TiO$_2$ is not usually disturbing but it can be removed to obtain a cleaned surface with a mechanical cleaning or with chemical solvents, but the high reactivity of Ti with the oxygen will quickly produce a new visible surface patina on the component. An interesting characteristic of NiTi alloys, associated with their superficial oxide protection, is their ability to resist corrosion and environmental agents. The surface appearance of an SMA component can be easily observed and used to acquire information on its physical behavior: the amber-brown oxide is hard and smooth, whereas its thickness is between 500 Å and 3000 Å; the black oxide is less hard, it is smooth and bright and its thickness is less than 4000 Å: it is used to produce special aesthetic characteristics but it can induce a dangerous flaked layer; the appearance of an etched surface is silver-gray and opaque with microscopical pitting because the oxide layer was removed with acid etching; mechanical polishing produces a bright silver aspect; ultrafine polishing is used to realize very smooth surfaces; many other surface treatments and coatings with different oxides, polyurethane, and other compounds are used to confer special characteristics to the external surface of a component.

After this list of surface treatments, referring to standard superelastic alloys, to identify some macroscopic mechanical properties that characterize the material, we should highlight the loading and unloading plateau, the residual strain, the ultimate tensile stress (UTS), and the maximal elongation. Typical mechanical properties of superelastic NiTi alloys are summarized in Table 2.10.

The high biocompatibility of NiTi alloys allows a wide range of applications in medical and orthodontic fields, where a technological demand is high; therefore, we can find surgical instruments, permanent orthopaedic and orthodontic implants. A significant medical device is the Simon Nitinol Filter: it is a microscopical umbrella-shaped filter and it is inserted, in closed configuration, into a vein; then, due to the shape memory effect, it is activated by the corporeal temperature resulting in an opened umbrella configuration that is able to filter eventual blood clots. An interesting orthopedic application is the Mitek system used to make a connection between tendons, ligaments, and bones; the device consists of a titanium structure connected with different NiTi wires, the titanium structure can be rigidly connected to

TABLE 2.10

Typical Mechanical Properties of Superelastic NiTi Alloys

	Code	C	N	S
Before the superelastic treatment	UTS [MPa]	1026	1026	950
	Breaking elongation %	6	7	7
After the superelastic treatment	UTS [MPa]	836	798	760
	Breaking elongation %	14	15	15
	Loading plateau [MPa]	361	323	285
	Unloading plateau [MPa]	190	114	76
	Residual strain %	< 0.25%	< 0.25%	< 0.25%

TABLE 2.11

Nitinol Properties

Density	6.45 g/cm³
Thermal conductivity	10 W/mK
Specific heat	322 J/KgK
Breaking stress	750–960 MPa
Breaking strain	15.5%
Austenite yield point	560 MPa
Austenite Young modulus	75 GPa
Martensite yield point	100 MPa
Martensite Young modulus	28 GPa

a bone through an artificially made hole, whereas the NiTi wires can be activated by the corporeal temperature to realize a permanent connection with another biological element.

As just said, the Nitinol was realized in 1962 at the U.S. Naval Ordinance Laboratory with a composition of 55% nickel and 45% titanium, resulting in an alloy exhibiting the shape memory effect and characterized by the principal properties listed in Table 2.11.

The common NiTi alloys are characterized by the the NiTi intermetallic compound, but often, there is also the presence of other intermetallic compounds, i.e., $NiTi_2$ and Ni_3Ti (Figure 2.80), because the temperature range of existence of NiTi is usually restricted. Furthermore, the behavior of the alloy can be influenced by additive elements or by contaminants, i.e., as just mentioned, the oxygen can exploit the reactivity of titanium to produce TiO_2; the result of an excess of external elements can result in difficulties during hot workings of the material, eventually causing fragility and cracks. The

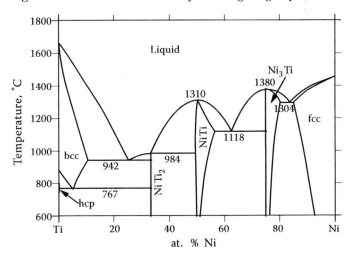

FIGURE 2.80
State diagram of NiTi.

hysteretic diagram of a NiTi alloy can be designed with a proper composition of the material, i.e., a nickel excess can drastically influence the value of the M_S, while the same nickel excess is able to influence drastically the A_S value for tempered alloys and less significantly for tempered and artificially aged alloys.

From a macroscopic viewpoint, as with other SMA alloys, Nitinol exhibits different mechanical behavior for austenitic and martensitic phases. As shown in Figure 2.81, the martensitic stress–strain diagram can be subdivided in three parts: an initial plateau, due to the SIM formation associated with a twinning deformation; a linear domain with a mixed elastic-twinning deformation; and an irreversible plastic domain. The austenitic behavior is influenced by temperature; between M_S and M_d, the elasticity is predominant, but with an increasing deformation, a SIM structure can appear due to twinning, while a further stress produces plasticity; over the M_d temperature, martensite cannot be produced; thus, the plastic behavior can be anticipated. An interesting phenomenon is the correlation between mechanical load and the transformation temperature, due to the energy density of the material. For this reason, when the yield performances of a NiTi alloy are improved with cold workings, the process can influence negatively the shape memory effect drastically reducing the maximal recoverable deformation; therefore, a cold working is usually followed by a thermal drawing. In fact, the work hardening produces a high density of dislocations that reduces the mobility of the geminated borders, whereas the thermal drawing produces a dislocation readjustment incrementing the geminate mobility. The thermal treatments are important to optimize the superelastic properties, i.e., incrementing the austenite yield point, the possibility of SIM formation are incremented without running into permanent deformations.

On the other hand, the transformation temperature can influence mechanical properties of the material through a change in the stress–strain relation (Figure 2.82).

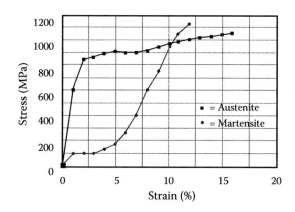

FIGURE 2.81
Strain–stress diagram of a NiTi alloy in the austenitic and martensitic phases.

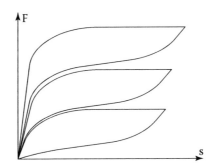

FIGURE 2.82
Effect of the temperature on the strain–stress relationship, the shift in the hysteresis cycle.

2.5.2 Some Observations for the Realization and the Selection of SMA Actuators

The principal advantages associated with the SMA technologies for the realization of actuation devices are the simple machine design and the favorable ratio between actuator weight and exertable mechanical power. Persuaded by these properties, different authors proposed a wide range of applications, i.e., manipulators or automatic clamps or more complex systems, whereas others concentrated their attention on the control problem, limiting the actuator stroke to reduce the hysteretic cycle with positive result on its predictability, or adopting a proper sensor, i.e., a thermal sensor, to realize a closed-loop control. Interesting simplified closed-loop systems are based on the variation of the electrical resistance of the shape memory alloy during the transformation and it is used to measure the martensite fraction and, indirectly, the deformation of the actuator (Ikuta, Tsukamoto, and Hirose 1998; Homna, Miwa, and Iguchi 1984).

As mentioned the hysteretic cycle influences negatively the predictability of the system; furthermore, the hysteresis depends on the applied mechanical load and on the internal temperature. Different authors have run experiments on shape memory alloys to highlight the decay of the SME or the fatigue phenomena (Reynaerts and Van Brussel 1998), obtaining relations between the mean deformation of an SMA component and its service life (Furuya et al. 1989), or incrementing the long-term performances with proper expedients (Beyer et al. 1990).

A crucial point for the use of these materials is the control of their temperature; the heating is directly induced on an energy insertion into the system, whereas the cooling depends highly on the environmental conditions and by the thermal exchange approach: conduction, convection, or irradiation. Normally, in our conditions, the conductive and the convective phenomena (only with a relative movement between the SMA component and an eventual environmental fluid) are predominant; therefore, the transferred thermal power can be expressed as in (2.134), where P_t is the transferred power,

h is the transmission coefficient, A is the exchange surface, and ΔT is the temperature difference.

$$P_t = h \cdot A \cdot \Delta T \tag{2.134}$$

The cooling time used to obtain a complete transformation to the martensitic phase from the temperature M_S to the temperature M_f can be obtained (2.139) from the simplified exchange equation (2.135), integrating an infinitesimal time interval (2.137) on the whole transformation (2.138).

$$m \cdot c' \frac{dT}{dt} + h \cdot A \left(T - T_0\right) = 0 \tag{2.135}$$

where m is the mass of the SMA component, c' is the equivalent specific heat (2.136), $T(t)$ is the mean temperature of the SMA, and T_0 is the external temperature.

$$c' = c + \frac{\Delta T}{\Delta h} \tag{2.136}$$

where Δh is the enthalpy of the transformation and c is the specific heat.

$$dt = -\frac{mc'}{h} \frac{dT(t)}{T(t) - T_0} \tag{2.137}$$

$$\int_0^t dt = -\frac{mc'}{h} A \int_{M_s}^{M_f} \frac{dt}{T(t) - T_0} \tag{2.138}$$

$$t = -A \frac{mc'}{h} \ln\left(\frac{M_f}{M_S}\right) \tag{2.139}$$

The term ΔT depends on the transformation temperature and on the environmental temperature, A is related to the shape of the component, whereas h is influenced by the exchange conditions (Table 2.12).

The results listed in Table 2.12 show how a low flow can improve the cooling performances, reducing the overall cooling time and, at the same time, the use of a cooling liquid can drastically increase the behavior of the system. Unfortunately, the choice of the environmental fluid is affected by the type of energy used for the heating process; when the heating is performed through a Joule effect, the presence of electric current is not directly associable with an explosive atmosphere or with a conductive liquid; therefore, explosive and conductive fluids should be avoided.

TABLE 2.12

Dependence of the Thermal Coefficient h
on the External Fluid

	Rarefied Fluid (Gas)	Dense Fluid (Liquid)
Immobile fluid	Low	Higher
Low flow	Higher	High
High flow	High	Very high

Furthermore, experimental results confirmed a reduction of the cooling
time with an increment of the applied mechanical load, due to the increment
of the transformation temperature that produces an increment of the differential temperature ΔT. Finally, the section effect on the thermal exchange can
be improved with the adoption of a square-shaped section, drastically reducing the cooling time with respect to a circular section but reducing the fatigue
performances, due to the presence of a locally reduced radius of curvature.

On the other hand, the heating process can be performed with an environmental heated fluid, but this technique is usually inconvenient and it is used
only in special measurement systems or in some biomedical applications, i.e.,
the Simons Nitinol filter, whereas the Joule effect, due to the electric current
flowing the natural resistance of the actuator, is a really controllable and
localized phenomenon; therefore, it can be an interesting option. The choice
of the Joule effect as heating method drives our selection of the proper alloy
to realize a suitable actuator: it should be an electric conductor and possess
a good electric resistivity, i.e., a NiTi alloy with a low copper tenor. The NiTi
alloys are usually cold-worked because this process favors the formation
of dislocations realizing desired starting points for the martensite enucle-
ations. Unfortunately, cold working is a difficult process and, therefore,
simple shapes are common to realize SMA actuators, such as wires, bars,
tubes, rings, or plates. The SMA wires can also be twisted to realize a spiral
spring, obtaining an amplification of the stroke with a reduced resistance to
mechanical loads; furthermore, the twisted configuration results in an increment of the volume of SMA alloy to realize an actuator with the same exertable force of a simple linear configuration, consuming more material and
degrading the dynamic speed of the system. NiTi wires exhibits a particular
efficiency for traction loads (Table 2.13), due to their geometrical shape; therefore, mechanical devices, using these active elements, should be designed to
produce only axial loads.

Different SMA wires can be used in to form a complex systems with series
or shunt configurations; a series of NiTi wire has a behavior of a long wire
and can be used to generate a stroke amplification, whereas a shunt configuration is able to increment the exerted force. Furthermore, the amplification system can be realized with external mechanical leverages and they
are used, generally, to increment the stroke decreasing the exerted force. The
connection between different SMA wires or between an SMA wire and the

TABLE 2.13

Mechanical Loads on SMA Wires

Mechanical Load	Efficiency	Energy Density
Traction	Good	High
Torsion	Low	Average
Flexion	Very low	Low

electrical feeding system can be realized with welds or with special clamps; the weldability of NiTi is really low and, therefore, the choice of clamps is the most common. These clamps should be realized with a conductive material characterized by a low resistivity to allow a suitable electrical flow and to reduce the local heating due to the contact resistance between clamp and SMA wire; furthermore, the material should be able to maintain a good pressure-level on the wire, in spite of its tendency to realize shear actions on the joint during the shape transformation process.

Due to these different phenomena and, as just mentioned, to hysteresis an open-loop feeding system can be realized only for low-speed applications and with a reduced positioning precision; furthermore, during the normal work of the actuator, uncompleted transformations can be observed with minor hysteretic loops within the principal loop associated with the complete transformation. Therefore, a precise feeding needs a closed-loop system that monitors in real time the behavior of the actuator and, with a proper controller (a PID or a simple PI can be adequate), it can realize the desired transformation. A common setup composed by an electric feeding can be realized with a pulse width modulated (PWM) system, whereas in a position control loop, the position signal is measured and compared with a reference value. Then the difference is used as input of a common proportional integral derivative (PID) or a proportional integral (PI) controller and, clearly, a high electric feeding power is associated with a fast movement (Figure 2.83), as expressed in (2.140), but a too high electric power can produce a damage of the SMA wire.

$$m \cdot c \cdot \frac{dT}{dt} = P + h \cdot A \left(T - T_0 \right) \qquad (2.140)$$

where m is the mass of the active element, c is the specific heat, $T(t)$ is the fictious temperature of the wire (the real temperature is a space distribution and assumes different values in different points of the wire), P is the input power, h is the transmission coefficient, A is the exchange surface, and T_0 is the external temperature.

The cooling power is generally passive, apart from Peltier cells that electrically produce a cooling, and the change of shape during cooling is due to an external recovery force. When the force is active, the controller can set this recovery force to obtain the desired motion profile; when the force is passive,

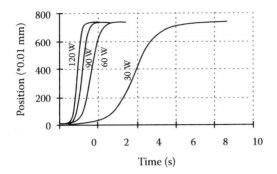

FIGURE 2.83
Time to realize the desired movement associated with the feeding power.

the controller can only produce a slowdown of the recovery motion through an electric feeding that contrasts with the passive cooling process.

2.5.3 Thermo-Electro-Mechanical Models of SMA Fibers

The characteristics of SMA components can be estimated using different methods. Some of them include: differential scanning calorimetry (DSC), liquid bath, resistivity measure, cycle with constant applied load, and traction test. DSC is a technique that is able to measure absorbed and released heat in an SMA specimen during crystallographic transformations. Liquid bath is a liquid with a controllable temperature, in which an SMA specimen is immersed and the imposed temperature is related to the macroscopic shape conversion. Resistivity measure is based on the change of resistivity during crystallographic transformation. An austenite–martensite–austenite cycle can be imposed with a constant mechanical load, allowing the measure of maximum and minimum deformations, which can be associated with critical points of the transformations. A traction test can be executed at a constant temperature to measure mechanical properties of the material and to relate them to the fixed temperature.

The numerical characterization of the alloy can be used to properly set a thermo-(electro)-mechanical dynamic model, which allows the open-loop control of the SMA used as an actuator. We will consider only dynamic/kinematic models of fibers; however, as we will see in the following, dynamic models of thin films are also interesting in SMA actuators based on the Peltier cell concept.

The literature on dynamic models of SMA fibers is an active field, and a classification of the different approaches could be helpful in choosing the correct modeling method. A first classification can be based on deriving principles; we can distinguish empirical models, micromechanical models, and thermodynamic models. Some models are directly structured to describe the behavior of SMA fibers (a 1-D problem) whereas other models are developed for more complex problems (2- or 3-D problems) and are then

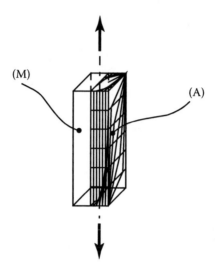

FIGURE 2.84
SMA Voigt model where (*M*) is the martensite phase and (*A*) is the austenite phase.

reduced to describe only the characteristics of a 1-D problem. Dynamic SMA models are usually composed by a stress–strain relationship and a kinematic law. We can distinguish models where these two laws are inseparable, separable but coupled, or separable and uncoupled. The majority of the models are numerical because of the high nonlinearity of the shape memory effect; however, we must mention some simplified analytic models. Despite their imprecision, analytic models are very useful for a first-dimensioning of a micro-actuator and in closed-loop applications. Many other distinctions can be considered, e.g., the presence of martensite variants in 3-D models or the set of independent variables of the model; however, the proposed taxonomy can be a good starting point to deepen the world of shape memory models. We will describe a famous numeric model that was started with a work of Tanaka, Kobayashi, and Sato (1986), which was first modified by Liang and Rogers (1990) and later by Brinson (1993), while Brailovski, Trochu, and Daigneault (1996) or Potapov and da Silva (2000) added some interesting simplified analytic models.

A simple micromechanical derivation of the first model was proposed by Brinson and Huang (1996). A parallel Voigt model of austenite and martensite phases is considered in a 1-D specimen as in Figure 2.84. The specimen is subject to an external stress, so an elastic strain appears and, due to the considered Voigt model, the relations between austenite and martensite stress and strain are as in (2.141).

$$\sigma = \left(1-\xi\right)\sigma_a + \xi\sigma_m \qquad \varepsilon_m = \varepsilon_a$$
$$\sigma_a = E_a\varepsilon_a \qquad\qquad\qquad \sigma_m = E_m\varepsilon_m \tag{2.141}$$

where σ is the stress of the specimen, ξ is the martensite fraction, σ_a and σ_m are, respectively, austenite and martensite stress, ε_a and ε_m are, respectively, austenite and martensite strain, whereas E_a and E_m are, respectively, austenite and martensite Young modulus. Relations in (2.141) can be combined to obtain (2.142).

$$\sigma = \left[\xi \cdot E_m + \left(1 - \xi\right) \cdot E_a\right] \cdot \varepsilon_a \tag{2.142}$$

We now consider a temperature increment able to generate a phase transformation. It results in an adjustment of the SMA Voigt model (Figure 2.85) and the consequential total strain is the sum of the elastic strain and of the transformation strain; thus, the stress–strain relation can be modified as in (2.143).

$$\sigma = \left[\xi \cdot E_m + \left(1 - \xi\right) \cdot E_a\right] \cdot \left(\varepsilon - \varepsilon_L \cdot \xi_S\right) \tag{2.143}$$

where ε_L is the maximum residual strain of the transformation and ξ_S is the detwinned martensite fraction.

The stress–strain relation (2.143) must be coupled with a kinetics equation, i.e., (2.144).

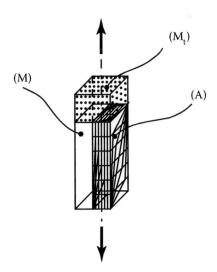

FIGURE 2.85
SMA Voigt model where (M) is the martensite phase, (A) is the austenite phase, and (M_t) is the transformed martensite.

$$\begin{cases} \xi = \dfrac{1-\xi_0}{2} \cos\left\{ \dfrac{\pi}{\sigma_S^{cr} - \sigma_f^{cr}} \left[\sigma - \sigma_f^{cr} - C_M \left(T - M_S \right) \right] \right\} \\ \text{with} \quad T > M_S, \quad \sigma_S^{cr} + C_M \left(T - M_S \right) < \sigma < \sigma_f^{cr} + C_M \left(T - M_S \right) \end{cases} \qquad (2.144)$$

where ξ_0 is the initial fraction, M_s is the critical martensite start temperature, σ_S^{cr} and σ_f^{cr} are, respectively, the martensite start and finish critical stress, C_M is the slope of the border line of the detwinned martensite field in the stress–temperature diagram. A complete explanation of the model can be found in (Brinson 1993).

2.5.4 Control Techniques

There are three main important control techniques based on three different physical phenomena: current control, resistance control, and Peltier effect control. We will show in this section some recent advances in the control of SMA actuators, especially of wire and rod actuators for linear movements. We will emphasize in particular the use of some physical properties to achieve the desired behavior of the considered system; and, we will neglect many interesting properties of the system in terms of "control theory." The scope is a brief mention of the control problems of SMA actuators, referring the reader to the specific bibliography for further detailed explanation. As mentioned above, the first type of control approach for SMA actuators considered is current control. SMA fibers are electrical conductors having particular values of electrical resistance. Because of this resistance and because of the Joule effect, a current going through the wire is able to generate heat, and consequently, the current can be used to control the temperature of the SMA wire.

Many authors reported studies of this type of SMA control; we can refer to Bhattacharyya, Faulkner, and Amalraj (2000) who considered an SMA wire subjected to an electric current, under no mechanical load, and under proper assumptions of the material's properties. Under their assumptions, the energetic balance can be formulated as in (2.145) referring to Figure 2.86.

$$\frac{\partial}{\partial x}\left[K(\xi)\frac{\partial T}{\partial x} \right] + \rho_E(\xi)J(t)^2 - \frac{2h}{R}\left(T - T_{amb} \right) = C_V(\xi)\frac{\partial T}{\partial t} - H\frac{\partial \xi}{\partial t} \qquad (2.145)$$

where ξ is martensite fraction (as usual, also Bhattacharyya et al. propose a cooling and a heating law, to be associated with the thermodynamic energetic equation); K, ρ_E, and C_V are, respectively, thermal conductivity, electrical resistivity, and heat capacity; h is the convection coefficient (Bhattacharyya et al. propose a linear definition of these properties); H is the transformation

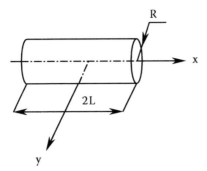

FIGURE 2.86
SMA wire.

latent heat; R is the cross-sectional radius of the wire; T and T_{amb} are, respectively, the temperature of the wire and of the environment; t is the time variable; and J is the current density, which is used to generate a forced heating of the wire. The boundary and the initial conditions of the Cauchy problem can be expressed by (2.146) if we consider a wire with a $2L$ length.

$$\begin{cases} T(x,t)\big|_{x=\pm L, \forall t \in \Re} = T_{amb} \\ T(x,t)\big|_{\forall x \in [-L,L], t=0} = T_{amb}, \quad \xi\big(T(x,t)\big)\big|_{\forall x \in [-L,L], t=0} = \xi_0 \end{cases} \tag{2.146}$$

The second control scheme taken into consideration is based on the change of wire resistance during the phase transformation (Airoldi, Besseghini, and Riva 1995). When the wire is subjected to a generic thermal load and a constant mechanical load, we can observe a linear relationship between the strain and the relative variation of the resistance of the wire (Wu, Fan, and Wu 2000). The slope of the relationship is dependent upon the specific transformation, and the relationship between the stress and the relative variation of resistance can be quite complex. This functional aspect can be used to control the movement of the micro-actuator with a simple control scheme. A quantitative work in this direction is reported by Sittner et al. (2000). The authors studied the behavior of an $Ni_{50}Ti_{45}Cu_5$ alloy and, utilizing the principles of (2.147) during martensitic transformation and some other theoretical and experimental observations, they developed a new, interesting behavioral model of this alloy.

$$\frac{\Delta R}{R_0} = \frac{\Delta \rho}{\rho_0} + (1+2v)\varepsilon + \frac{C^R\left(T-T^R\right)}{\rho_0} \tag{2.147}$$

where R_0 and ρ_0 are, respectively, the resistance and resistivity in the austenitic phase at the reference temperature T^R, ΔR, and $\Delta \rho$ are, respectively, the absolute change of resistance and resistivity during the transformation at the generic T temperature, ε is the resulting strain, whereas v and C^R are two material parameters. The authors, according to other researchers (De Araujo, Morin, and Guenin 1999), observed a linear relation between relative resistivity and strain in martensitic and austenitic transformation (2.148).

$$\frac{\Delta \rho}{\rho_0} = K\varepsilon \tag{2.148}$$

where K represents the constant slopes associated with the shape memory transformations.

The third, and most promising, control scheme considered is based on the Peltier effect (Lagoudas and Kinra 1993), which focuses on the ability to increment the working frequency of SMA micro-actuators. This special phenomenon is produced using a thermoelectric element, obtained by sandwiching an SMA layer between two semiconductor layers, a positively doped (P) and a negatively doped (N). Alternating the current direction, the SMA layer becomes the hot or the cold junction of a thermoelectric couple, because the Peltier effect generated by electric current causes a temperature differential at a junction of the dissimilar metals. The general thermal transfer model for this control scheme is proposed by Ding and Lagoudas (1999) and is presented in (2.149).

$$K_i \frac{\partial^2 T_i}{\partial x^2}(x,t) + \rho_i J^2(t) - H \frac{P_C}{A_C}\left(T_i(x,t) - T_0\right) = C_v^i \frac{\partial T_i}{\partial t}(x,t),$$

$$x \in I_i, t > 0, i \in \{P, S, N\} \tag{2.149}$$

$$I_P = \left]-L-\frac{d}{2}, -\frac{d}{2}\right[, I_S = \left]-\frac{d}{2}, \frac{d}{2}\right[, I_N = \left]\frac{d}{2}, L+\frac{d}{2}\right[$$

where P, S, and N are, respectively, a positively doped semiconductor, an SMA layer, and a negatively doped semiconductor (Figure 2.87); ρ_i is the electrical resistivity, J is the current density, C_v^i is the heat capacity per unit volume, T is the temperature with t is its time variable, P_C and A_C are the perimeter and the area of the cross section, and H is the heat convection coefficient.

The interface conditions consist in the temperature continuity and the equal exchange of net heat flux across the interfaces (2.150).

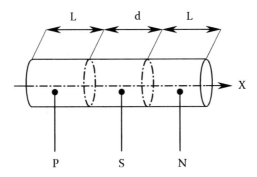

FIGURE 2.87
SMA Peltier actuator.

$$T_S\left(-\frac{d}{2},t\right)=T_P\left(-\frac{d}{2},t\right), \quad T_S\left(\frac{d}{2},t\right)=T_N\left(\frac{d}{2},t\right)$$

$$-K_S\frac{\partial T_S}{\partial x}\left(-\frac{d}{2},t\right)=-K_P\frac{\partial T_P}{\partial x}\left(-\frac{d}{2},t\right)+\alpha_P T_P\left(\frac{d}{2},t\right)J(t) \qquad (2.150)$$

$$-K_S\frac{\partial T_S}{\partial x}\left(\frac{d}{2},t\right)=-K_N\frac{\partial T_N}{\partial x}\left(\frac{d}{2},t\right)+\alpha_N T_N\left(\frac{d}{2},t\right)J(t)$$

where α_N and α_P are the Seeback coefficients. The end boundary conditions are defined in (2.151) and the initial conditions consist in the thermal equilibrium of the system at the T_0 environmental temperature.

$$T_P\left(-L-\frac{d}{2},t\right)=T_0, \quad T_N\left(L+\frac{d}{2},t\right)=T_0 \qquad (2.151)$$

The solution of the differential problem provides us a model to understand the behavior of the SMA system under the presence of the current density J.

3

Finite Element Analysis

The design and realization of micro- and meso-actuators benefit from simulation methods based on finite elements that allow a high precision level. Particularly, different integrated and visual metaphysics—freeware or commercial—instruments are available to developers' and users' communities, utilizing the experience of the former, the examples proposed by the latter, and the performing computational power, in a single environment. The advantages of this type of instrument consist, primarily, in the robustness of the computational process and the reasonable development time, whereas the disadvantages can be associated with a certain complexity to realize simple operations, which implies a sort of entry barrier for occasional users; also, a high complexity in realizing very simple custom operations can be a limitation for exigent and specialized users. After an evaluation of the advantages and disadvantages, we believe that finite element simulators are very interesting instruments to perform the most common design operations and realize a virtual prototype, and, after a good experience of the designer, they can be really useful also for special situations. Therefore, in this chapter, we examine the use of a finite element model (FEM) multiphysic environment for our specific interests relative to micro- and meso-actuators, particularly the ANSYS multiphysics software for its wide diffusion in technical and scientific circles, which, hence, can be used as a reference solution for comparison with other similar software. This deveopment environment is provided with a user-friendly graphic interface that allows a rapid learning of the first rudiments; further, a textual interface remains in the ANSYS Parametric Design Language (APDL). We refer to the APDL approach rather than the graphical input because the textual approach allows a rapid elaboration of repetitive actions and a consciousness of the simulative process, at least initially, and is suitable for an exposition in a textual monograph. On the other hand, the graphical interface can be learned more easily by the user without special preliminary explanations. We must specify that the proposed description regarding some commands and examples of ANSYS' use are not intended to replace the user manual but is only an overview of how this instrument is suitable for useful analyses of actuators or, in general, of microsystems. Therefore, before producing technical results, their validity should be accurately reviewed by an expert familiar with the user manual, with proper experimental calibration and comparison with the scientific/technical literature (Gyimesi M., Wang J. S., Ostergaard D., 2001), (Gyimesi M., Ostergaard D., 1999).

An APDL code is composed, normally, of a rather standard sequence of sections. First, we should choose the type of simulation to individuate the proper basic element that is used as the elementary block to constitute the body or the set of bodies that we desire to analyze. Every elementary component is characterized by specific degrees of freedom and constraints, and proper physical properties that allow us to represent only some physical behaviors; therefore, the choice of the element should be performed wisely. The body, or the system of bodies, submitted to a simulation is represented by a specific spatial geometry with assigned material properties. Hence, it is subjected to external loads and constraints that identify completely with the Cauchy problem. The next step consists in generating a mesh automatically, semiautomatically, or manually to subdivide the continuous geometry in a series of finite elements for the proper numerical solver. Finally, postprocessing allows a numerical or graphical output of results produced by the solver to permit a correct technical interpretation by the user.

3.1   Introduction to APDL

As mentioned, APDL is the ANSYS Parametric Design Language, and it is a scripting language used to build the model, eventually using parameters, and automate some tasks; further, it is a powerful instrument that can be used to modify a simulation realized with a graphical interface, or design the simulation on a slow computer. APDL is a real programming language with if-then-else branching and do-loops; therefore, it can also be used to implement custom algorithms or perform mathematical computations, as shown in the following sections.

3.1.1   Parameters

APDL parameters are really variables with associated values that can be changed during the execution of the program; the name of a variable must begin with a letter, must be composed only of alphanumerical and underscore characters, and must contain less than 32 characters. Some examples of correct names are shown in (3.1).

$$RHO \_ WIRE$$

$$RHO \_ TEMP \_ \tag{3.1}$$

$$THETA9$$

Further, the names of the ANSYS labels or, generally, reserved ANSYS words, should be avoided; therefore, a useful trick is to use long names or add a custom prefix or suffix to every custom parameter.

An interesting function can be associated with the variables with names ending with the underscore character: they can be hidden with the command *STATUS. This aspect is really interesting with complex programs, or during debugging; whereas the undocumented function *STAT,_PRM can show the parameters used by the ANSYS environment, the function *STAT,PRM_ shows all the trailing underscore parameters and the function *DEL,,PRM_ can be used to delete this group of parameters.

Different types of parameters can be defined for general or specific purposes: a scalar is an integer or floating point number with double precision, a character is a sequence of alphanumeric symbols, an array is an organized array of numeric values, a char is an array of characters, a table is a special numeric array implementing linear interpolation, and a string is an organized list of characters. Also, parameters including more than one element can be dimensioned with the *DIM function; an example of declarations is shown in (3.2).

$$h \_ input = 2$$

$$x \_ input = 3.45$$

$$Title \_ input = 'Cramer'$$

$$*DIM, C \_ char \_ input, CHAR, 6 \qquad (3.2)$$

$$*DIM, A \_ array \_ input, ARRAY, 5, 4, 4$$

$$*DIM, T \_ table \_ temp \_, TABLE, 3, 2, 2$$

$$*DIM, S \_ string \_ temp \_, STRING,$$

As mentioned, the *STATUS,_PRM function can be used to obtain the environmental variables, producing a result in the output window similar to those in Table 3.1; the function *STATUS shows the user parameters as in Table 3.2 and *STATUS,PRM_ lists the voluntary hidden parameters as in Table 3.3.

The value assumed by a parameter can be assigned during execution with the *SET command or, as shown in (3.2), with the symbol "=", where, as usual, the left operand is the name of the parameter and the right operand is the assumed value; the correct syntax is shown in (3.3); an example of its use is described in (3.4).

$$Parameter = Value$$
$$\qquad (3.3)$$
$$*SET, Parameter, Value$$

TABLE 3.1

Output Result of the Command *STATUS,_PRM

Name	Value	Type	Dimensions
_GUI_CLR_BG	systemButtonFace	Character	
_GUI_CLR_FG	systemButtonText	Character	
_GUI_CLR_INFOBG	systemInfoBackground	Character	
_GUI_CLR_SEL	systemHighlight	Character	
_GUI_CLR_SELBG	systemHighlight	Character	
_GUI_CLR_SELFG	systemHighlightText	Character	
_GUI_CLR_WIN	systemWindow	Character	
_GUI_FNT_FMLY	Arial	Character	
_GUI_FNT_PXLS	16.0000000	Scalar	
_GUI_FNT_SLNT	r	Character	
_GUI_FNT_WEGT	medium	Character	
_RETURN	0.00000000	Scalar	
_STATUS	0.00000000	Scalar	
_UIQR	1.00000000	Scalar	

TABLE 3.2

Output Result of the Command *STATUS,

Name	Value	Type	Dimensions
A_ARRAY_INPUT	—	Array	5 4 4
C_CHAR_INPUT	—	Char array	6 1 1
H_INPUT	2.00000000	Scalar	—
TITLE_INPUT	Cramer	Character	—
X_INPUT	3.45000000	Scalar	—

TABLE 3.3

Output Result of the Command *STATUS,PRM_

Name	Value	Type	Dimensions
S_STRING_TEMP_		String array	8 1 1
T_TABLE_TEMP_		Table	3 2 2

$$h\_bar = 3.25$$

$$*SET, h\_bar, 3.25$$

(3.4)

The *SET function can also be used to delete a parameter with a blank right-hand side: (3.5).

$$h \_ bar =$$

$$*SET, h \_ bar, \quad 3.25 \tag{3.5}$$

The value of a parameter can also be set at start-up as an argument, launching ANSYS from the command line, as shown in (3.6).

$$ansys100 \quad -h \_ bar \quad 3.25 \tag{3.6}$$

The value of a parameter can also be assigned by extracting it from an ANSYS item. ANSYS works on different types of entities, nodes, lines, volumes, and areas, and every entity is characterized by different properties called items, which assume specific values. These values can be stored in APDL parameters with the *GET function, as shown in (3.7) and in the example in (3.8) with the extended and short forms.

$$*GET, Parameter, Entity \_ Type, Entity \_ Num, Item, Item \_ Label \tag{3.7}$$

$$*GET, Position \_ x, NODE, 24, LOC, X$$

$$Position \_ x = NX(24) \tag{3.8}$$

The example in (3.8) associates the x location of node 24 with the parameter Position_x; also this specific*GET function has an inverse implementation, as shown in (3.9), which gives the node nearest to the position:

$$NODE(3.24, 4.26, 5.34) \tag{3.9}$$

The assignments of array values can be performed using the *SET function or the shorthand symbol, "=" but no more than ten elements can be assigned at the same time; therefore, the statement includes the starting element of the assignment in the array, as shown in (3.10), that produces output results as in Table 3.4:

$$*DIM, A \_ Array, ARRAY, 11, 1$$

$$A \_ Array(1) = -3.4, 2.3, 5.6, 3, 67, 2.5, 6.3, 9.8, 0, 34$$

$$A \_ Array(11) = -8.2 \tag{3.10}$$

$$*STATUS, A \_ Array$$

TABLE 3.4

Array Assignment

Location	Value
1 1 1	-3.40000000
2 1 1	2.30000000
3 1 1	5.60000000
4 1 1	3.00000000
5 1 1	67.0000000
6 1 1	2.50000000
7 1 1	6.30000000
8 1 1	9.80000000
9 1 1	0.00000000
10 1 1	34.0000000
11 1 1	-8.20000000

TABLE 3.5

VFILL Assignment

Location	Value
1 1 1	3.93971306
2 1 1	2.12790961
3 1 1	1.84456374
4 1 1	3.10490839
1 2 1	22.5610025
2 2 1	20.5519790
3 2 1	14.8037957
4 2 1	35.9167772
1 3 1	2.17588758
2 3 1	1.55129477
3 3 1	3.60664129
4 3 1	2.63189744
1 4 1	18.6213358
2 4 1	34.2498857
3 4 1	27.3179957
4 4 1	33.1888780

An interesting function for array assignment is *VFILL, which can be used to assign specified data, as *SET, or extract values from a ramp or randomly from a uniform, Gaussian, triangular beta or gamma distribution. That is, the expression (3.11) assigns random values to the matrix M, extracting them from a Gaussian distribution characterized by a mean value equal to 2.3 and standard deviation equal to 1.6, and from a uniform distribution with a lower bound equal to 11 and an upper bound equal to 36, corresponding to the output result in shown in Table 3.5:

$$*DIM, M, ARRAY, 4, 4$$

$$*VFILL, M(1, 1), GDIS, 2.3, 1.6$$

$$*VFILL, M(1, 1), RAND, 11, 36 \qquad (3.11)$$

$$*VFILL, M(1, 1), GDIS, 2.3, 1.6$$

$$*VFILL, M(1, 1), RAND, 11, 36$$

$$*STATUS, M$$

An interacting version of the *SET assignment can be performed with the *VEDIT function, producing a visual interface between the APDL program and the user, who is able to fill the visual mask manually. Further, the values can be assigned in array form, extracting them from ANSYS entities with the *VGET function. This option can enhance the readability of the APDL code,

TABLE 3.6

Output of the *VGET Function

Location	Value
1 1 1	3.74165739
2 1 1	1.00000000
3 1 1	4.58257569

reducing the use of the "for" or "do-while" loops, with an improvement in the performance of the simulation. With proper selection, the *VGET function can transfer the corresponding properties of sequences of entities or properties of the same entity; that is, the input in (3.12) generates free key points, then it connects the three key points with three lines, and, finally, it inserts the lengths of the three lines in an array with the *VGET function, as shown in the output Table 3.6.

```
*DIM, Out, ARRAY, 3, 1

/prep7

K, 1, 2, 0, 1

K, 2, 3, 2, 4

K, 3, 3, 2, 5                                        (3.12)

L, 1, 2

L, 2, 3

L, 3, 1

*VGET, Out, LINE, 1, LENG, , , , 2

*STATUS, Out
```

APDL allows also the use of files, where parameters or entities can be easily saved. Equation (3.13) shows a simple example of file usage through the functions *CFOPEN, *VWRITE, and *VREAD. The array A_source is dimensioned and filled with four values, then the file Medium.txt is then opened with the *CFOPEN function, the next step consists in a writing operation of A_source on Medium.txt through the *VWRITE function, and, finally, the

TABLE 3.7

Output of the *VWRITE/*VREAD Process

Location	Value
1 1 1	0.100000000
2 1 1	2.10000000
3 1 1	3.50000000
4 1 1	6.00000000

values in Medium.txt are read and assigned to the array B_receiver through the *VREAD function, as shown in the output file listed in Table 3.7.

```
*DIM, A _ source, ARRAY, 4, 1

*DIM, B _ receiver, ARRAY, 4, 1

A _ source(1) = 0.1, 2.1, 3.5, 6

*CFOPEN, Medium, txt

*VWRITE, A _ source(1)

(F10.8, 2X)

*VREAD, B _ receiver(1)

(F10.8, 2X)

*STATUS, B _ receiver
```
(3.13)

3.1.2 Entities

During the construction of the geometry, different types of entities can be adopted—key points, lines, areas, and volumes—while the determination of the FEM can be performed through just two types of entities: nodes and elements. The command K can be used to produce a key point in a Cartesian location, as shown in (3.14), with the corresponding result as shown in Figure 3.1.

```
/prep7

K, 1, 2, 0, 1

K, 2, 3, 2, 4

K, 3, 3, 2, 5

K, 4, 5, 5, 5

L, 1, 2

L, 1, 3

L, 1, 4

L, 2, 3

L, 2, 4

L, 3, 4
```
(3.14)

The command L is used to generate lines connecting different key points, as in (3.15), where six lines connect the four key points realized with the command sequence in (3.14), corresponding to the result shown in Figure 3.2.

FIGURE 3.1
Four key points generated with the command K.

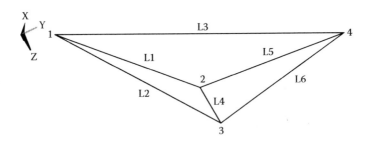

FIGURE 3.2
Six lines generated with the command L.

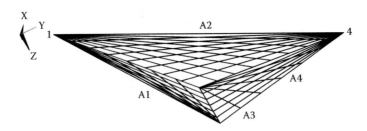

FIGURE 3.3
Four areas generated with the command AL.

By analogy, the traced lines can be considered as boundaries of corresponding areas; therefore, the command AL can be used to generate these entities (3.15), as shown in Figure 3.3.

$$
\begin{aligned}
&AL, 1, 2, 4\\
&AL, 1, 5, 3\\
&AL, 2, 6, 3\\
&AL, 4, 5, 6
\end{aligned}
\tag{3.15}
$$

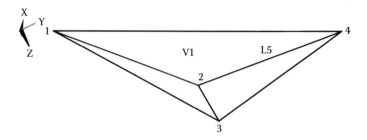

FIGURE 3.4
Volume generated with the command VA.

Also, the traced areas can be used as boundaries for a volume using the command VA as in (3.16), with the result shown in Figure 3.4

$$VA, 1, 2, 3, 4 \tag{3.16}$$

The geometrical definition can also be performed with visual approaches or by operating directly on solid entities with Boolean or other three-dimensional transformations after description of the geometrical model. The next step consists in the selection of a fundamental element according to the desired simulation, as shown in (3.17), with the command ET, with the proper material parameters not specified for the sake of simplicity. Finally, with the VMESH command, the meshing operation associates volumetric elements and nodes to the geometrical body. These elements and nodes, shown in Figure 3.5 and in Figure 3.6, constitute the FEM, which, with boundary conditions and applied load, is the starting point for the simulation of the desired micro- or meso-actuator.

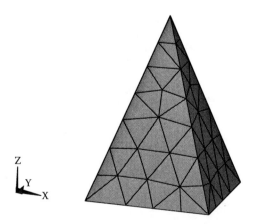

FIGURE 3.5
Elements of a finite element model.

FIGURE 3.6
Nodes of a finite element model.

```
/prep7
K, 1, 1, 1, 0
K, 2, 2, 2, 0
K, 3, 2, 1, 0
K, 4, 1.5, 1.5, 1.4
L, 1, 2
L, 1, 3
L, 1, 4
L, 2, 3
L, 2, 4
L, 3, 4
AL, 1, 2, 4
AL, 1, 5, 3
AL, 2, 6, 3
AL, 4, 5, 6
VA, 1, 2, 3, 4
ET, 1, SOLID92
SMRTSIZE, 6
VMESH, ALL
```

(3.17)

3.1.3 Controlling Program Flow

The previous expressions and examples of APDL codes are characterized by a sequential execution of a list of commands. The flow of the program can be controlled with appropriate control structures that allow a conditioned execution of instruction blocks, resulting in a powerful programming tool.

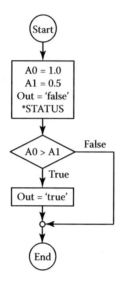

FIGURE 3.7
Example of "if" statement use.

The *IF command allows execution of different blocks according to the value assumed by a conditional variable. The conditional variable is a Boolean expression that can be false or true and is evaluated by comparing the values of two parameters with the help of a comparison operator, as shown in (3.18) and in the corresponding diagram block in Figure 3.7.

$$
\begin{aligned}
&\text{A0 = 1.0}\\
&\text{A1 = 0.5}\\
&\text{Out = 'false'}\\
&\text{/prep7}\\
&\text{*STATUS}\\
&\text{*IF, A0, GT, A1, THEN}\\
&\qquad\text{Out = 'true'}\\
&\text{*ENDIF}\\
&\text{*STATUS}
\end{aligned}
\tag{3.18}
$$

The flowchart in Figure 3.7 shows how the result of the first *STATUS command is a "false" value for the parameter *Out*, whereas the second *STATUS command produces a "true" value; this is because the value assumed by parameter A0 is greater than that assumed by parameter A1. Other comparison operands can be used to produce the desired Boolean result, as listed in Table 3.8; More than one condition can be connected to evaluate a Boolean expression, as listed in Table 3.9, allowing a compact evaluation, as shown in (3.19) and in the corresponding flowchart in Figure 3.8, that produces three output values: Out1, Out2, and Out3, assuming, respectively, the values true, false, and true.

TABLE 3.8

Comparison Operators

Operator Name	Meaning (left OPERATOR right)
EQ	Left equal to right
NE	Left not equal to right
GT	Left greater than right
LT	Left less than right
GE	Left greater than right or equal to right
LE	Left less than right or equal to right
ABGT	Absolute value of left greater than the absolute value of right
ABLT	Absolute value of left less than the absolute value of right

TABLE 3.9

Boolean Operands

Operator Name	Meaning (left OPERATOR right)
AND	True if left and right are true; false otherwise
OR	True if left or right is true; false otherwise
XOR	True if left or right is true; false if left and right are true or otherwise

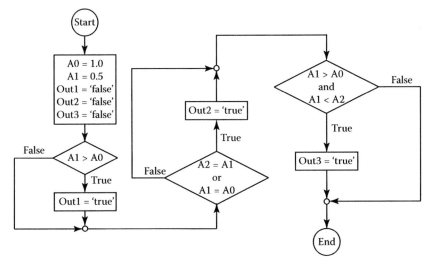

FIGURE 3.8
Example of "if" statement used with Boolean expressions.

```
A0 = 0.5
A1 = 1.0
A2 = 1.7
Out1 = 'false'
Out2 = 'false'
Out3 = 'false'
/prep7
*IF, A1, GT, A0, THEN                                    (3.19)
        Out1 = 'true'
*ENDIF
*IF, A2, EQ, A1, OR, A1, EQ, A0, THEN
        Out2 = 'true'
*ENDIF
*IF, A1, GT, A0, AND, A1, LT, A2, THEN
        Out3 = 'true'
*ENDIF
```

The "if-then" structure can be completed with the ELSE statement, which allows execution of an alternative block of instructions when the Boolean expression evaluated in the *IF command is equal to false, as shown in (3.20) and in the corresponding flowchart shown in Figure 3.9, where a simple program generates a random key point with the abscissa greater than the ordinate.

```
*DIM, Coo, ARRAY, 2
/prep7
*VFILL, Coo(1) , RAND, 1, 3
*IF, Coo(1) , GT, Coo(2) , THEN
        K, , Coo(1) , Coo(2) , 0            (3.20)
        *ELSE
        K, , Coo(2) , Coo(1) , 0
*ENDIF
```

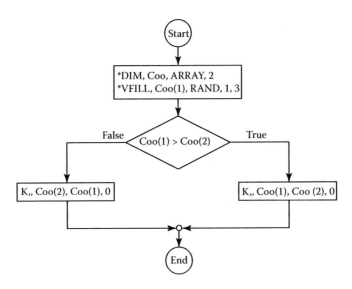

FIGURE 3.9
Example of "if-then-else" structure.

The instruction block, or the alternative instruction block, of the "if-then-else" structure does not have a limitation on the number or the type of instructions; therefore, this structure can be freely nested into a similar one, as shown in (3.21), (3.22) with the contracted form of the *ELSEIF command, and in the associated flowchart in Figure 3.10, where a modified version of the previous program generates random key points with the abscissa greater than the ordinate and discards key points with the abscissa equal to the ordinate.

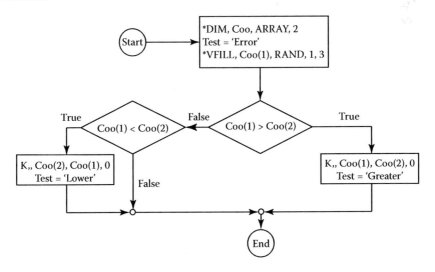

FIGURE 3.10
Nested "if-then-else" structure.

```
*DIM, Coo, ARRAY, 2

Test = 'Error'

/prep7

*VFILL, Coo(1) , RAND, 1, 3                          (3.21)

*IF, Coo(1) , GT, Coo(2) , THEN

        K, , Coo(1) , Coo(2) , 0

        Test = 'Greater'

        *ELSE

        *IF, Coo(1) , LT, Coo(2) , THEN

                K, , Coo(2) , Coo(1) , 0

                Test = 'Lower'

        *ENDIF

  *ENDIF


*DIM, Coo, ARRAY, 2

Test = 'Error'

/prep7

*VFILL, Coo(1) , RAND, 1, 3                          (3.22)

*IF, Coo(1) , GT, Coo(2) , THEN

        K, , Coo(1) , Coo(2) , 0

        Test = 'Greater'

        *ELSEIF, Coo(1) , LT, Coo(2) , THEN

                K, , Coo(2) , Coo(1) , 0

                Test = 'Lower'

        *ENDIF

  *ENDIF
```

Further APDL can also loop single instructions or block of instructions to reduce drastically the length of the code iterating repetitive operations. An easy way is the *REPEAT command, which executes the previous instructions

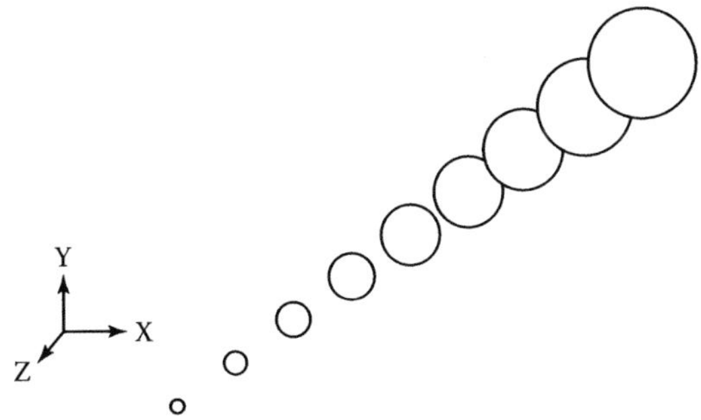

FIGURE 3.11
Repeat command used to realize a sequence of spheres.

the desired number of times, incrementing at every iteration the value of the input parameters with a fixed step, as shown in (3.23), where the *SPH command is repeated nine times, incrementing the coordinates of the center on the working plane with a step value equal to one, and the radius, with a step value equal to 0.1. The result of this iteration is a sequence of nine spheres with their centers on the same line and with increasing radii, as shown in Figure 3.11.

$$/\text{prep7}$$

$$*\text{SPH4, 1, 1, 0.1} \qquad (3.23)$$

$$*\text{REPEAT, 9, 1, 1, 0.1}$$

An interesting looping structure can be obtained with the *DO command, which can iterate a block of instructions, incrementing or decrementing a parameter from an initial value to a final value with a desired increment step. This powerful method is sufficient to improve readability and reduce the length of different codes, but it should be used with caution compared with the array approach, because it can impact the computational performance of the simulation. The next expression (3.24), as explained in the corresponding flowchart in Figure 3.12, produces a strip between two logarithmic spirals (shown in Figure 3.13), creating 200 key points, connecting them with interpolating lines and, then, drawing the areas bounded by these lines.

```
Theta = 0

J = 0

/prep7                                    (3.24)

K, 1, 1, 0, 0

K, 101, 2, 0, 0
```

```
L, 1, 101

*DO, I, 2, 100, 1

        Theta = I/10

        K, 1, cos (Theta) * exp (Theta) , sin (Theta) * esp(Theta) , 0

        K, 1 + 100, 1.1 * cos (Theta) * exp (Theta) + 1, 1.1 *
        sin (Theta) * esp(Theta) , 0

        L, I − 1, I

        L, I + 100 − 1, I + 100

        L, I, I + 100

        J = I + 2 * (I − 2)

        AL, J − 1, J, J + 1, J + 2

*ENDDO
```

A compact instruction for loop programming is the implied colon do-loop, which consists in a "for loop", where the loop variable is not explicitly declared, but only initial value, final value, and step value are expressed in the APDL instruction. A simple descriptive example can be obtained by modifying the program in (3.24)—precisely, by declaring the array variable lengths and filling it with an implied colon do-loop that measures the distances between corresponding key points on the two spirals (3.25):

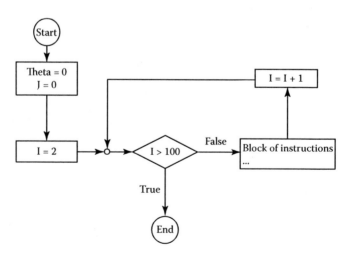

FIGURE 3.12
Flowchart of a DO-loop.

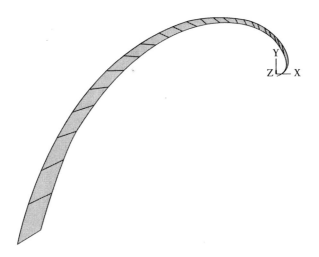

FIGURE 3.13
Strip between two logarithmic spirals.

```
Theta = 0
J = 0
*DIM, Lengths, ARRAY, 100
/prep7
K, 1, 1, 0, 0
K, 101, 2, 0, 0
L, 1, 101
*DO, I, 2, 100, 1
        Theta = I/10

        K, 1, cos (Theta) * exp (Theta) , sin (Theta) * esp(Theta) , 0

        K, 1 + 100, 1.1 * cos (Theta) * exp (Theta) + 1, 1.1 *
        sin (Theta) * esp(Theta) , 0

        L, I − 1, I

        L, I + 100 − 1, I + 100

        L, I, I + 100

        J = I + 2 * (I − 2)

        AL, J − 1, J, J + 1, J + 2
*ENDDO
*GET, Lengths(1 : 100) , LINE, 1 : 300 : 3, LENG
```

$$(3.25)$$

Another useful loop is implemented with the *DOWHILE command, which can be used with an unknown number of cycles; the loop is interrupted when a condition is satisfied, realized on, usually, a Boolean parameter. The loop repeats until the looping expression becomes false, i.e., it is less than or equal to 0.0. Also, the do-while loop needs an initialization of the looping parameters involved in the logical expression, and this initialization must be external to the loop. In (3.26), a do-while block, as explained in the corresponding flowchart in Figure 3.14, generates random spheres with their centers on the work plane, and the sum of their radii is lower than a fixed value, producing the result shown in Figure 3.15.

```
*DIM, Coo, ARRAY, 2

*DIM, R _ Rad, ARRAY, 1

M = 1

RR = 0

/prep7

*DOWHILE, M

        *VFILL, Coo(1) , RAND, 0, 10

        *VFILL, R _ Rad(1) , RAND, 0.1, 0.5

        SPH4, Coo(1) , Coo(2) , R _ Rad(1)

        RR = RR + R _ Rad(1)

        *IF, RR, GT, 9, THEN

                M = 0

        *ENDIF

*ENDDO
```

(3.26)

APDL also allows anticipated exits from loops using special functions such as STOP, which terminates the execution of the program; EXIT, which exits completely from the cycle; and CYCLE, which bypasses the current iteration of the loop, continuing the execution of the program with the next iteration.

3.1.4   Functions and Macros

During programming activity, there are, commonly, repetitive blocks of instructions used to solve a specific problem or a range of problems; to avoid rewriting the same set of commands, APDL provides compiled functions and procedures to realize user-defined macros. An often-used set comprises

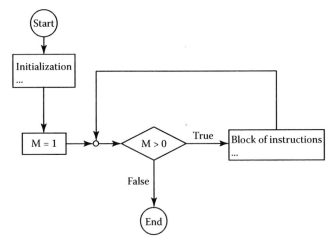

FIGURE 3.14
Flowchart of a do-while structure.

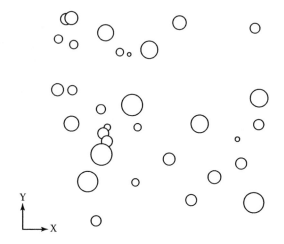

FIGURE 3.15
Random spheres with their centers on the work plane.

mathematical functions (Table 3.10), allowing a simplified approach to computation in ANSYS programming.

Different functions are implemented for a wide range of purposes, so the official ANSYS documentation should be consulted to finds the desired function, and verify the correct syntax and proper application field. User can also realize their own macros with simple approaches, using customized ANSYS commands to avoid repetition, or simply improve the readability of the program source. A macro can be produced with the *CREATE command or with a text editor, and can be saved in a macro file, with a MAC extension, or in a

TABLE 3.10

Mathematical Functions

Name	Meaning	Sample
ABS(x)	Absolute value of x	ABS(-3.2)=3.2
EXP(x)	Exponential of x	EXP(2.3)=7.34
LOG(x)	Natural log of x	LOG(2.0)=0.69
LOG10(x)	Base 10 log of x	LOG10(2.0)=0.30
SQRT(x)	Square root of x	SQRT(2)=1.41
MOD(x,y)	Reminder of x/y	MOD(7,3)=1
SIN(x), COS(x), TAN(x)	Trigonometric functions	SIN(2)=0.91
SINH(x), COSH(x), TANH(x)	Hyperbolic functions	SINH(2)=3.63
ASIN(x), ACOS(x), ATAN(x)	Inverse trigonometric functions	ASIN(1)=1.57
ASINH(x), ACOSH(x), ATANH(x)	Inverse hyperbolic functions	ASINH(1)=0.88

macro library file, which can contain a set of macros; e.g., the program (3.17) can be written in a schematic form (3.27), using the macro file Macro_Lib shown in (3.28) and called with the command *ulib. The command *use is employed to call a specific function in the ANSYS search path, in a specified path, or, as in this situation, in a macro library; Also, this command can be omitted when the macro is saved in a MAC file.

```
/Prep7

*ulib, Macro _ Lib, mlib

*use, Create _ Ks

*use, Create _ Ls                                    (3.27)

*use, Create _ As

*use, Create _ V

*use, Create _ Mesh


    Create _ Ks
    K, 1, 1, 1, 0
    K, 2, 2, 2, 0
    K, 3, 2, 1, 0                                    (3.28)
    K, 4, 1.5, 1.5, 1.4

    /EOF
```

```
                        Create _ Ls
                        L, 1, 2
                        L, 1, 3
                        L, 1, 4
                        L, 2, 3
                        L, 2, 4
                        L, 3, 4
                        /EOF

                        Create _ As
                        AL, 1, 2, 4
                        AL, 1, 5, 3
                        AL, 2, 6, 3
                        AL, 4, 5, 6
                        /EOF

                        Create _ V
                        VA, 1, 2, 3, 4
                        /EOF

                        Create _ Mesh
                        ET, 1, SOLID92
                        SMRTSIZE, 6
                        VMESH, ALL
                        /EOF
```

Further, a list of parameters can be passed to the macro with commas after the call. These parameters, within the macro, have specific names composed of the prefix "arg" and a sequential numeric suffix. As an example, the macro library in (3.28) can be modified as in (3.30) using input arguments for the Create_Ks macro, while the principal program (3.27) implements the input call shown in (3.29).

```
        /Prep7

        *ulib, Macro _ Lib, mlib                                    (3.29)

        *use, Create _ Ks, 1, 1, 0, 2, 2, 0, 2, 1, 0, 1.5, 1.5, 1.4
```

```
*use, Create _ Ls
*use, Create _ As
*use, Create _ V
*use, Create _ Mesh

Create _ Ks
K, 1, arg 1, arg 2, arg 3
K, 2, arg 4, arg 5, arg 6
K, 3, arg 7, arg 8, arg 9
K, 4, arg 10, arg 11, arg 12
/EOF

Create _ Ls
L, 1, 2
L, 1, 3
L, 1, 4
L, 2, 3
L, 2, 4
L, 3, 4
/EOF

Create _ As
AL, 1, 2, 4
AL, 1, 5, 3
AL, 2, 6, 3
AL, 4, 5, 6
/EOF

Create _ V
VA, 1, 2, 3, 4
/EOF

Create _ Mesh
ET, 1, SOLID92
SMRTSIZE, 6
VMESH, ALL
/EOF
```

(3.30)

Only the value of the variable is passed, not the entire variable; this value is stored in a temporary variable within the macro and denominated with the term ARGXX, where XX is an integer number between 1 and 19. Also, within the macro, some local variables can be defined and denominated with ARXX, where XX is an integer number between 20 and 99, and these variables are not visible outside of the macro. To explain previous assertions, the main program in (3.31) calling the macro in (3.32) can be observed, and their debug can be described with STATUS commands.

$$a = 6$$

$$/prep7$$

$$*STATUS$$

$$*STATUS, ARG \qquad (3.31)$$

$$MyMacro, a$$

$$*STATUS$$

$$*STATUS, ARG$$

$$*STATUS$$

$$*STATUS, ARG$$

$$b = 3$$

$$AR30 = 4 \qquad (3.32)$$

$$ARG1 = ARG1 + 1$$

$$*STATUS$$

$$*STATUS, ARG$$

The first *STATUS command in the main program (3.31) shows the value of the only global parameter of the system, which is equal to 6 (Table 3.11), and the second *STATUS, ARGX command does not show any result, because every local parameter of the main program assumes a null value. Then, the program proceeds, entering into the macro MyMacro and assigning the same value of "a" to the first argument of the macro; within the macro, the first command, *STATUS, always shows only one global parameter, as in Table 3.11, while the second command, *STATUS, ARGX, shows that the argument ARG1 assumes a value equal to 6, received by the global parameter "a".

TABLE 3.11

Output of the *STATUS Command

Parameter	Value
A	6.00000000

TABLE 3.12

Output of the *STATUS Command

Parameter	Value
A	6.00000000
B	3.00000000

TABLE 3.13

Output of the *STATUS,argx Command

Parameter	Value
ARG1	7.00000000
AR30	4.00000000

Subsequently, within the macro, a second global parameter "b" is defined with an assigned value equal to 3, and the non-null local parameter AR30 appears with an assigned value equal to 4. Then, the value of the input argument ARG1 is modified, and it assumes an assigned value equal to 7. These changes are highlighted by the successive *STATUS commands within the macro, which show the global parameter "a" equals 6 and the global parameter "b" equals 3 (Table 3.12).

The next *STATUS,ARGX command, always within the macro, indicates a change—the input argument ARG1 assumes a value equal to 7, and the local parameter AR30 assumes a value equal to 4 (Table 3.13).

Exiting from the macro and reentering the main program, the next *STATUS command indicates the values assumed by the global parameters "a" and "b" as, respectively, 6 and 3 (Table 3.12), whereas the last *STATUS, ARGX command does not show the presence of non-null local parameters. After exiting the macro, the local parameters ARG1 and AR30 are not visible, but the global parameter "b" defined within the macro remains visible outside of the macro.

3.2   ANSYS Simulations

An ANSYS program is usually subdivided into a sequence of standard activities: the model definition, the constraints/loads application, the numerical solution of the problem, and the postprocessing of results. Therefore, an example is used all along this section to illustrate the programming structure and some specific statements. The first task consists in the declaration of the title for the simulation with the command /TITLE (3.33). To preserve a proper order in the program source, different comments can be inserted according to user practice, i.e., the name of the author, a concise description of the simulative functionalities, the date of realization, or the version:

$$/\text{title,   Basic   Simulation} \tag{3.33}$$

Further, a coherent unit system must be selected to express numerically the value of input parameters. A coherent system allows us to forecast the measurement unit of the output results of the simulation. This choice can also be implemented without the command /UNIT, which explicitly declares the unit system, associating a label with a physical property; in fact, the choice can be implicit in the value defined for every input parameter. In this situation, the software simulation does not provide for a measurement unit, but the user can properly forecast it. The SI, or, in restricted form, the MKS, is a widely used system employed to derive specific unit systems for MEMS, i.e., the μMKSV or the μMSVfA, which permits expression of the physical properties with numerical values characterized, in scientific notation, by an exponent that is as much as possible near zero, as shown in appendix B. After a measurement is selected, numerical values can be assigned to the parameters of the simulation, with the associated measurement unit reported, for convenience, in the corresponding comment (3.34).

```
Exx = 2.06E11          ! YoungModulus[Pa]
Nxy = 0.33             ! PoissonModulus
Pressure = −1.04E4     ! AreaPressure[Pa]              (3.34)
RCyl = 0.001           ! CylinderRadius[m]
HCyl = 0.01            ! CylinderHeight[m]
```

The next task consists of the definition of the geometry that represents the domain of the simulation; this geometry can be imported from an external file or can be realized within the ANSYS environment. An example of (3.35) realizing a cylindrical element is shown:

```
/prep7                                                              (3.35)

CYL4, 00, RCyl, , , , HCyl   ! Cylinder, Center = 0, r = RCyl, h = HCyl
```

Then, according to the type of geometry—linear, planar, or spatial—and according to the associated physical properties, one or more types of fundamental elements can be imported into the simulation. In the proposed example, which can be surely solved in a simple and intelligent way, a solid block is used—precisely, a tetrahedral solid structure furnished with ten nodes and denominated as SOLID92 (see official ANSYS documentation for a detailed description of its characteristics); thus, the SOLID92 element is used to define the material of the cylinder with the associated Young and Poisson moduli (3.36):

```
ET, 1, SOLID92      ! Material1
MP, EX, 1, Exx      ! YoungModulus                     (3.36)
MP, PRXY, 1, Nxy    ! PoissonModulus
```

Loads and constraints can be applied on the geometry or directly on the nodes. In our example, we propose the application of a kinematic constraint on any face of the cylinder to avoid movements along the z-axis while a mechanical load, represented by a pressure, is applied on the opposite face; therefore, a traction of the solid system is realized, producing constant stress in the direction of the axis of the cylinder.

$$
\begin{array}{ll}
\texttt{DA, 1, UZ, 0} & \texttt{! AvoidMovements} \\
\texttt{SFA, 2, , PRES, Pressure} & \texttt{! ConstantPressure}
\end{array}
\tag{3.37}
$$

Further, the solid geometry is subdivided into a wide set of SOLID92 elements, realizing a volumetric mesh (3.38); the meshing step can be controlled in a semiautomatic way, or manually, to decide the mesh subdivision and gain interesting results.

$$
\begin{array}{ll}
\texttt{TYPE, 1} & \texttt{! ActivateMaterial1} \\
\texttt{SMRTSIZE, 1} & \texttt{! HighestSmartPrecision} \\
\texttt{VMESH, ALL} & \texttt{! VolumetricMesh} \\
\texttt{FINISH} &
\end{array}
\tag{3.38}
$$

At this point, after the finite element system is defined, it can be examined by setting the type of analysis—a simple static structural analysis for us—and launching the proper solver (3.39). Different types of analysis can be performed, as shown in the next paragraph, while the solver can be deeply controlled, as shown in the official ANSYS documentation.

$$
\begin{array}{ll}
\texttt{/SOLU} & \\
\texttt{ANTYPE, STATIC} & \texttt{! StaticAnalysis} \\
\texttt{SOLVE} & \\
\texttt{FINISH} &
\end{array}
\tag{3.39}
$$

Finally, the results of the analysis can be postprocessed; i.e., the output numerical values can be observed and used to realize graphics, tables, or sequences of data for further elaboration. In our example, a simple contour diagram is realized to show the stress in the z direction (3.40), depicted in Figure 3.16.

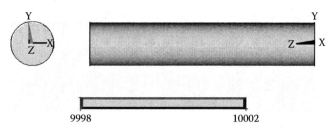

FIGURE 3.16
Stress in the z direction for a cylindrical solid in traction.

$$/POST1$$

$$(3.40)$$

$$PLNSOL, S, Z$$

The complete source of the illustrative example used during this explanation of the structure of an ANSYS program is expressed in the following expression (3.41) for the sake of clarity.

```
/title, Basic Simulation

Exx = 2.06E11              ! YoungModulus[Pa]
Nxy = 0.33                 ! PoissonModulus
Pressure = -1.04E4         ! AreaPressure[Pa]
RCyl = 0.001               ! CylinderRadius[m]
HCyl = 0.01                ! CylinderHeight[m]
/prep7
CYL4, 00, RCyl, , , , HCyl ! Cylinder, Center = O, r = RCyl,
                             h = HCyl
ET, 1, SOLID92             ! Material1
MP, EX, 1, Exx             ! YoungModulus
MP, PRXY, 1, Nxy           ! PoissonModulus
DA, 1, UZ, 0               ! AvoidMovements
SFA, 2, , PRES, Pressure   ! ConstantPressure
TYPE, 1                    ! ActivateMaterial1
SMRTSIZE, 1                ! HighestSmartPrecision
VMESH, ALL                 ! VolumetricMesh
FINISH
/SOLU
ANTYPE, STATIC             ! StaticAnalysis
SOLVE
FINISH
```

$$(3.41)$$

3.2.1 Types of Simulations

The ANSYS environment places a wide range of possible simulations at the disposal of the user, as summarized in Table 3.14.

Structural analysis is applied on mechanical objects to calculate, under desired boundaries and initial conditions, the displacements, strains, stresses, and reaction forces. Depending on these conditions, different types of structural analyses can be performed within the ANSYS environment, as listed in Table 3.15. Static analysis allows the determination of the structural

TABLE 3.14

Type of ANSYS Simulations

Type	Description
Structural	Mechanical actions, reactions
Contact	Mechanical interactions
Thermal	Heat transfer
Fluid	Simple and complex fluid simulations
Low-frequency electromagnetic	Electromagnetic field analysis
High-frequency electromagnetic	H-F and resonators
Coupled field	Interaction of different fields

TABLE 3.15

Type of Structural Simulations

Type	Description
Static	Static loadings
Modal	Natural frequencies
Harmonic	Harmonic loadings
Transient dynamic	General time-variable loadings
Spectrum	Spectrum response
Explicit dynamic	Fast solutions for complex dynamics

TABLE 3.16

Special-Purpose Features for Structural Simulations

Type	Description
Fracture mechanics	Cracks and flaws occurring in structures
Composites	Bonded materials
Fatigue	Cyclic load application
p-method	Specified degree of accuracy
Beam analyses	One-dimensional idealization

configuration under linear or nonlinear static loadings, employing a number of analytical techniques: modal analysis is performed to compute the natural frequencies and the mode shapes of a structure; the response of a structure to harmonically time-varying loads can be determined with harmonic analysis; transient dynamic analysis is employed to determine the response of a structure to general time-varying loads; spectrum analysis can be performed to calculate stresses and strains caused by a response spectrum or random vibrations; buckling analysis can forecast the buckling loads and the buckling mode shapes; and explicit dynamic analysis allows fast solutions for complex dynamic problems.

Finally, some special-purpose features can be used, as shown in Table 3.16.

TABLE 3.17

Fluid Simulations

Type	Description
Computational fluid dynamics	Fluid flow
Acoustics	Sound waves
Thin films	Small gap of fluid

TABLE 3.18

Computational Fluid Dynamics

Type	Description
Laminar flow	Ordered velocity field
Turbulent flow	Turbulent fluctuating velocity field
Thermal	Convection
Compressible flow	Density changes
Non-Newtonian fluid flow	Viscosity changes with the strain rate
Multiple species transport	Two or more fluids
Free surface	Unconstrained gas–liquid surface

Contact analyses can be subdivided into two main classes: rigid-to-flexible and flexible-to-flexible. The contact is between two bodies and, in rigid-to-flexible analysis, one of the bodies is characterized by a higher stiffness, whereas the other is compliant. On the contrary, in flexible-to-flexible analysis, the two bodies possess comparable stiffnesses. Also, these problems are highly nonlinear and can be associated with self-excited frictional vibrations, or they can take into account multifield effects, such as heat transfer or electromagnetic flux through the contact. Therefore, convergence difficulties can be experienced during these simulations.

Thermal analyses can be steady-state or transient, depending on loading conditions, and is used to determine the temperature distribution and other thermal quantities for the desired physical system. The heat transfer can be due to conduction, convection, or radiation. Also, special-purpose options, such as change of phase or internal heat generation, can be activated when necessary.

Fluidic analyses pertain to general fluid dynamics, acoustic problems, or special situations in thin fluidic films (Table 3.17); general fluid simulations involving computational fluid dynamics can be associated with different physical phenomena (Table 3.18).

Laminar flow analysis is associated with ordered and smooth velocity fields, but, in turbulent flow analysis, turbulent fluctuations appear in the velocity fields. Also, if fluid properties depend on temperature, a thermal analysis can be performed considering heat transfer through the system. Compressible flow analysis is associated with changes in density caused by pressure gradients. To describe complex relations between stress and the rate of strain, a non-Newtonian fluid flow analysis can be realized. Multiple species transport analysis is used to study systems composed of two or

TABLE 3.19

Low-Frequency Electromagnetic Analyses

Type	Description
Magnetic	Permanent magnets or electromagnets
Electric	Steady-state, harmonic, or transient fields
Electrostatic	Electrostatic fields
Electric circuit	Lumped parameters circuits

more fluids. Finally, free surface analyses involves an unconstrained gas–liquid surface. These types of analyses are not mutually exclusive but can be combined to realize a real-world simulation, i.e., a turbulent compressible analysis.

An acoustic analysis involves physical properties associated with acoustic wave transfer: the pressure waves move into a compressible fluid, eventually interacting with solid objects; further, the fluidic medium is not flowing, and its viscosity is neglected during the acoustic simulation.

Thin-film analysis is devoted to the study of fluidic thin layers between moving surfaces. The associated phenomena especially affects microstructure behavior owing to damping and stiffening of air or other fluids between the elements of microelectromechanical systems.

Low-frequency electromagnetic analysis is based on Maxwell's equations, employing different formulations, i.e., magnetic vector potential (MVP), magnetic scalar potential (MSP), or edge-based formulations. These approaches can be used to solve magnetic, electric, electrostatic, or electric circuit problems, as listed in Table 3.19.

Magnetic analysis can be static, if devoted to simulating fields caused by direct current (DC) or permanent magnets; harmonic, if when studying fields produced by AC currents or voltages; or, generally, transient, if associated with arbitrary electric or magnetic sources. Analogously, an electric analysis can be performed to study steady-state current conduction, harmonic quasi-static behavior due to AC sources, or transient quasi-static simulations associated with arbitrary time-dependent sources. Electrostatic field analysis is associated with electrostatic phenomena and can be performed with h-method or p-method; the former uses many simple elements, whereas the latter uses few complex elements. Finally, electric circuit analysis can be harmonic, static, or transient and can be associated with AC, DC, or arbitrary time-varying sources.

High-frequency electromagnetic field analysis is associated with electromagnetic phenomena with wavelengths comparable with the dimensions of the geometrical structure; energy transfer occurs through electromagnetic wave transport.

A coupled-field analysis is performed through the interaction of different physics fields, i.e., mechanical and electrical, or thermal and electrical, to realize complex simulations; different approaches can be used according to simulative situations or user abilities (Table 3.20).

TABLE 3.20

Coupled-Field Analysis

Type	Description
Sequentially coupled physics analysis	Different sequential analyses
Multifield single-code coupling	Single run/code
Multifield multiple-code coupling	Different runs/codes
Unidirectional load transfer	Transfer to fluid dynamics analysis
Reduced-order modeling	Finite number of eigenmodes
Direct coupled-field analysis	Coupled-field elements
Coupled physics circuit simulation	Circuital elements

Sequentially coupled physics analysis is performed with a sequence analysis involving different physical fields transferring partial results to the succeeding simulation; multifield single-code coupling is implemented with a single APDL code transferring the result of a single-field analysis to the next analysis implicitly at every iteration of the simulation. Multifield multiple-code coupling is a different iterative coupler that can be used to study complex problems, because it solves separately every field block of the equation; the unidirectional transfer method is employed to transfer unidirectionally the results of a field analysis to a fluid dynamics analysis; the reduced-order modeling uses a reduced number of eigenmodes of the system to simplify analysis; the direct coupled-field analysis is performed through special finite elements that incorporate the coupling into their basic properties; finally, coupled physics circuit simulation involves circuital elements, enabling interesting simplified simulations also for very complex systems.

3.3 Examples of Actuator Analysis

3.3.1 Analysis of an Electrostatic Actuator

An analysis of an electromechanical system is proposed as follows, where an electrostatic actuator moves an inertial mass connected to the frame with a spring, which exerts a return force on the actuator and, with a damper, reduces the vibratory effects produced by the contemporaneous presence of the elastic and inertial elements; therefore, a circuital scheme can be drawn as in Figure 3.17.

The proposed APDL program for the simulation of the transient behavior of the system starts with some generic instructions, such as the definition of the number of microprocessors used on the hardware machine for the simulation, the maximum number of results, the title associated with the simulation, and a reminder on the measurement unit system used:

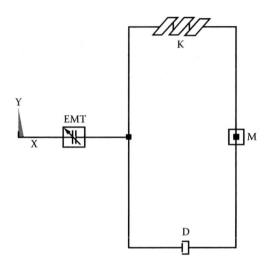

FIGURE 3.17
Circuital scheme: EMF is the ideal electromechanical actuator, M is the lumped mass, K is the spring, and D is the damper.

$$
\begin{aligned}
&\text{/batch, list} \\
&\text{/CONFIG, NPROC, 2} \qquad \text{! NumberCPU = 2} \\
&\text{/CONFIG, NRES, 10000} \qquad \text{! NumberResults < 10000} \\
&\text{/show, file} \\
&\text{/title, Electrostatic} \\
&\text{/com, μMKSV   Units}
\end{aligned}
\tag{3.42}
$$

After an introductory phase, the parameters of the system, listed in Table 3.21, are defined to express the electric tension applied on the electromechanical actuator and the temporal characteristics of the simulation during the transitory phase. This electric action on the electromechanical actuator is converted into mechanical work during the simulation to move the mass M in the horizontal direction x.

TABLE 3.21

Parameters of the Simulation

Name	Description
End	Number of steps used to define the input electric square wave
SubEnd	Number of substeps of a single step, where a random disturbance is added to the input tension, with mean value equal to zero
SubStep	Temporal extension of a single substep
VLoad	Vector of constant electric tensions during every temporal substep
TLoad	Vector of temporal substeps

The vector of electric tension on the actuator is constructed by superimposing a square wave of amplitude 5 V and a random disturbance in the range –0.5 to 0.5 V; the temporal scale is produced with subsequent constant temporal substeps (3.43). The result is similar to Table 3.22.

```
End = 4

SubEnd = 10

SubStep = 0.0025

*DIM, VLoad, ARRAY, End * SubEnd

*DIM, TLoad, ARRAY, End * SubEnd

*DO, I, 1, End, 1

    *IF, MOD(I, 2) , GT, 0, THEN

        *DO, J, 1, SubEnd, 1

            VLoad( (i − 1) * SubEnd + J) = RAND(−0.5, 0.5)

        *ENDDO

    *ELSE

        *DO, J, 1, SubEnd, 1

            VLoad( (i − 1) * SubEnd + J) = RAND(4.5, 5.5)

        *ENDDO

    *ENDIF

*ENDDO

*VFILL, TLoad, RAMP, SubStep, SubStep

*STATUS

*STATUS, VLoad

*STATUS, TLoad
```

$$(3.43)$$

After the definition of the system parameters, some nodes can be generated to serve as basic points for the connection of circuital elements. The first node is fixed and constitutes the electric ground, the second node moves with respect to the first one because it is the basic point for the mobile mass and, finally, the third mode represents another fixed point for the connection of a return spring:

TABLE 3.22

VLoad and TLoad Values

Location	VLoad	TLoad E-02	Location	VLoad	TLoad E-02
1	0.347174960	0.25	21	0.272046570	5.25
2	-4.281461000E-02	0.50	22	5.462490000E-02	5.50
3	-0.112006490	0.75	23	-2.243365000E-02	5.75
4	0.192367430	1.00	24	0.304237820	6.00
5	3.755990000E-02	1.25	25	0.336442640	6.25
6	0.117920840	1.50	26	-0.381753190	6.50
7	0.347848170	1.75	27	0.258927270	6.75
8	-0.496671090	2.00	28	0.213630290	7.00
9	-3.090208000E-02	2.25	29	9.080437000E-02	7.25
10	-0.179711430	2.50	30	-0.436638070	7.50
11	5.29272897	2.75	31	4.86779659	7.75
12	5.08212903	3.00	32	4.52197646	8.00
13	5.19514657	3.25	33	4.95314855	8.25
14	4.57000457	3.50	34	5.34293517	8.50
15	4.84728017	3.75	35	4.81807375	8.75
16	4.61244488	4.00	36	5.11415747	9.00
17	5.02974319	4.25	37	5.27517996	9.25
18	5.48948181	4.50	38	5.20741118	9.50
19	5.13166988	4.75	39	4.59674861	9.75
20	4.65089488	5.00	40	5.07933975	10.0

$$
\begin{aligned}
&\texttt{/prep7} \\
&\texttt{/N, 1} \quad\quad \texttt{! Origin} \\
&\texttt{/N, 2, 0.1} \quad \texttt{! x = 0.1} \\
&\texttt{/N, 3, 0.2} \quad \texttt{! x = 0.2}
\end{aligned} \tag{3.44}
$$

The element types used for the simulation can be categorized as circuital concentrated elements, particularly, the element TRANS126, which represents the ideal electromechanical transducer; an inertial element MASS21; and two mechanical elements COMBIN14, which constitute the return spring and a damper, respectively:

$$
\begin{aligned}
&\texttt{/ET, 1, TRANS126} \quad \texttt{! EM - Transducer} \\
&\texttt{/ET, 2, 21, , , 4} \quad\quad \texttt{! Mass(DOF = X, Y)} \\
&\texttt{/ET, 3, 14, , 1} \quad\quad\quad \texttt{! Spring} \\
&\texttt{/ET, 4, 14, , 1} \quad\quad\quad \texttt{! Damper}
\end{aligned} \tag{3.45}
$$

One or more constants are associated with every element to influence its behavior during the simulation: the electromechanical transducer is characterized by an initial gap between the faces equal to 150 µm, and an electric capacitance proportional to 800 pF · µm and inversely proportional to the relative position of the faces; the mobile mass is equal to 10^{-4} kg; the linear stiffness assumes the value 250 µN/µm; and the damping constant is equal to $30 \cdot 10^{-3}$µN s/µm (3.46).

```
R, 1, , 1, 150      ! Gap
RMORE, 8.00E2       ! C0, C = C0/x
R, 2, 1E - 4        ! Mass                    (3.46)
R, 3, 250           ! Stiffness
R, 4, , 30E - 3     ! Damper
```

The properties defined in (3.46) are assigned to circuital elements that are positioned between two nodes or in a single node, i.e., the ideal electromechanical actuator is connected to nodes 1 and 2, the mass is placed in node 2, and the spring and the damper are arranged between nodes 2 and 3 (3.47).

```
TYPE, 1    ! ElementType = 1
REAL, 1    ! ConstantSet = 1
E, 1, 2    ! Position
TYPE, 2    ! ElementType = 2
REAL, 2    ! ConstantSet = 2
E, 2       ! Position                         (3.47)
TYPE, 3    ! ElementType = 3
REAL, 3    ! ConstantSet = 3
E, 2, 3    ! Position
TYPE, 4    ! ElementType = 4
REAL, 4    ! ConstantSet = 4
E, 2, 3    ! Position
```

Different constraints are imposed on the electromechanical circuit, and they remain in place during the entire simulation. From a mechanical viewpoint, the transducer, the spring, and the damper are subjected only to horizontal movements in the direction of the x-axis; therefore, the impediments to the horizontal movements of nodes 1 and 3 must be set. But, as the mass can move also along the vertical direction, for the sake of clarity, its movements in the direction of the y-axis are prevented with the appropriate constraint. Furthermore, from an electric viewpoint, the ground is set in node 1, annulling the conventional value of its electric potential (3.48).

```
D, 1, UX, 0      ! Fixed  left  transducer
D, 2, UY, 0      ! Avoided  y  mass  movements
D, 3, UX, 0      ! Mechanical  ground                    (3.48)
D, 1, VOLT, 0    ! Voltage  ground
finish
```

The solution of the transition problem is achieved through successive steps. At every step, the value of the electric tension at node 2, i.e., the potential difference between the faces of the actuator, is set in accordance with the respective values assumed by the parameter VLoad (3.49).

```
/solu
ANTYP, TRANS                      ! Transient  analysis
KBC, 1                            ! Step  boundaries
DELTIM, .00005, .00001, .0001     ! Time  increments
AUTOS, ON                         ! Auto  time-step
AUTOS, ON                         ! Save  intermediates
CNVTOL, F                         ! Force  convergence     (3.49)
*DO, I, 1, End * SubEnd, 1
    D, 2, VOLT, VLoad(I)          ! Transducer  voltage
    TIME, TLoad(I)                ! Time  step
    SOLVE
*ENDDO
finish
```

Finally, during the postprocessing, the position of the mobile mass is monitored at every instant (3.50), as shown in Figure 3.18.

```
/post26
NSOL, 2, 2, U, X                  ! Mass  displacement
/XRANGE, 0, TLoad(End * SubEnd)
/YRANGE, −.008, .004
/AXLAB, X, Time[s]                                         (3.50)
/AXLAB, Y, Displacement[m]E − 6
PLVAR, 2                          ! Plot  displacement
finish
```

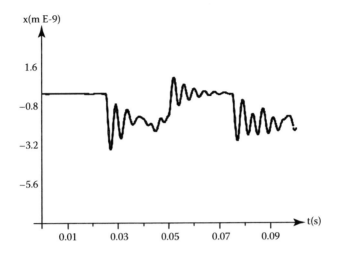

FIGURE 3.18
Movement of the mobile mass.

3.3.2 Analysis of a Piezoelectric Actuator

Another example is shown to highlight the use of a circuital approach for the analysis of a piezoelectric actuator. In this section, the desired geometry is assigned to a deformable body with piezoelectric properties (Figure 3.19), and the feeding system is modeled with a simple lumped parameter model (Figure 3.20).

The source of the APDL program starts with the configuration of the system, i.e., setting the CPU's number of the hardware machine, incrementing the standard number of results, or associating the output to a graphical file (3.51).

$$
\begin{array}{lll}
\texttt{/CONFIG, NPROC, 2} & \texttt{! CPU\ \ number} & \\
\texttt{/CONFIG, NRES, 10000} & \texttt{! Result\ \ number} & (3.51) \\
\texttt{/SHOW, FILE} & \texttt{! File\ \ output} &
\end{array}
$$

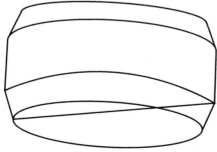

FIGURE 3.19
Geometrical configuration of the piezoelectric body.

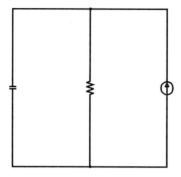

FIGURE 3.20
Electrical feeding circuit.

The properties of the piezoelectric material are then assigned, referring to the classical PZT-5A; thus, its mass density is identified with a scalar variable (3.52).

$$/\text{prep7}$$
$$\text{Density} = 7750 \tag{3.52}$$

From an electrical viewpoint, the material can be characterized through the relative permittivity matrix that, for anisotropic materials, is symmetric, as shown in (3.53); therefore, only some elements of the matrix can be defined (3.54), and the undefined components are automatically set to zero.

$$\boldsymbol{\varepsilon} \equiv \begin{bmatrix} \varepsilon_{11} & \varepsilon_{12} & \varepsilon_{13} \\ \varepsilon_{21} & \varepsilon_{22} & \varepsilon_{23} \\ \varepsilon_{31} & \varepsilon_{32} & \varepsilon_{33} \end{bmatrix} = \begin{bmatrix} \varepsilon_{11} & \varepsilon_{21} & \varepsilon_{31} \\ \varepsilon_{12} & \varepsilon_{22} & \varepsilon_{32} \\ \varepsilon_{13} & \varepsilon_{23} & \varepsilon_{33} \end{bmatrix} \equiv \boldsymbol{\varepsilon}^T \tag{3.53}$$

$$\begin{aligned} p11 &= 916 \\ p22 &= 830 \\ p33 &= 916 \end{aligned} \tag{3.54}$$

The mechanical properties of the piezoelectric material are summarized by the stiffness matrix, which is always symmetric for elastic anisotropic materials (3.55).

$$\boldsymbol{c} \equiv \begin{bmatrix} c_{11} & c_{12} & c_{13} & c_{14} & c_{15} & c_{16} \\ c_{21} & c_{22} & c_{23} & c_{24} & c_{25} & c_{26} \\ c_{31} & c_{32} & c_{33} & c_{34} & c_{35} & c_{36} \\ c_{41} & c_{42} & c_{43} & c_{44} & c_{45} & c_{46} \\ c_{51} & c_{52} & c_{53} & c_{54} & c_{55} & c_{56} \\ c_{61} & c_{62} & c_{63} & c_{64} & c_{65} & c_{66} \end{bmatrix} \tag{3.55}$$

$$
= \begin{bmatrix}
c_{11} & c_{21} & c_{31} & c_{41} & c_{51} & c_{61} \\
c_{12} & c_{22} & c_{32} & c_{42} & c_{52} & c_{62} \\
c_{13} & c_{23} & c_{33} & c_{43} & c_{53} & c_{63} \\
c_{14} & c_{24} & c_{34} & c_{44} & c_{54} & c_{64} \\
c_{15} & c_{25} & c_{35} & c_{45} & c_{55} & c_{65} \\
c_{16} & c_{26} & c_{36} & c_{46} & c_{56} & c_{66}
\end{bmatrix} \equiv c^T
\tag{3.55}
$$

The stiffness matrix can be organized in different ways according to different notations that influence the position of the components of the matrix. Therefore, the elastic relation between stress and strain can be represented to establish the proper position of every component (3.56). Under the hypothesis of symmetry for stiffness matrix, the necessary components can be assigned to perform the desired simulation (3.57).

$$
\begin{bmatrix}
\sigma_{11} \\
\sigma_{22} \\
\sigma_{33} \\
\sigma_{12} \\
\sigma_{23} \\
\sigma_{13}
\end{bmatrix}
=
\begin{bmatrix}
c_{11} & c_{12} & c_{13} & c_{14} & c_{15} & c_{16} \\
c_{21} & c_{22} & c_{23} & c_{24} & c_{25} & c_{26} \\
c_{31} & c_{32} & c_{33} & c_{34} & c_{35} & c_{36} \\
c_{41} & c_{42} & c_{43} & c_{44} & c_{45} & c_{46} \\
c_{51} & c_{52} & c_{53} & c_{54} & c_{55} & c_{56} \\
c_{61} & c_{62} & c_{63} & c_{64} & c_{65} & c_{66}
\end{bmatrix}
\begin{bmatrix}
\varepsilon_{11} \\
\varepsilon_{22} \\
\varepsilon_{33} \\
\varepsilon_{12} \\
\varepsilon_{23} \\
\varepsilon_{13}
\end{bmatrix}
\tag{3.56}
$$

$$
\begin{aligned}
c11 &= 12.1E10 \\
c21 &= 7.52E10 \\
c31 &= 7.54E10 \\
c22 &= 11.1E10 \\
c32 &= 7.52E10 \\
c33 &= 12.1E10 \\
c44 &= 2.11E10 \\
c55 &= 2.11E10 \\
c66 &= 2.26E10
\end{aligned}
\tag{3.57}
$$

The last characteristics of the piezoelectric material are represented by the piezoelectric matrix that allows coupling between mechanical and electrical properties. Also, in this case, the explicit form of the piezoelectric relation should be shown to dispel all doubts about the notation adopted for its construction (3.58). The terms with a nonnegligible absolute value can then be assigned, as shown in (3.59).

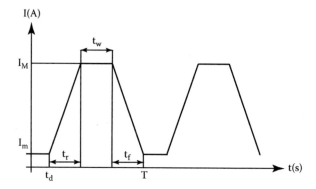

FIGURE 3.21
PWM-supplied current, where I_M and I_m are, respectively, the maximal and the minimal currents, and $t_d, t_r, t_f, t_p,$ and T are, respectively, the delay time, the rise time, the fall time, the pulse time, and the period.

$$
\begin{bmatrix}
\sigma_{11} \\
\sigma_{22} \\
\sigma_{33} \\
\sigma_{12} \\
\sigma_{23} \\
\sigma_{13}
\end{bmatrix}
=
\begin{bmatrix}
e_{11} & e_{12} & e_{13} \\
e_{21} & e_{22} & e_{23} \\
e_{31} & e_{32} & e_{33} \\
e_{41} & e_{42} & e_{43} \\
e_{51} & e_{52} & e_{53} \\
e_{61} & e_{62} & e_{63}
\end{bmatrix}
\begin{bmatrix}
E_1 \\
E_2 \\
E_3
\end{bmatrix}
\tag{3.58}
$$

$$
\begin{aligned}
e12 &= -5.4 \\
e22 &= 15.8 \\
e32 &= -5.4 \\
e41 &= 12.3 \\
e53 &= 12.3
\end{aligned}
\tag{3.59}
$$

After the proper identification of the proposed piezoelectric material with a set of parameters, the properties of the electric circuit depicted in Figure 3.20 can be defined. The resistor is simply represented by its electrical resistance (3.60), and the electric source is a PWM (pulse width modulated) current, as shown in Figure 3.21, and can be identified by the sequence of parameters listed in (3.61).

```
Resist = 8.00E6   !Electric   resistance
```
(3.60)

```
Imax = 8.00E6              ! Maximal  current
Imin = 0.0                 ! Minimal  current
Tdelay = 1.0E − 4          ! Delay  Time
Trise = 1.0E − 4           ! Rise  Time          (3.61)
Tfall = 1.0E − 4           ! Fall  Time
Tpulse = 2.0E − 4          ! Pulse  Time
Tperiod = Trise + Tfall + 2 * Tpulse
```

The geometry of the piezoelectric body depends only on the characteristic parameter H, and the simulation time is set to 2 ms (3.62).

$$
\begin{aligned}
H &= 5.0E − 4 \\
T &= 2.0E − 3
\end{aligned}
\tag{3.62}
$$

The finite element types employed are a SOLID226, which identifies the piezoelectric material, and two CIRCU94's, which represent the current source and the electric resistor (3.63).

```
ET, 1, SOLID226, 1001    ! Piezoelectric   transducer
ET, 2, CIRCU94, 0        ! Electric   resistor          (3.63)
ET, 3, CIRCU94, 3, 2     ! Current   source
```

Then, the electric properties defined in (3.60–3.61) are transferred to the respective parametric blocks (3.64) and used to identify correctly the associated CIRCU94 elements.

```
R, 1, Resist
                                                         (3.64)
R, 2, Imin, Imax, Tdelay, Trise, Tfall, Tpulse, Tperiod
```

The properties of the piezoelectric material, as just shown, are identified by the mass density (3.65), the relative permittivity matrix (3.66), the mechanical stiffness matrix (3.67), and the piezoelectricity matrix (3.68); these properties are then associated with the SOLID226 element (3.69).

```
MP, DENS, 1, Density    ! Mass   density          (3.65)
```

```
TB, DPER, 1

TBDATA, 1, p11
                                                   (3.66)
TBDATA, 2,       p22

TBDATA, 3,             p33
```

```
TB, ANEL, 1

TBDATA, 1,  c11,  c21,  c31
TBDATA, 7,        c22,  c32
TBDATA, 12,             c33                              (3.67)
TBDATA, 16,                   c44
TBDATA, 19,                         c55
TBDATA, 21,                               c66


TB, PIEZ, 1

TBDATA, 2,        e12
TBDATA, 5,        e22
TBDATA, 8,        e32                                    (3.68)
TBDATA, 10, e41
TBDATA, 15,             e53



MAT, 1   $   TYPE, 1                                     (3.69)
```

After the definitions of the employed material, the geometry of the piezoelectric body is identified by a cylindrical volume surmounted by a truncated cone, corresponding to another symmetrical truncated cone (Figure 3.19). The three-dimensional primitives are traced with extrusions starting from a two-dimensional figure on the working plane. Therefore, to construct a tower of primitives, the working plane should be moved with the command WPAVE (3.70), or different working planes can be defined directly in the APDL program.

```
CON4, 0, 0, 9 * H / 10, H, H / 4    ! Truncated   cone
WPAVE, 0, 0, H / 4                  ! Move   working   plane
CYL4, 0, 0, H, , , , H / 2          ! Cylinder
WPAVE, 0, 0, 3 * H / 4              ! Move   working   plane (3.70)
CON4, 0, 0, H, 9 * H / 10, H / 4    ! Truncated   cone
WPAVE, 0, 0, 0.0                    ! Move   working   plane
VADD, ALL                          ! Add   all   volumes
```

The geometry identified in (3.70) is subdivided into tetrahedral finite elements by a three-dimensional mesh, and the dimensional control of the mesh is achieved through the semiautomatic instruction SMRTSIZE (3.71), obtaining the result shown in Figure 3.22.

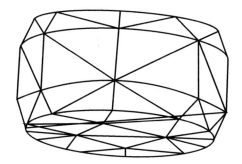

FIGURE 3.22
Three-dimensional piezoelectric mesh.

$$
\begin{array}{lll}
\text{SMRTSIZE, 8} & \text{! Smart size} & \\
\text{MSHAPE, 1, 3D} & \text{! Tetrahedral shape} & (3.71) \\
\text{VMESH, ALL} & \text{! Volumetric mesh} &
\end{array}
$$

The resistor and current source are identified by lumped elements, and, when they are generated, an identification number is assigned to them according to sequential order. Therefore, one way to know the identification number of each circuital element consists in saving the number of the last piezoelectric element and using it as a reference (3.72).

$$
*\text{GET, PiezoElem, ELEM, , COUNT} \qquad (3.72)
$$

The boundary conditions for the simulation are set by identifying two electrodes represented by the minor bases of the truncated cones, which constitute the piezoelectric body. An electrode is formed by different mesh nodes subjected to the same electric tension. The lower electrode is conventionally considered a ground element, setting its voltage level at zero; from a mechanical viewpoint, its movements along the symmetry axis are avoided (3.73).

```
NSEL, S, LOC, Z, 0                    ! Select nodes
CP, 1, VOLT, ALL                      ! Define electrode
*GET, Electrode0, NODE, 0, NUM, MIN   ! Get label
NSEL, S, LOC, Z, H                    ! Select nodes
CP, 2, VOLT, ALL                      ! Define electrode      (3.73)
*GET, Electrode1, NODE, 0, NUM, MIN   ! Get label
NSEL, S, LOC, Z, 0
D, ALL, UZ, 0                         ! Mechanical frame
D, Electrode0, VOLT, 0                ! Electrical ground
```

The two circuital elements are then generated and connected to the electrodes of the piezoelectric body (3.74).

```
TYPE, 2                        ! Element   type
REAL, 1                        ! Parametric   set
E, Electrode1, Electrode0      ! Connect   resistor
TYPE, 3                        ! Element   type
REAL, 2                        ! Parametric   set
E, Electrode1, Electrode0      ! Connect   generator
NSEL, ALL
FINISH
```
$$(3.74)$$

After completing the preprocessing task, the finite element problem can be solved, proposing a transitory analysis along the temporal period T subdivided into 100 temporal steps (3.75).

```
/SOLU
ANTYPE, TRANS        ! Transient   analysis
NSUBS, 100           ! Substeps   number
TIME, T              ! Time   period
KBC, 1               ! Step   loads
OUTRES, ESOL, ALL
SOLVE
FINISH
```
$$(3.75)$$

The solution of the problem allows postprocessing, which consists of the realization of different graphical or textual outputs to reproduce the temporal evolution of some solution variable. In particular, every node defined during the preprocessing activity is characterized by an associated sequence of degrees of freedom with their respective variables, which can be saved and shown with a proper combination of APDL commands (3.76). For example, see the evolution of the electric tension between the ends of the resistor and the evolution of the elastic strain of an element on the upper electrode shown in Figures 3.23–3.24.

```
/POST26
ESOL, 2, PiezoElem + 1, , SMISC, 1, V _ Res    ! Electric   tension
ESOL, 3, 157, , EPEL, Z, Z _ Strain            ! Strain
PRVAR, V _ Res
PLVAR, V _ Res
PRVAR, Z _ Strain
PLVAR, Z _ Strain
FINISH
```
$$(3.76)$$

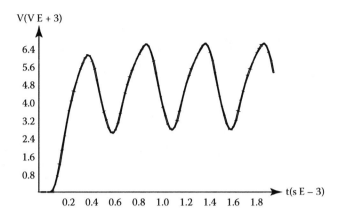

FIGURE 3.23
Electric tension on the resistor.

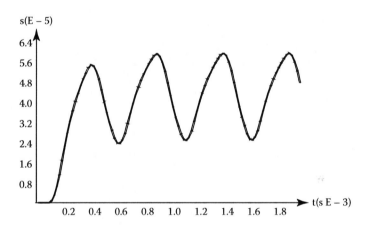

FIGURE 3.24
Elastic strain of an element on the upper electrode.

3.3.3 Multiphysic Analysis of an Electrostatic Actuator System

In the following section, the last example of finite element analysis is proposed, involving an improved simulative complexity with respect to the previous situations. The observed physical phenomenon is the movement of a perforated steel plate caused by an electrostatic action (Figure 3.25). The plate is rigidly connected to the frame by two beams that comply with the effect of the electrostatic action, allowing movement of the plate in a direction perpendicular to its wide surface. The electrostatic action is due to an electric potential difference between the low surface of the plate and the base of the system, which is set at a different potential level (Figure 3.26).

A thin layer of fluid material, air, is present between the two surfaces at different electric potential levels; thus, due to the particular truncated cone shape

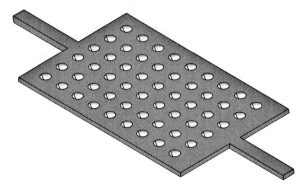

FIGURE 3.25
Perforated plate moved by an electrostatic action.

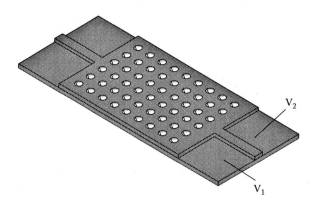

FIGURE 3.26
Surfaces at different potential levels.

of the hole on the plate (Figure 3.27), a fluid flux can be generated from the low to the high surface, depending on the vibratory speed of the perforated plate.

Although fluidic flux is present, in this example we are interested only in the electromechanical interaction between the plate and the generator of potential difference, neglecting the interaction between fluid and structure or the proper choice of the number of holes and their dimensions; in fact this is only an illustrative simulation. The examined electromechanical problem is treated with a multifield single-code simulation, where the solution is obtained by coupling two physical fields through the reciprocal transfer of generated loads. Thus, the resulting simulative process shows a robust behavior for complex problems also.

The simulation starts with the configuration of the system declaring the number of hardware CPUs, setting the maximal dimension of the results, and forcing the program to keep the memory allocated during the analysis phase (3.77). This last option can favor the acceleration of the computation when the used hardware machine is endowed with sufficient physical memory; it

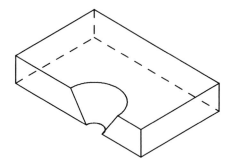

FIGURE 3.27
Truncated cone hole on the moving plate.

can reduce the number of allocations and disallocations of memory in some solutions that employ temporary variables.

$$
\begin{aligned}
&\text{/CONFIG, NPROC, 2}\\
&\text{/CONFIG, NRES, 10000}\\
&\text{MEMM, KEEP, ON}
\end{aligned}
\tag{3.77}
$$

After configuring the simulator and choosing the μMKS measurement system, some parameters are used to define the geometry of the problem (3.78). The electrostatic plate is conceptually shaped from a parallelepiped, with dimensions bl, bh, and bw; removing four extreme parallelepipeds, with dimensions tl, th, and tw; and, finally, producing a series of holes with a truncated cone shape positioned using the parameters hl and hw.

$$
\begin{aligned}
\text{bl} &= 100.0\\
\text{bh} &= 1.5\\
\text{bw} &= 40.0\\
\text{bw2} &= 4.0\\
\text{tl} &= 20.0\\
\text{th} &= \text{bh}\\
\text{tw} &= (\text{bw} - \text{bw2})/2\\
\text{hl} &= 2.0\\
\text{hw} &= 2.0\\
\text{hh} &= \text{bh}\\
\text{gap} &= 2.0
\end{aligned}
\tag{3.78}
$$

The adopted type of finite element for the plate is SOLID95, which presents a standard three-dimensional shape characterized as a deformable structure with 20 nodes, 8 in the vertices of the figure and 12 in the median points of the segments connecting the vertices. The material associated with the elements SOLID95 is described by the Young modulus, Poisson's coefficient, and mass density (3.79).

$$
\begin{aligned}
&\text{ET, 1, 95} \\
&\text{MP, EX, 1, 931E3} \\
&\text{MP, NUXY, 1, 0.1} \\
&\text{MP, DENS, 1, 1.74E} - 15
\end{aligned}
\tag{3.79}
$$

The geometry of the plate is defined using the parameters shown in (3.78) to generate the principal volume and four smaller parallelepipeds, which are detracted from the first one (3.80).

$$
\begin{aligned}
&\text{BLOCK, 0, bl, 0, bh, 0, bw} \\
&\text{BLOCK, 0, tl, 0, th, 0, bw} \\
&\text{BLOCK, 0, tl, 0, th, tw + bw2, bw} \\
&\text{BLOCK, bl} - \text{tl, bl, 0, th, 0, tw} \\
&\text{BLOCK, bl} - \text{tl, bl, 0, th, tw + bw2, tw}
\end{aligned}
\tag{3.80}
$$

The holes with a truncated cone shape are generated in the plate perpendicular to the assigned reference plane, where a point with coordinates x,y represents the intercept between the reference plane and the axis of the hole. From this axis, two circles are drawn with two declared radii at a defined distance, producing the truncated cone. Therefore, if we want to position the hole on a different plane from the actual reference (O,x,y), a new reference plane should be defined, or the actual reference plane should be relocated; if we opt for the command WPROTA, producing a rotation of the reference frame of 90° around the x-axis, then the holes are generated, and the reference frame is driven to the original configuration (3.81).

$$
\begin{aligned}
&\text{WPROTA, 0, } -90, 0 \\
&\text{*DO, J, 0, 8, 1} \\
&\quad\text{*DO, I, 0, 5, 1} \\
&\qquad\text{CON4, tl} + \left(5\ /\ 2 + 3 * \text{J}\right) * \text{hl}, -\left(5\ /\ 2 + 3 * \text{I}\right) \\
&\qquad\text{*hw, 0.5, 1.5, hh} \\
&\quad\text{*ENDDO} \\
&\text{*ENDDO} \\
&\text{WPROTA, 0, 90, 0}
\end{aligned}
\tag{3.81}
$$

The generated volumes are detracted from the original plate (3.82), producing the desired electrostatic plate (Figure 3.28).

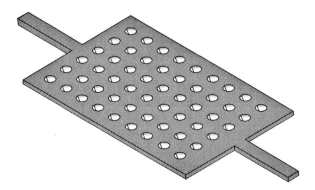

FIGURE 3.28
Electrostatic plate.

$$\text{VSEL, ALL}$$
$$\text{VSBV, 1, ALL}$$
(3.82)

The next phase consists of the realization of a volumetric mesh on the electrostatic plate associating the material 1 with the generated volume, setting the dimensions of the finite elements semiautomatically with the command SMRTSIZE and, finally, selecting the shape of the mesh elements with the command MSHAPE (3.83). We opted for a tetrahedral mesh because of the presence of holes with a truncated cone shape in a parallelepipedal plate. Hence, the SOLID95 elements degenerate into a tetrahedral deformable structure with ten nodes, four in the vertices of the finite volume, and six in the median points of the segments that connect the vertices. Thus, the mesh shown in Figure 3.29 is obtained.

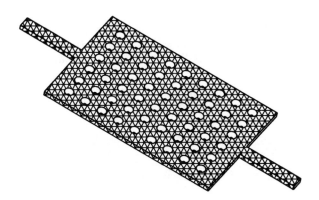

FIGURE 3.29
Volumetric mesh on the electrostatic plate.

```
VATT, 1, , 1
SMRTSIZE, 1
MSHAPE, 1, 3D
MSHKEY, 0
VMESH, ALL
```
(3.83)

The boundary conditions are set on the mesh to define the Cauchy problem, blocking the external extremities of the flexible bars. Also, the multifield approach demands additive boundary conditions represented by the surface interface. On this surface, the electric and the mechanical fields exchange information to couple the two physical fields with a single code (3.84).

```
ALLSEL
ASEL, S, LOC, Y, bh/2
ASEL, R, LOC, Z, bw/2
NSLA, S, 1
DA, ALL, UX
DA, ALL, UY
DA, ALL, UZ
ALLSEL
ASEL, S, LOC, Y, bh/2
ASEL, R, LOC, Z, 0
NSLA, S, 1
DA, ALL, UZ
ALLSEL
NSEL, S, LOC, Y, 0
SF, ALL, FSIN, 1
```
(3.84)

As mentioned, under the plate there is a thin gap of fluidic material that separates it from the other face of the generator of the electric potential difference. Hence, to describe this fluidic material, we have selected the element SOLID123, which is made up of a tetrahedral structure with ten nodes, four in the vertices and six in the median points of the segments connecting the vertices, and every node is provided with an electric degree of freedom identified by its potential level. Also, the use of the µMKS measurement system requires the command EMUNIT to set the permittivity of the material, defining its electric properties (3.85).

```
ET, 2, 123
EMUNIT, EPZRO, 8.8541878E - 6
MP, PERX, 2, 1
```
(3.85)

The movement of the electrostatic plate produces a deformation of the fluidic film; therefore, to implement these transformations of the interface surface, the morphing option is activated (3.86). Then, the geometry of the material is represented, for the sake of simplicity, by a parallelepiped (3.87).

$$\text{MORPH, ON} \qquad\qquad (3.86)$$

$$\text{BLOCK, 0, b1, } -\text{gap, 0, 0, bw} \qquad\qquad (3.87)$$

Hence, the mesh is applied on the volume; setting the dimensions of the elements with the semiautomatic command SMRTSIZE associated with the tetrahedral shape (3.88), which gives a volumetric electric mesh subject to the morphing of the interface surface, owing to displacements produced by the mechanical field (Figure 3.30).

$$
\begin{aligned}
&\text{VSEL, S, VOLU, , 1}\\
&\text{SMRTSIZE, 1}\\
&\text{MSHAPE, 1, 3D}\\
&\text{MSHKEY, 0}\\
&\text{VATT, 2, 2}\\
&\text{VMESH, ALL}
\end{aligned}
\qquad (3.88)
$$

As we are not interested in the mechanical interaction between the fluid and the structure, or in the dynamics of the fluid through the holes on the plate, we avoid the specifications of the mechanical properties of the fluid, and we simply set the boundary conditions on the domain so as to annul the

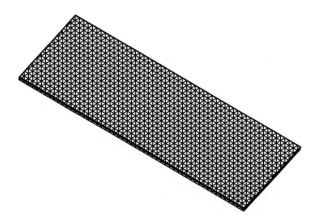

FIGURE 3.30
Mesh of the electric material.

displacements along the external surfaces x (3.89), z (3.90), and y (3.91) of the parallelepiped, respectively.

$$
\begin{aligned}
&\text{ASLV, S}\\
&\text{ASEL, R, LOC, X, 0}\\
&\text{DA, ALL, UX, 0}\\
&\text{ASLV, S}\\
&\text{ASEL, R, LOC, X, bl}\\
&\text{DA, ALL, UX, 0}
\end{aligned}
\tag{3.89}
$$

$$
\begin{aligned}
&\text{ASLV, S}\\
&\text{ASEL, R, LOC, Z, 0}\\
&\text{DA, ALL, UZ, 0}\\
&\text{ASLV, S}\\
&\text{ASEL, R, LOC, Z, bw}\\
&\text{DA, ALL, UZ, 0}
\end{aligned}
\tag{3.90}
$$

$$
\begin{aligned}
&\text{ASLV, S}\\
&\text{ASEL, R, LOC, Y, -gap}\\
&\text{DA, ALL, UY, 0}
\end{aligned}
\tag{3.91}
$$

The surface of the dielectric parallelepiped, which coincides with the low face of the electrostatic plate, represents the interface toward the mechanical field, and its geometrical configuration is identified, through morphing, by the movement of the electrostatic element; therefore, an explicit boundary condition cannot be set for this surface (3.92).

$$
\begin{aligned}
&\text{ASLV, S}\\
&\text{ASEL, R, LOC, Y, 0}\\
&\text{NSLA, S, 1}\\
&\text{SF, ALL, FSIN, 1}
\end{aligned}
\tag{3.92}
$$

Finally, the electric boundary conditions are identified by the electric tension on the low and high surfaces of the electric domain, assuming a positive value on the interface and a conventional null value on the opposite face (3.93), which represents the electric ground.

Therefore, the union of the electric and mechanical meshes with their boundary and interface conditions represent the physical problem (Figure 3.31), to be solved by a suitable numerical algorithm.

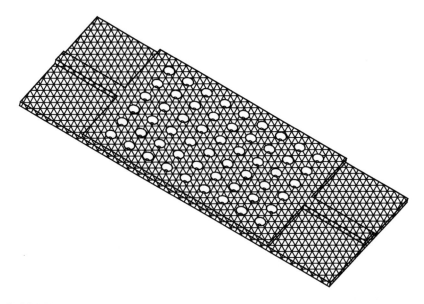

FIGURE 3.31
Electrical and mechanical meshes.

$$
\begin{aligned}
&\text{D, ALL, VOLT, 100}\\
&\text{NSEL, S, LOC, Y, −gap}\\
&\text{D, ALL, VOLT, 0}\\
&\text{ALLSEL, ALL}\\
&\text{FINISH}
\end{aligned}
\qquad (3.93)
$$

The instruction block for the solution starts with the activation of the multi-physic solver, then the physic fields are defined and associated with the materials used for the meshes, and finally, the solution order of the problems within the multiphysic solver is identified and the desired tolerance set (3.94).

$$
\begin{aligned}
&\text{/SOLU}\\
&\text{MFANALYSIS, ON}\\
&\text{MFELEM, 1, 1}\\
&\text{MFELEM, 2, 2}\\
&\text{MFORDER, 2, 1}\\
&\text{MFCONV, ALL, 1.0E − 6}
\end{aligned}
\qquad (3.94)
$$

The solution option relies on a static analysis employing the sparse direct solver and forcing an in-core variable solution (3.95).

```
ANTYPE, STAT
EQSLV, SPARSE
BCSOPTION, , INCORE
MORPH, ON
MFCMMAND, 2
ANTYPE, STAT
NLGEOM, ON
MORPH, OFF
KBC, 1
MFCMMAND, 1
```
(3.95)

The two physic fields exchange loads in a conservative way: the structural field receives forces from the electrostatic field, returning kinematic displacements (3.96).

```
MFINTER, CONS
MFSURFACE, 1, 2, FORC, 1
MFSURFACE, 1, 1, DISP, 2
SOLVE
SAVE
FINISH
```
(3.96)

After the solution, the next step is the postprocessing, where we plot the displacements along the x-axis (3.97), y-axis (3.98), and z-axis (3.99); the electric field in the fluid (3.100); and the intensity of the mechanical stress on the electrostatic plate (3.101), corresponding, respectively, to the results shown in Figures 3.32–3.36.

```
/POST1
FILE, FIELD1, RST
SET, LAST
ESEL, S, TYPE, , 1
NSLE, S
PLNS, U, X
FINISH
```
(3.97)

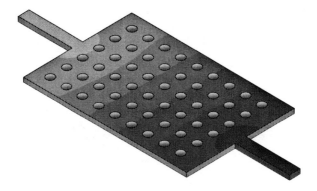

FIGURE 3.32
Displacements along the *x*-axis.

```
/POST1
FILE, FIELD1, RST
SET, LAST
ESEL, S, TYPE, , 1
NSLE, S
PLNS, U, Y
FINISH
```
(3.98)

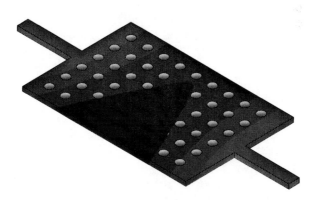

FIGURE 3.33
Displacements along the *y*-axis.

```
/POST1
FILE, FIELD1, RST
SET, LAST
ESEL, S, TYPE, , 1                    (3.99)
NSLE, S
PLNS, U, Z
FINISH
```

,

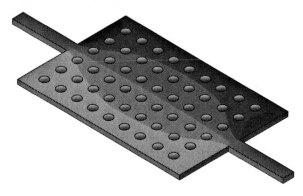

FIGURE 3.34
Displacements along the *z*-axis.

```
/POST1
FILE, FIELD2, RST
SET, LAST
ESEL, S, TYPE, , 2                    (3.100)
PLNS, EF, SUM
FINISH
```

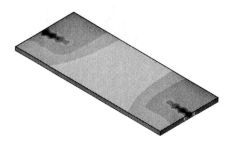

FIGURE 3.35
Intensity of the electric field.

```
/POST1
FILE, FIELD1, RST
SET, LAST
ESEL, S, TYPE, , 1
NSLE, S
PLNS, S, INT
FINISH
```

(3.101)

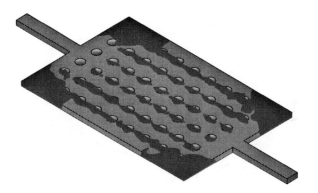

FIGURE 3.36
Intensity of the mechanical stress.

4

Design and Characterization
of a Pneumatic Mesovalve

As described in chapter 1, the shape memory effect can be exploited to develop thermal actuators through the Joule effect; therefore, we introduce the description of a simple, pneumatic, normally closed (NC), two-way, monostable mesovalve activated by a threadlike micro-actuator realized with an NiTi alloy. A preliminary prototype allows a wide range of experimental analyses to highlight the most important design defects and, thus, to improve them with an optimized solution. A pneumatic test bench was employed to properly manage some functional parameters and input–output variables of the prototype device. The electric feeding circuit was designed to generate a square wave train with adjustable amplitude, frequency, and duty cycle. Experimental results on the life of the device and different functional tests showed correlations between the behavior of the valve and some physical quantities, i.e., the mass flow rate and the input pressure.

Standard and recognizable parameters shall be proposed to quantify the real behavior of the valve. As described in section 4.5, the International Standard suggests, as comparison parameter, the flow capacity C for the quasi-static behavior (CEI EN 60534-2-1 Industrial-process control valves—Part 2.1: Flow capacity—Sizing equations for fluid flow under installed conditions; CEI EN 60534-2-3 Industrial-process control valves—Part 2.3: Flow capacity—Test procedures) and the main times for the dynamic behavior of the valve (ISO 12238 Pneumatic Fluid Valves, Directional Control Valves, and Measurement of Shifting Time).

Different tests are described with calm air and compressed air in the last part of the chapter to permit the characterization of the mesovalve. The static tests permit the experimental determination of the flow rate through the valve during different working conditions and permit us to estimate the loss of load and flow behavior. On the other hand, the dynamic tests permit us to establish the response time and the maximum activation frequency, controlling the step-response velocity and changing the activation frequency of the device.

The use of shape memory wires is due to wide international research and design activity, resulting in improvements of functional and mechanical properties; then, we choose, as operating principles, assembly simplification and the use of economical materials.

4.1 Introduction

4.1.1 Shape Memory Alloys

In this section some key concepts of shape memory alloys (SMAs) are briefly mentioned. The SMA considered for our application exhibits two distinct stable crystallographic phases, depending on the temperature and stress state; then, martensitic transformations occur from a crystallographic more-ordered parent phase, the austenite (A), to a crystallographic less-ordered product phase, the martensite (M). In general, the austenite is stable at high temperatures and low values of stress, whereas the martensite is stable at low values of temperature and high values of stress.

Thus, the SMA in the martensitic phase can be deformed in an irreversible manner until the breaking point of the chemical bonds. To return to the austenitic phase, or rather the previous form, the alloy has to be heated. Although NiTi is soft and easily deformable at a low temperature, it resumes its original shape and rigidity when heated to a high temperature. This is called the *one-way shape memory effect*. The ability of SMAs to recover a preset shape upon heating above the transformation temperature and to return to a certain alternate shape upon cooling is known as the *two-way shape memory effect*. The temperatures at which the SMA changes its crystallographic structure are characteristic of the alloy and can be tuned by varying the elemental ratios. Typically, M_S denotes the temperature at which the structure starts to change from austenite to martensite upon cooling, and M_F is the temperature at which the transition is completed; accordingly, A_S and A_F are the temperatures at which the reverse transformation from martensite to austenite starts and finishes, respectively. All these main points of temperature depend on the material composition as well as on the thermomechanical treatment. If the volume percentage of the transformed martensite is drawn as a function of the temperature, hysteresis line can be observed, characterizing the phase transformation. Moreover, the shape change is a thermodynamically irreversible process of the internal structure during the phase change; in other words, there are energy losses due to internal friction. These phenomena are depicted in the hysteresis diagram in Figure 4.1.

The material is completely martensite at the start point 1; the variation between martensite and austenite follows the lower line during heating. The austenitic phase starts when the temperature reaches A_S and finishes at A_F, because the material is completely austenite; whereas if the temperature gets down under M_S, the martensite begin to form, following the highest line until the point M_F.

The singular crystallographic structure of the martensitic phase consists of a thick disposition of crystallographic planes with a high relative mobility. When the material is deformed by an external force, these crystallographic planes change their configuration, distributing the total deformations without suffering the effects of atomic displacements (Pitteri, Zanzotto, 2002).

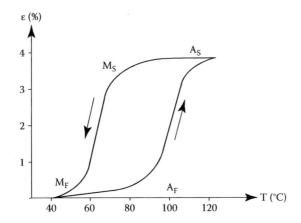

FIGURE 4.1
Hysteresis of a shape memory alloy (ε is a two-way strain).

Another important property of the SMA is its capability to transform not only with the temperature but also with stress, or with a combination of these effects in suitable conditions. Phenomenologically, the material immediately recovers the original shape with remarkable elasticity, that is, pseudoelasticity or superelasticity, because the material is significantly deformed and immediately recovers its original shape. During the transformation, the material forms the martensite progressively and is quasi-instantly deformed to rearrange the crystallographic structure without damage. The martensite produced because of the stress effect is not in the stable-phase range and, when the stress is removed, reverts to austenite, the material recovering its original shape. This effect, which makes the alloy appear extremely elastic, is known as *superelasticity* or *pseudoelasticity.*

In general, the shape memory effect can happen in three ways: free recovery, forced recovery, and actuation system. The first process occurs freely and, the material is subjected only to a displacement. The forced recovery is caused by a noncompliant constraint, so the material cannot recover its original shape. Hence, it produces stress that could be used for the realization of some connections. The third way, the actuation system, can be realized with an SMA linked to a compliant constraint, made of a structural constraint (section 1.2.6). The valve described in this chapter is an actuation system. The mechanical work is obtained by contraposition between the structural constraint and the recovery of the SMA, where a return spring or, simply, the pressure of the compressed air, is the structural constraint.

The SMA is made of different elementary metals. The composition and the type of the alloy depend on the properties that the SMA is to have; hence, the manufacturer can fix the transition temperature by adjusting the alloy ratio; i.e., an NiTi alloy can be compounded by similar quantities of nickel and titanium. A difference less than 1% in this ratio determines a variation of the transition temperature in a range between −100°C and +100°C. The range of the transition temperature of the used alloy is between 70°C and 90°C, and one of the stable phases is at the ambient temperature. Considering all the SMAs, the NiTi alloys are particularly interesting for some good mechanical properties (Reynaerts, Van Brussel, 1998), i.e., the ductility, good movement recovery, excellent corrosion resistance, stable transition temperature, perfect biocompatibility (hence, important medical applications such as dental braces or artery dilatators), and the capability of shape recovery on being electrically heated. Although the NiTi alloy is expensive, hard to realize, and exhibits a remarkable hysteresis, there are many reasons to prefer the SMAs when the actuator is realized. It has an electrical resistance higher than copper, improving heating performance through the Joule effect, as in the case described in this section, where in (4.4) a phase change is obtained with an electrical current.

The NiTi alloy is a binary intermetallic compound and exhibits a good solubility toward nickel or titanium excesses. This property allows us to add a lot of elements to change the mechanical characteristics and properties of the martensitic transformation. Nickel (1%) is the element that is most commonly added, permitting us to reduce the transformation temperature and to increase the breaking load in the austenitic phase. Similarly, iron and chromium reduce the transition temperature; copper reduces the hysteresis cycle; niobium permits us to reduce the start temperature of the martensitic phase and to widen the gap between A_S and M_S, and platinum or gold with hafnium and zirconium increase the transformation temperature of the martensite considerably.

4.1.2   Pneumatic Valve

Using the functional scheme of a mechanical machine described in chapter 1, the device can be represented as a sequential structure of an actuator A, a transmission T, and a user u (Figure 4.2). The actuation system is due to the mechanical energy E_{mecc} produced by the NiTi wires after the phase change is accomplished by the conversion of supplied electrical energy E_{el} to heat E_{term} (the Joule effect). In this case the transmission is direct, and the user is associated with the plug.

This industrial device is identified with the pneumatic attribute because the controlled media is the air. The pneumatic elements can have different characteristics whether for their functional properties, which identify the ability of elaborating the air flux, or for the working pressure, which defines the class of application. The power components directly joined to the actuator are subjected to a standard pneumatic pressure, namely, in a range between 4 and 10 bar. If the aim is the elaboration of a logic pneumatic signal without power functionality, the pressure can be reduced; therefore, there are pneumatic logic elements working at 3–4 bar and micropneumatic elements made of membranes that work in the range 1.4–2.5 bar. Finally, there are also some elements that can operate at less than 1 bar. In addition, the pneumatic components can be active or passive—the former has only a piloting

FIGURE 4.2
Functional definition of a mechanical machine. A is the actuator, T is the transmission, U is the user, E is energy, m is a mechanical function, u is a usable mechanical function, and p is an environmental perturbation.

aim, and their elements are permanently connected with the compressed air source and the control signal; the latter do not possess feeding connections. The active elements can have an output signal higher than the control signal, because they work only on the commutation of power flow, i.e., a signal amplification can be realized with the energy supplied by the compressed air source. On the other hand, the passive elements do not require energy, because there are no feeding connections.

The pneumatic valves are used to control the airflow in a pneumatic circuit; the function of the valve depends on the relative quantity of the flow; hence, there are power valves and control valves—the former operate directly on the compressed air, whereas the latter control other valves regulating the flow. Pneumatic valves can perform different operations, so we focus attention on the direction control valves. The main aim of these is to change the logical connections between different pipes, and they are characterized by the number of flow paths and valve positions. The simplest valve is 2/2 two-way, having two possible positions. As shown in Figure 4.3, the valve is made up of an external body and a moving internal plug (O), forming a parabolic shape. The plug is moved with a suitable drive to open or close the valve. In the Figure 4.3, the valve symbol is drawn according to the standard ISO 1219-1 Fluid power systems and components—Graphic symbols and circuit diagrams—Part 1: Graphic symbols, where the square representation is used; a square represents the working position of a valve.

In this chapter a new 2/2 valve will be described from the design to test and analysis phase as a base study for other typologies of valves. After defining the main parameters and solving the problems due to the first prototype, the realization of the other typologies is easy. The next step is the 3/2 valve shown in Figures 4.4 and 4.5. In this valve, line 2 can be alternatively

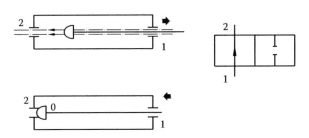

FIGURE 4.3
Two-way valve—functional scheme and symbol according to ISO 1219-1 standard.

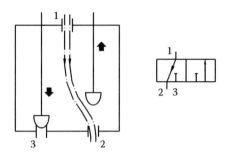

FIGURE 4.4
Three-way valve—functional scheme according to ISO 1219-1 standard.

connected with line 1 or with line 3 really with a suitable electrical drive, as described; moreover, lines 1 and 3 can be simultaneously opened or closed.

The activation of the valves can be accomplished in different ways, according to the desired function of the valves; hence, the different typologies of activation can be listed: manual control; mechanical control with cams, levers, and comparisons of limit switches; pneumatic control used for realization of the pneumatic circuits; and electric control, which represents the interfaces between the electrical and the pneumatic systems. The displacement of the mobile elements is due to various reasons: direct mechanical actuation, spring action, push of a pneumatic piston, or electromagnetic attraction.

As a rough estimation, the valve prototype can enter into the electrovalves group. Although there are no electromagnets to move the plug, the plug has an electrical connection. Properly stated, the activation is thermal; in fact, the supplied electrical energy heats the shape memory alloy, and at transformation temperature the SMA contracts and the output hole is opened or closed, depending on the valve configuration. Then, the recovery of the initial length is due to the spring and cooling (Figure 4.6).

Some positions of the valve can be transitory or, rather, the position is maintained by a signal; the number of stable positions characterizes the valve (Figure 4.7). Therefore, the two-way valve could have a stable position (monostable valve) and only one command, or two stable positions (bistable valve) and two control signals. The monostable valve has some components

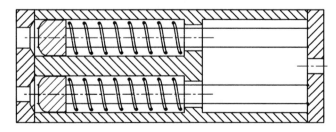

FIGURE 4.5
Three-way valve activated by shape memory wires.

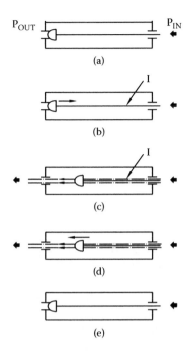

FIGURE 4.6
Functional scheme of a 2/2 valve: (a) normally closed valve; (b) electrical activation and heating of the wires; (c) contraction of the NiTi wires and opening of the valve; (d) electrical deactivation, cooling of the NiTi wires, and closing of the valve due to the spring force, (e) complete closing of the valve.

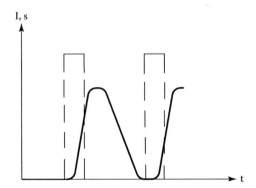

FIGURE 4.7
The phases of the valve and the displacement of the plug(s) in function of current activation I: the broken and the continuous lines represent, respectively, the activation and the displacement.

that restore the plug to the original position when the signal is turned off. These components could be springs, a small inline piston, or simply the force exerted by the flow.

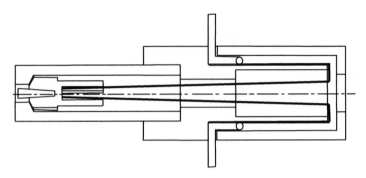

FIGURE 4.8
Pneumatic mesovalve with shape memory wires.

As already described in the section 4.1, the topic of this chapter is a pneumatic valve with the following characteristics: normally closed, two-way, monostable, active, and activated by threadlike SMA micro-actuators (Figure 4.8). The valve is designed to work at a range of overpressure between 0 and 3 bar, as the logical elements and different mesopneumatic elements.

The valve is "meso" when it is compared to the existing devices for low pressure and limited flow rate, whereas it is defined in the strict sense "meso" when its size is of the order of millimeters. Mesopneumatic valves are widely used for industrial applications because of the good flow characteristics and the possibilities of realizing a compact assembly; i.e., they are common in automatic assembly lines where mesoelectrovalves are used for flow control with a low response time. Hence, they are comparable to the proposed mesovalve actuated by an SMA. The sizes of the SMA mesovalve depend on the manufacturing technology and the design choices, whereas the miniaturization possibilities are high.

Microvalves and microfluidic elements are used in biomedical applications, in the devices for microdosage, for chemical applications, and for microsensors. Bioengineering is another wide field for the development of sophisticated and accurate devices.

The most interesting applications for SMA valves are associated with reduced electric power and the low pressure of the pneumatic circuit, as in piloting systems for pneumatic industrial valves. Furthermore, their reduced size is suitable for compact solutions, and the actuation thermal effect can be produced with a simple electric circuit.

The standard EN 60534-2-1 specifies which operative data shall be supplied for the proper choice of a valve and some of the main properties are the working pressure field, working flows (i.e., the compressed dry and filtered air), possible presence of lubrication, mass, geometrical sizes and used materials, valve type (in this case, 2/2 NC with a parabolic plug and activation by the shape memory wires), admitted temperature range, flow properties, and dynamic tests.

TABLE 4.1

Technical Data for NiTi Wire (Obtained in Calm Dry Air at 20°C)

Typology of NiTi Alloy	A	B	C	D	E	F
Diameter size (μm)	25	37	50	100	150	250
Minimum radius of curvature (mm)	—	1.9	2.5	5	7.5	12.5
Linear resistance (Ω/m)	1770	860	510	150	50	20
Recommended current (mA)	20	30	50	180	400	1000
Recommended power (W/m)	—	0.8	1.3	4.9	8	20
Maximum recovery stress at 600 MPa (g)	—	64	117	469	1056	2933
Off time 70°C wire						
Contraction time (s)	0.5	0.5	0.5	0.5	0.5	0.5
Relaxation velocity (s)	0.7	0.7	0.8	1.3	2.5	6.2
Cycle speed (cycle/min)	55	52	46	33	20	9
Off time 90°C wire						
Contraction time (s)	—	0.5	0.5	0.5	0.5	0.5
Relaxation velocity (s)	—	0.4	0.4	0.7	1.5	4.1
Ratio topic cycle (cycle/min)	—	68	67	50	30	13

4.1.3 Wire Choices

When the plug of a normally closed (NC) valve is subjected to feeding pressure, it attains the corresponding force to open the valve. This type of valve allows us to obtain an important pass section with short movements, because section and displacement are independent. Hence, these characteristics drive the design of the plug valve toward the actuation with shape memory wires. In fact, the NiTi wires generate high forces and limited displacements; furthermore, the wire shape supplies the necessary force with reduced usage of NiTi material.

The actuation is realized with shape memory fibers with diameter size of 50–100 μm exerting a contraction when electrically stressed; then the wires are cooled by air and stretched by a small force to return to starting conditions. The choice of a particular wire diameter is made by estimating the response time and the maximum stretch force, as shown in Tables 4.1 and 4.2; the wires should be thin for better dynamic performance, and on the other hand, the diameter of the wires should be large for high input pressure.

The movement realized by the shape memory effect is caused by a reallocation of the solid state in the NiTi alloy, without acoustic emissions and with good fatigue resistance; furthermore, it should be possible to make a physical movement in a narrow space with a reasonable cycle velocity (Table 4.3).

Hence, the benefits offered by these wires are as follows:

- They have a small footprint compared to electric motors or solenoids.
- They are very light compared to small motors or solenoids.

TABLE 4.2

Property of an NiTi Alloy

Property of an NiTi Alloy	Value
Density (g/cc)	6.45
Heating capacity (J/g°C)	0.32
Latent heat (J/g)	24.2
Maximum recoverable force (MPa)	560
Recoverable strain recommended (MPa)	187
Deformation force recommended (MPa)	35
Ultimate tensile strength (MPa)	1000
Work density (J/g)	5
Ratio energy-efficiency (%)	8

TABLE 4.3

Technical Data of NiTi Wire: Low (70°C) and High (90°C) Activation Temperature

	70°C Wire	90°C Wire
Start activation temperature (°C)	68	88
Finish activation temperature (°C)	78	98
Start relaxation temperature (°C)	52	72
Finish relaxation temperature (°C)	42	62
Tempering temperature (°C)	300	300

TABLE 4.4

Technical Data for NiTi Wire: Phase

Phase	Martensite	Austenite
Resistivity ($\mu\Omega$cm)	76	82
Young's modulus	28	75
Magnetic susceptility (μemu/g)	2.5	3.8

- As shown in the section 4.1.1, because crystalline realignment takes place as a result of internal temperature changes, it also occurs with ambient temperature change (Table 4.4). The Joule effect can be used in these applications, so the state change of SMA can be driven with a simple electrical circuit.
- The wires do not produce grinding sounds, clicks, or tickings.

4.2 Layout

A preliminary layout and its assemblability are described, underlining the problems and the reasons for further improvements. The choice of the valve size is associated with the opportunity to insert it into the standard line of the pneumatic systems. The first prototype of the valve is made up of 12 elements, as shown in Figure 4.13, which shows an exploded view. In particular, the valve body (Figure 4.9 and Figure 4.10) is realized with an NC lathe in two different parts for easy unassembly and reassembly.

The input of the flow is realized in Teflon from a 12 mm diameter bar, the hole is Φ 3.8 mm and depth 2.2 mm, and two metallic clamps are fixed on the internal boundary as shown in Figure 4.12, whereas the external diameter

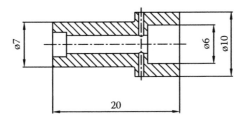

FIGURE 4.9
Valve body, input of airflow (dimensions in mm).

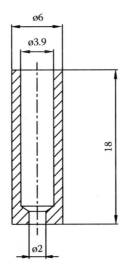

FIGURE 4.10
Valve body, output of airflow (dimensions in mm).

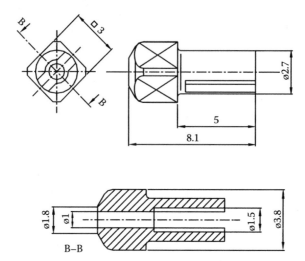

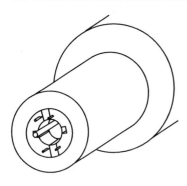

FIGURE 4.11
Plug to open and close (dimensions in mm, section view according to the European standard).

Φ 6 mm is jointed with a quick coupling or directly to the pneumatic system piping. The copper wires for the electrical connection are fixed to one side on the metallic clamps and the other side through the hole Φ 0.8 mm on the electrical circuit. This type of connection is the weak point of the valve; it presents some assembly difficulties and does not give a long-life electrical connection.

The second half of the valve (Figure 4.10) is mechanically machined or molded, if necessary, using Plexiglas. This material is selected so that valve behavior during the working process can be observed.

The plug used to open and close the valve (Figure 4.11) is directly moved by the NiTi wire, and its shape is determined by different requirements. The central hole (Φ 1 mm) is fixed to the SMA wires with a conical metallic pin to close the electrical circuit. The filleted area of the plug (Φ 2.7 mm) houses the spring, whereas the chamfered area is airtight when the valve is closed. The plug is made in brass with a lathe. An improvement would be to use different plastic materials as the Polyamide PA66+30%FV, which can be molded or machined during the prototyping phase.

On the inlet hole (Figure 4.12) there are the two metallic clamps that represent the electrical positive and negative poles; therefore an insulating material should be inserted between the two clamps to avoid short circuit.

FIGURE 4.12
Copper and NiTi wires on the metallic clamps.

The described fixing system of the wires on the metallic clamps can be imprecise; hence,

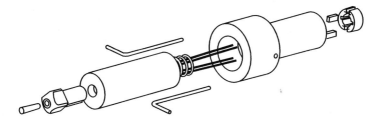

FIGURE 4.13
Exploded view of the preliminary prototype of the valve.

the clamps could be removed in a second proposed prototype, in which pneumatic load losses and manufacturing costs are also reduced.

One of the main problems of the preliminary prototype shown in Figure 4.13 is its assemblability. The two metallic clamps are mortised to the internal face of the valve, and the plastic blocks are inserted between these clamps, insulating the positive and negative poles. The next operation consists of inserting the two copper wires (common electric monocable wires) into the dedicated holes and fixing these wires on the metallic clamps. The difficulty of this operation lies in obtaining the ideal electric contact, although this contact can be broken during the tests. The hole should be closed with glue for a good air seal. Afterward, the four NiTi wires are inserted through the brass plug, through the spring, and then through the Teflon body to the metallic clamps, where they will be fixed into the dedicated sites. Then, the NiTi wires are stretched and fixed with a conical pin on the brass plug. The fixing of the four wires on the metallic clamps is not always perfect because NiTi wires are extremely thin and flexible, and the wires easily come out of their sites. Besides, if the wires are not adequately stretched, the plug cannot be controlled. The last operation consists of placing the spring into the body of the valve and closing the two bodies of the valve. A further observation is that these operations are not particularly suitable for automated production and, obviously, a flexible automation with standard components is preferred for different reasons, but the most important reason is the reduction of production costs.

When the valve was tested operationally without compressed air, some problems appeared. Some of them are due to the design of the hole deformation and nonideal fixings of the NiTi wires, and others are due to reduced assemblability. Finally, all these operations are not sustainable, not only for the complexity of the assembly but also for the low dynamic performances and for the short reliability; hence, the number of components can be reduced and the method of assembly can be improved.

4.2.1   Development of a First Series of Improvements

The tests without compressed air have underlined some problems due to the nonideal fixings, as just described. Furthermore, the displacement of the plug

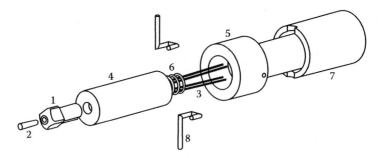

FIGURE 4.14
Valve components: (1) the brass plug; (2) steel conical pin; (3) NiTi Φ 50 mm wire; (4) body of valve with output hole; (5) internal body of valve; (6) spring; (7) cap with hole Φ 4 mm for the air input; and (8) copper wires.

should be incremented, increasing the length of wires to reduce the number of wires and to change the method of fixing. In the second prototype there are no metallic clamps, and there are two wires that are 100 mm long. Seven components of the new prototype with a global weight of 3 g are shown in Figure 4.14.

The assemblability is improved with a time reduction and reliability increment following these steps. Insert two NiTi wires with a needle through the hole of the plug, through the spring, and then through the body of the valve (5). This last component, shown in Figure 4.15, replaces the component of the first layout (Figure 4.9). Then, the NiTi wires are joined around the relative copper wires, and the heads of the NiTi wires are fixed on the brass plug with the conical pin, as shown in Figure 4.16. Finally, the valve is closed with the Plexiglas body (4) and with the cap (7), while air-loss points are glued as shown in Figure 4.17.

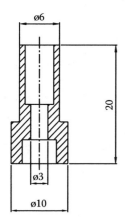

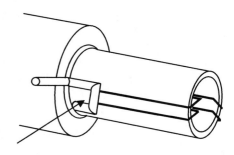

FIGURE 4.15
Internal body of the valve.

FIGURE 4.16
Connection and fixing of NiTi wires and copper wires.

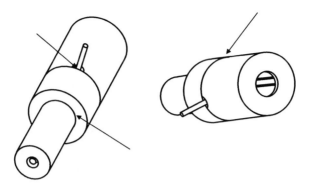

FIGURE 4.17
Assembled valve and point sealed with glue.

4.3 Forces

The NiTi wires will exert a force F to move the plug of the valve overcoming the spring force and the force due to the air pressure. The maximum force exerted by the spring can be calculated with equation (4.1), and force due to the pressure with equation (4.2), where Δl is the extension, k is the elastic constant of the spring, F_m has a value between 0.91 N and 1.27 N, and F_p is the force due to the pressure P on the area A_{ott} of the size of the plug.

$$F_m = k \cdot \Delta l \tag{4.1}$$

$$F_p = P \cdot A_{ott} \tag{4.2}$$

$$F = k \cdot \Delta l + P \cdot A_{ott} \tag{4.3}$$

Therefore, the minimum force that the NiTi wires shall realize, while the alloy phase changes from martensitic to austenitic, is calculated with equation (4.3). On the other hand, the maximum applicable force is usually declared by the NiTi producers or can be measured with experimental tests and depends especially on the wire diameter. Therefore, the pressure is the only functional variable in (4.3) that drives the design choices, whereas the limit condition, as shown in Figure 4.57, is reached when the pressure is too high or, better, when the wires gain the yield point.

4.3.1 Extension and Contraction Operation

The stretch of the wires is due to preloading of a passive mechanical component, a return spring that exerts a return force to close the valve; furthermore, the spring, due to its compliance, allows the opening of the valve when the NiTi wires are activated. In the various tests, four types of spring are used with different stretch properties, but with the same internal diameter and length, because the geometrical size depends on the internal housing in Plexiglas and on the housing of the closing-pin. Examining equation (4.3), the area of the plug A_{ott} and the spring stretch are fixed. The maximum force that the wires can tolerate is known from Tables 4.1–4.2; hence, the elastic coefficient k is chosen as low as possible to obtain a wide pressure range and, with a compromise between the spring stability and the linearity of force-length ratio, lateral flexions can be avoided. Then, when the valve is activated, the NiTi wires contract and move the plug. Although the use of the spring is essential for the return to the starting condition of the NiTi wires, this could produce some problems. As a matter of fact some tests, with a work frequency higher than 1 Hz, show lateral flexions when the valve is opened, and this phenomenon increases when the pressure differential rises.

The cooling of the NiTi wires happens almost instantly when the SMA wires are subjected to forced cooling with compressed air; furthermore, the spring force could be replaced with an opportune recovery pressure P0 applied on the area of the internal surface of the plug, obtaining a design simplification, due to the elimination of the spring. Finally, the plug can be redesigned to suit the effects of inlet pressure (Figure 4.18).

$$E_{el} = R \cdot I^2 \cdot t \tag{4.4}$$

The phase change of the SMA wires is associated with a thermal phenomenon, which produces a contraction at 88°C. The temperature increment is produced with a supplied energy E_{el} due to a PWM (pulse width modulated) current, as shown in Figure 4.4 where t is the time, I the simulated constant current, and R the electrical resistance of the wires.

These valves can also be used as passive components, that is, as flux thermostats, without electrical activation of the wires, obtaining directional control valves (Figure 4.19 and Figure 4.20) that change the flow direction

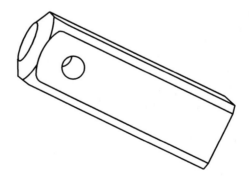

FIGURE 4.18
Plastic plug.

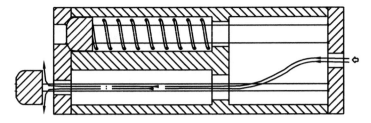

FIGURE 4.19
Configuration with flow at temperature lower than the activation of SMA wires' temperature.

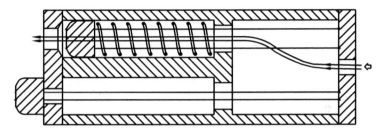

FIGURE 4.20
Configuration with flow at temperature higher than the activation of SMA wires' temperature.

depending on the temperature. As a matter of fact, if a normally closed and a normally open plug are fixed at appropriately chosen SMA wires, opening 1 is closed and the opening 2 is opened simultaneously when the flow reaches the activation temperature T_a. In this way, the flow direction is changed, but the valve remains in this stable configuration as long as the temperature of the flow does not fall below T_a. Therefore, the valve is 3/2 bistable passive.

4.4 Instruments: Test Bench

A wide range of tests is a basic operation, which precedes further investigations on the real behavior of the proposed valve; hence, a pneumatic test bench was planned to properly manage some functional parameters and input–output variables of the prototypical device. This test bench can be separated into different parts for a convenient description: the activation circuit of the valve, the fluid control system, the measurement instruments and, finally, the power supplies. Therefore, some specific aspects of the test bench are examined in depth in the following subsections.

4.4.1 Valve Activation: Electrical Circuit

An analog feeding circuit for the valve can be designed for ease of implementation and can be improved afterward with more performing devices in terms of both precision and controllability. Consequently, an analog circuit was assembled that rectifies the signal starting from the electrical grid and produces constant output current with an adjustable amplitude used to feed the valve, allowing control of the behavior of the pneumatic device. Then, the proposed feeding circuit was replaced by an improved still analog circuit that was able to generate a square wave train with adjustable amplitude, frequency, and duty cycle. It was used for life tests and for functional tests showing correlations between the behavior of the on–off valve and some parameters, i.e., the mass flow rate and the input pressure.

After an approximate setup of the system parameters, the last step results in a digital electric circuit with a microchip that permits a high resolution and counts the number of square waves. This digital circuit can be simply interfaced with other measurement instruments, and it can be turned into a closed-loop system for real-time dynamic control.

The use of a wire geometry for the NiTi actuator simplifies the heating method, because the NiTi alloy can be activated simply by feeding it with a DC or AC current and, because of the Joule effect, the shape memory wire changes its physical phase from martensitic to austenitic within the nominal temperature domain. Therefore, the heat quantity produced in the time unit matches the current power and the generated thermal energy can be approximately described with the simplified expression (4.5), where E is the dispersed energy converted into heat during the time interval t, R is the electrical resistance of the NiTi wire, and I is the current through the wire.

$$E = R \cdot I^2 \cdot t \tag{4.5}$$

The current heats the wire until the final austenitic temperature, whereas the wire contracts, conserving a constant volume and the section increases with a proportional reduction of the length. From an electric viewpoint, during the thermomechanical transformation, we consider a constant electric resistance for the sake of simplicity.

The current and the exposition time should be accurately controlled to avoid excessive thermal stresses on the wire. Particular attention should be paid in calm air tests because there is no convective cooling due to airflow. Then, fixing the exposition time t, the Joule equation (4.5) can estimate a maximum current amplitude I avoiding a loss of memory property.

After a series of tests on the assemblability of the first prototype (Figure 4.21), the experimental results suggest that feeding with copper wires inserted into holes of the valve body is too time consuming and does not guarantee electrical contact. In fact, the first layout constrains us to take the two copper wires, pass them through dedicated holes, bend and fix them with the NiTi wires on metallic demilunes; however, in this way, the wires cannot remain stretched; hence, the internal contact cannot be permanent. It sometimes caused the breakdown of the feeding circuit in calm air tests, owing to absence of contact or short circuit. Therefore, a set of three tests was realized

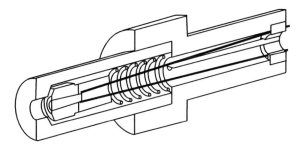

FIGURE 4.21
Section drawing of the valve with copper wires passed through and fixed on metallic blocks.

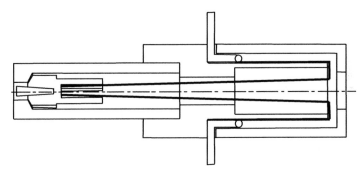

FIGURE 4.22
Section drawing of new prototype of the valve.

on the first prototype (Figure 4.21) in calm air with a 240 mA feeding current, respectively, with an on-off 1 Hz frequency for 100 cycles, an on–off 2 Hz frequency for 100 cycles, and an on–off 1 Hz frequency for 90 cycles, and it always shows the presence of sparks in the neighborhood of the electrical contact. This effect can be caused by local oxidation in the NiTi driving to overvoltage or, simply, short-circuit by wire-twisting production. The presence of sparks is not dangerous unless the fluid passing through the valve is dry, filtered air, but this situation cannot be accepted with noninert gases; besides, the connection between copper and NiTi wires (Figure 4.22) cannot be assured after almost 300 cycles. For these reasons, the valve was redrawn to improve the assemblability and to obtain more efficient electrical connections. The NiTi wires can be connected directly to the copper wires; hence, in the second prototype of the valve, the metallic blocks (Figures 4.23 and 4.24) are removed.

The improved prototype is realized with only two long NiTi wires connected directly to the copper wires, as shown in Figure 4.25. The NiTi wires in the new prototype are twice the length of the NiTi wires in the old prototype: this preserves the stroke of the valve pin.

As already described, the first step for valve activation consists in the plan of an analog circuit for feeding (Figures 4.26–4.27). This receives the

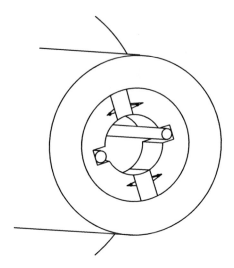

FIGURE 4.23
Particular of metallic blocks and fixing of the wires.

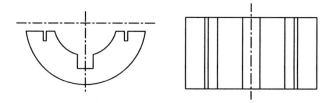

FIGURE 4.24
Metallic clamp.

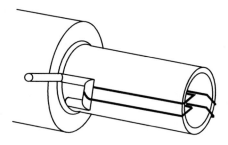

FIGURE 4.25
The new connection between copper and NiTi wires.

electrical grid (AC 230 V with 50–60 Hz frequency) as input and supplies constant current I_{cost}, changing the amplitude of the current step with an appropriate potentiometer. In this way, the proper current is fed for the activation of NiTi wires.

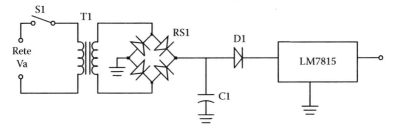

FIGURE 4.26
Constant current feeding.

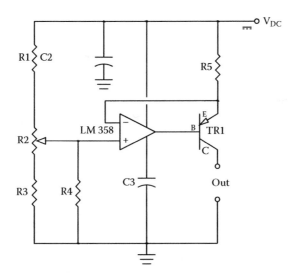

FIGURE 4.27
Representation of the second part of the constant current circuit.

The complete representation of the electric circuit is reproduced at the end of this section in Figure 4.36. The first part of the circuit rectifies the alternating voltage wave. After the transformer T1, a complete Wheatstone bridge with four 1N4148 diodes rectifies the half-waves (Figure 4.26). After the diode bridge there is a pulse voltage 50 Hz that is leveled with an electrolytic high-K capacitor C1; the rectified voltage V_{cc}, at the pins of the electrolytic capacitor C1, is described approximately by equation (4.6), where 0.7 is the mean value of the rectifier-diode forward voltage, and V_a is the net voltage.

$$V_{CC} = V_a \cdot 1.41 - 0.7 \tag{4.6}$$

The second part of the analog circuit is able to change the gain of the amplifier stage using a constant-current source, as shown in Figure 4.27. The pin in the noninverting input of the operational amplifier LM358 is connected

TABLE 4.5

Electrical Characteristics of Some NiTi Alloys from Different Producers

Typology of NiTi Wires	Diameter (µm)	Resistance (Ωm)	Typical Current (mA)	Force (g)
B	37	860	30	20
C	50	510	50	35
D	100	150	180	150
E	150	50	400	330
F	250	20	1000	930

with the resistance cursor R2, whereas the inverting pin is connected with the emitter of the transistor TR1. The constant current detected by this circuit depends on the feeding voltage and on the resistance R5 applied on the transistor emitter, as described in (4.7), where V_{CC} is the feeding positive-voltage value, V_{in} is the voltage applied on noninverting input of the operational amplifier and can be controlled by the R2 trimmer. The voltage V_{CC} depends on voltage V_a, as shown in (4.6), and the component LM7815 levels the voltage to the fixed value equal to 15 V.

$$I = \left(V_{CC} \cdot V_{in} \right) / R5 \qquad (4.7)$$

Some characteristics of different NiTi wires are listed in Table 4.5. These data are affected by environmental conditions, different factors associated with the crystal structure of the alloy, alloy impurities, and the type of electrical feeding. Hence, some tests with the analog feeding circuit are required to try a parameter adjustment.

Hence, after a series of preliminary tests, acceptable behavior was obtained using a constant current I_{cost} equal to 240 mA for 500 ms on a wire with a length of 10 cm and a diameter equal to 50 µm. With this current amplitude, the valve opens immediately because the NiTi wire heats instantly until the temperature of the phase changes and it contracts; whereas the martensitic phase cannot be easily recovered by cooling the NiTi wires, because the recovery time is at least 500 ms with these parameters in calm air at ambient temperature; consequently, the activation frequency is almost 1 Hz. The next set of trials is devoted to increasing the activation time of the wire under dry and filtered airflow conditions, with a square wave generator (changing the current gain) focusing the optimization of the frequency. Further, the compressed air effect is positive on the recovery and cooling of the wire owing to convective phenomena.

Referring to Figure 4.28, the square wave generator has a duty cycle DC equal to the ratio between T_{ON} and the sum of T_{OFF} and T_{ON}, and is adjustable to permit the activation and deactivation of the feeding circuit with a desired frequency f and a fixed current I_{cost}.

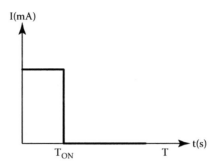

FIGURE 4.28
Square waveform with zero logical level.

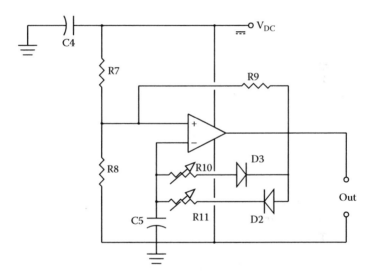

FIGURE 4.29
Scheme of analog square wave generator.

The two resistances $R10$ and $R11$ with the two diodes put in reversed polarity, connected between output and the inverting operational pin (Figure 4.29), permit us to obtain a generator that supplies the square wave with a non-symmetrical duty cycle (Figure 4.28) with the positive half-wave different from the negative half-wave. Hence, the following equations (4.8–4.9) can be used to estimate the time T_{ON} and T_{OFF} in milliseconds, where the resistance values are expressed in $k\Omega$ and the capacitors' value in μF.

$$T_{ON} = \ln 2 \cdot \left(R11 \cdot C5 \right) \tag{4.8}$$

$$T_{OFF} = \ln 2 \cdot \left(R10 \cdot C5 \right) \tag{4.9}$$

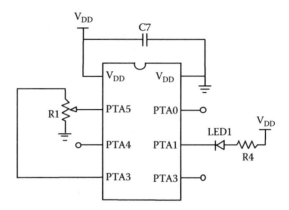

FIGURE 4.30
Motorola MC68HC908-QT4 microchip.

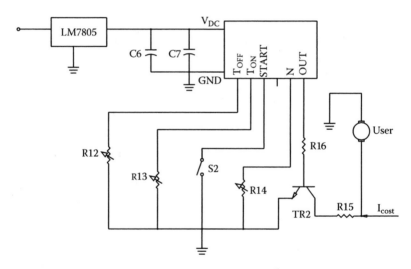

FIGURE 4.31
Digital circuit scheme with Motorola MC68HC908-QT4 (*N* is the number of pulses).

The output square wave takes a value equal to zero when the logical state is OFF and is equal to the feeding voltage when the logical state is ON.

With the analog circuit, the frequency and the duty cycle adjustments are not linear. As a result, two groups of trimmers, $R10$ and $R11$, are set to change the T_{ON} and T_{OFF} values with proper calibration, whereas the $C5$ capacitor should be removed and changed to obtain reasonable accuracy on the duty cycle DC. Therefore, a digital circuit (Figure 4.31) with a Motorola chip QT4 (Figure 4.30) is assembled to improve the accuracy of the generated signal and to systematically manage the test parameters.

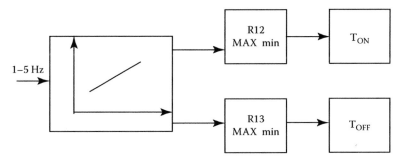

FIGURE 4.32
Part of block diagram of the main program.

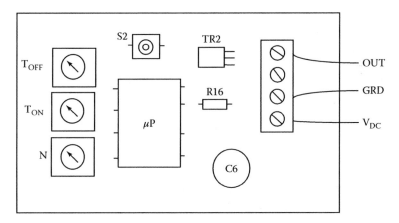

FIGURE 4.33
Electronic card with components listed in Table 4.6.

The duty cycle can be controlled independent of the frequency with a dedicated program for the Motorola microchip; besides, the cycle number (or maximum number of pulses) can be easily set, the maximum and minimum values of time T_{ON} and T_{OFF} are calculated and allocated to bounds of the trimmers $R12$ and $R13$, which represent the two variables. Also, the frequency can be controlled by fixing the bound values between 1 and 5 Hz and then allocating the values of the two variables T_{ON} and T_{OFF} (Figure 4.33). By default, the bound values of the two variables are 100 and 500 ms; they can then be modified by the resistances $R12$ and $R13$ without recompiling the flash memory (Table 4.6).

In the example (Figure 4.32), the cycle number lies between a minimum value of 10 and a maximum value equal to the value assumed by the variable *CH_NR_CICLI* multiplied by 4: the trimmer ($R14$), as shown in Figure 4.31, permits fixing the cycle number in the range defined by the chip QT4. The program puts the circuit on hold until switch $S2$ is activated. In this case the

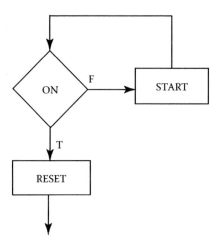

FIGURE 4.34
Particulars of block diagram of the main program.

function of *S2* is a starter, whereas with a working circuit, *S2* acts as a reset, which is necessary for an emergency stop (Figure 4.34).

The chip is programmed by a 9-pin serial gate connected to the computer through a serial-USB converter cable. The digital circuit gives some important advantages; i.e., it is easily interfaceable with measurement, so that the circuit can be integrated with a feedback system for dynamic real-time control to improve the accuracy of the device (Figure 4.35).

TABLE 4.6

List of Used Components in the Feeding Electrical Circuit

Symbol	Value	Symbol	Value
T1	Transformator 15 W (nTN01.32) sec.18 v 0.8 A	D2, D3	Diode 1N4148
S1, S2	Switch	R10, R11	From equation
RS1	Diode bridge 1N4148	R8, R9, R7	100 kW
D1	Diode led (red)	Op. Amp. 2	LM358
C1	100 nF	C4	100 nF
LM7815	Chip 15 V	C5	From equation
R1	1kW	C2, C3	100 nF
R2	Trimmer	R12, R13, R14	Trimmer
R3	1kW	R15	100 W
R4	10^3 kW	R16	10 W
R5	From equation	C7	0.1 μF
C2, C3	100 nF	C6	10 μF
Op. Amp. 1	LM358	LM7805	Chip 5 V
TR1	Power transistor pnp type BDX54C	TR2	BD237

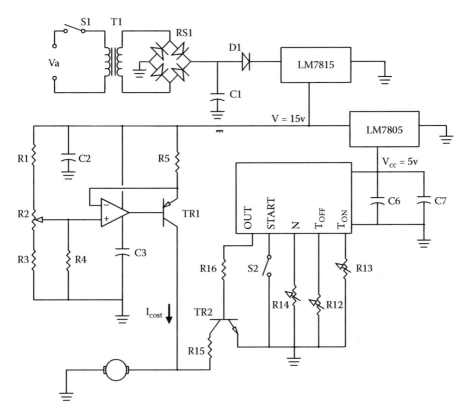

FIGURE 4.35
Complete scheme of the electrical circuit.

4.4.2 Measurement

The distinction between static (or quasi-static) and dynamic measures is fundamental: the first is realized by the system at regime conditions, when the measured physical quantity does not change (or changes very slowly), whereas the latter is realized during the transitory when the measured physical quantity changes its value. The static tests are performed to find the output flow and pressure of the valve, when it is opened by the electric circuit for a time that permits to establish a regime condition, whereas the variables of the dynamic test are specified later. The difference between the real value of quantity (input signal) and the measured value depends on the behavior of the instrument; hence, proper technical data are specified for every instrument that is part of the test bench for the valve, that is, the nonlinearity (representing the maximum difference between the indication of measure and the straight line that linearizes the curve of the instrument, expressed as a percentage of the measurable maximum value), the hysteresis (the effect of the load history), the repeatability, the accuracy, and the sensibility threshold.

4.4.3   Experimental Setup

The test bench is assembled in four main blocks, as shown in Figure 4.36: the electric feeding circuit of the valve, already described in the previous section, permits supply of a constant current step I_{cost} to the SMA wires with frequency f and an adjustable duty cycle for a fixed number of cycles; the pneumatic feeding permits generation of compressed air and regulation of the input pressure P_1 of the dry and filtrated air; the measuring instruments, the pressure transducer and the mass flow meter, respectively, for the pressure P_2 and the flow Q at the output of the valve; and, finally, the logic system for data elaborations.

The pneumatic equipment used in the test bench works with compressed air produced by a generator group. The air should have specific characteristics to determine the static and dynamic characteristics of the valve and to avoid damage to the instruments. The air is filtered and dehumidified as required by the technical norms. A volumetric compressor (P_{max} = 8 bar maximum regime pressure, W = 1.5 kW regime power) is used to compress the air, whereas an *FRL* group (filter–reducer–lubricator) cleans, filters, and lubricates the air (Figure 4.37).

Then, a manometer is fixed at a maximal pressure equal to 6 bar to avoid damage to the instruments and to maintain a constant pressure at the input regulator; hence, compressed air is supplied from the feeding pneumatic block at 6 bar, dehumidified, and filtered at 20 μm (maximal dimension of the particulate in the air).

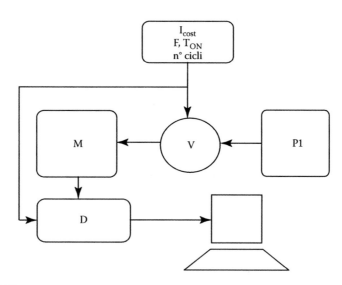

FIGURE 4.36
Block schema of the test bench.

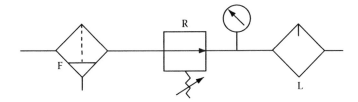

FIGURE 4.37
FRL group, symbols from norm ISO 1219-1; F = filter, R = reducer, L = lubricator.

4.4.4 Pressure Measurement

Pressure is one of the most important variables for pneumatic systems and, as described in the following, when the value of pressure is compared and related to the value of local ambient pressure, it is called relative or effective pressure and is usually measured by manometers, whereas the absolute pressure is associated with the zero-conditions pressure (the absolute pressure is used for computing the flow through the valve and the flow capacity). In the industrial field, the airflow is expressed qualitatively as standard pressure (Figure 4.38), referring to the atmospheric pressure as described in ISO 8778 (Pneumatic fluid power—Standard reference atmosphere).

The spring force F_m and the pressure P operating on the plug act on the valve system, as described in equation 4.3, against the force F exerted by SMA wires. The spring force can be estimated as a linear function of the spring elongation, whereas the force due to phase transition is valued experimentally. A proportional pressure regulator can change the input pressure P_1 of the air to realize static or dynamic tests. This instrument permits us to control the pressure with a 3% accuracy, within a 0.2–10 bar pressure range and with a 0.1 bar resolution (Table 4.7). The input pressure can be considered

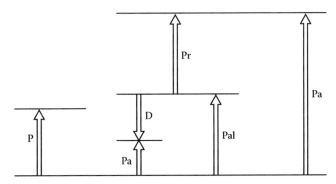

FIGURE 4.38
Pressure levels: P is the normal pressure, Pa is the absolute pressure (the biggest value of Pa is associated with a pressure in the plant, whereas the lowest value is associated with a depression in the plant), D is the depression, Pr is the relative pressure, Pal is the absolute local pressure.

TABLE 4.7

Functional Characteristic of the Proportional Regulator of Pressure

Technical Data	
Fluid	Dehumidified and filtered air at 20 µm
Minimum input pressure	P_{OUT} + 1bar
Maximum input pressure	10 bar
Ambient temperature	$-5° - +50°C$
Output pressure	0.2 – 9 bar
Nominal flow rate (6 bar Δp 1 bar)	1100 Nl/min
Output flow	1300 Nl/min
Air consumption	< 1 Nl/min
Weight	360g
Feeding voltage	24 VDC ± 10%
Signal reference (voltage)	0–10 VDC
Linearity	≤ 1%
Hysteresis	≤ 1%
Repeatability	≤ 1%
Sensitivity	≤ 1%
Resolution	0.1 bar

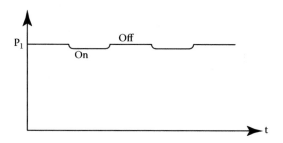

FIGURE 4.39

Qualitative loss of input pressure load due to opening of the valve.

approximately constant, whereas, when the valve is opened, there is a deflection of the feeding pressure, as shown qualitatively in Figure 4.39. If the deflection is excessive, a pneumatic capacitor can be inserted in the system.

The output pressure of the valve is measured by a transducer with high accuracy and stability (accuracy class equal to 0.05%) (Table 4.8).

4.4.5 Flow Measurement

The flow rate of compressed air into the valve represents the quantity of gas that passes through a reference section in unit time. Two kinds of flow rates are commonly used: the mass flow and the volume flow. The mass flow rate G represents the mass quantity that passes through a reference section in unit time (kg/s), whereas the effective volumetric flow rate Q (m^3/s or l/s)

TABLE 4.8

Functional Characteristics of the Transducer

Technical Data	
Relative pressure	10 bar
Absolute pressure	50 bar
Linearity and hysteresis	≤ 0.05%
Temperature effect (1°C)	≤ 0.005%
Feeding voltage	24 VDC ± 10%
Signal reference (voltage)	0–10 VDC

represents the volume of compressed air and is related to the mass flow rate *G* and the compressed air density ρ as shown in equation 4.10.

$$Q = G/\rho \tag{4.10}$$

The volume normal flow rate Q_N (4.11) represents the air quantity that passes through a reference section in unit time expressed in standard conditions of the air: 20°C and 65% of relative humidity as defined by the standard ISO 8778.

$$Q_N = G/\rho_N \tag{4.11}$$

The normal quantities are measures in normal units denoted by the symbol *N*, (i.e., the normal-liter per minute is *Nl/min*, whereas the dual normal-cube meter per minute is Nm^3/min). A mass flow with a low differential pressure is employed for measuring the flow rate (Table 4.9) and the result is converted into a digital RS232 signal, or into an analog 0–10 V signal.

This measurement process is based on the anemometer principle, and the effective volume flow is measured by a thermal sensor. As shown in Figure 4.40, the heater produces the known pulses of heat, and when the flow passes through, the heaters measure the variations of temperature,

TABLE 4.9

Functional Characteristics of Mass Flow

Property	Value
Full scale range	0.02 to 50 Nl/min (N_2 equivalent)
Operating media	Neutral
Maximum operating pressure	10 bar
Linearity	± 1% FS
Accuracy	± 0.3% FS
Repeatability	± 1% FS
Power supply	24 VDC ± 10%
Output signal (digital)	RS232

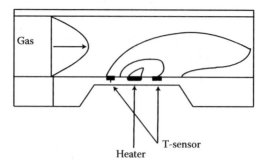

FIGURE 4.40
Flow measurement.

TABLE 4.10

Average Values of the Flow Rate and
Flow Coefficient C for Different Pressures
with a Data Acquisition Time of 30 s

Input Pressure (bar)	Average of the Flow Rate Q_{med} (Nl/min)	C
1.2	10.0	78.8
1.3	11.5	73.6
1.4	12.8	70.4
1.5	16.2	79.1
1.6	17.3	76.4
1.7	17.5	71.0
1.8	20.5	77.0
1.9	23.1	81.1
2.0	24.5	80.8

computing the flow speed v. Therefore, the flow rate Q can be calculated using (4.10) or (4.11) with the characteristics of the fluid, the section A, and the density ρ, whereas the mass flow can be described by (4.12).

$$G = \rho \cdot v \cdot A \tag{4.12}$$

The advantage of this method is that it is unaffected by variations of pressure or temperature; besides, it offers high accuracy and sensitivity as described in Table 4.10. The flow-measuring instrument may be located upstream or downstream of the test section and is used to determine the true time-averaged flow rate with a 2% accuracy.

4.4.6 Installation of Test Specimen

A correct installation of the instruments in the test bench is a required step to obtain the proper measurement results. The instruments are connected

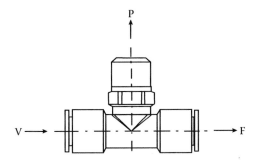

FIGURE 4.41
Pneumatic output connections of the valve: P is the pressure gauge, F the mass flow, and V the valve.

to the pneumatic system by quick couplings, limiting the variations of section as much as possible. The international standard EN 60534-2-3 advises considering all couplings using the piping geometry factor F_p into the equations for the determination of the flow capacity C (see EN60534 norm for accurate definitions).

As shown in the international standards, the test section should consist of two straight lengths of pipe. The upstream and downstream piping adjacent to the test specimen should conform to the nominal size of the test specimen connection. The inside surface should be free from obstructions, which may cause excessive flow disturbance. Moreover, the alignment between the centerline of the test section piping and the centerline of the inlet and outlet of the test specimen should be within a tolerance of 1 mm. For the static tests, the valve shall be oriented so that the flow pattern does not produce a velocity head at the pressure tap, as shown in Figure 4.41, and the valve stroke shall be fixed with an accuracy ± 0.5 during any flow test.

4.4.7 Data Acquisition System

In section 4.4.1 the parameters of the digital circuit, numbers of cycles, T_{ON} and T_{OFF}, are set electronically; now, for a graphical comparison with the output pressure, we acquire the data from mass flow and the square wave from the valve-feeding circuit. A data acquisition card is used for the acquisition of analog signals from the mass flow, the transducer, and feeding circuit, whereas a dedicated software package is required for the analysis. The connections between the card and the three signals are highlighted in the following patterns Figures 4.42–4.44.

The acquired signal is unsettled because of noise and interference. Thus, the highest frequencies are filtered (Figure 4.45) and the signal is reconstructed with a fast Fourier transform (FFT) function (Figures 4.46–4.47; Numerical Recipes, 2002). The Johnson noise or white noise cannot be filtered, but it does not change the physical meaning of the sensing result.

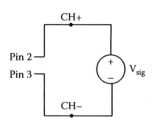

FIGURE 4.42
Connection card–transducer signal.

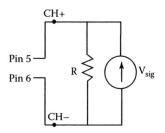

FIGURE 4.43
Connection card–feeding-step signal.

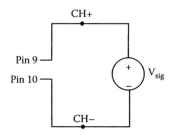

FIGURE 4.44
Connection card–mass flow signal.

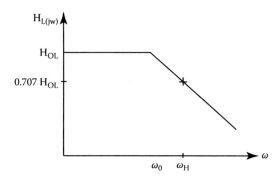

FIGURE 4.45
Low-pass filter.

4.5   Valve Characterization

Standard and recognizable parameters should be selected to study the real behavior of the valve; the equations listed in the following text identify the relationships between flow rates, flow coefficients, related installation factors, and service conditions, allowing the computation of characteristic times for dynamic tests.

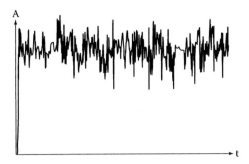

FIGURE 4.46
FFT function and noise peaks.

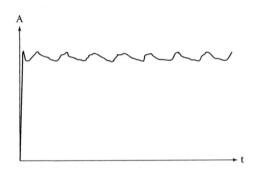

FIGURE 4.47
Nonfiltered signal and signal with lower cut off at 10 Hz.

4.5.1 Flow Capacity C

The International Standard suggests the flow capacity C as a comparison parameter to characterize the prototype and to compare it with commercial valves. It can be expressed in two ways: as K_V if the reference standard is the European Standard EN and the units belong to S.I., or as C_V if the reference standard is the National Bureau of Standard NBR. Different factors are shown in the following to identify the relationship between flow rate, flow coefficients, related installation conditions, and service state. The ratio x between differential pressure and inlet absolute pressure (4.13) is the ratio between ΔP and upstream pressure P_1, where ΔP is the differential pressure between upstream and downstream pressure taps. The differential pressure ratio factor of a control valve without attached fittings for choked flow is deduced approximately from standard tables and is calculated with (4.25). Then, the specific heat ratio factor $F\gamma$ is the ratio between the specific heat ratio γ and the value 1.40.

$$x = \frac{\Delta P}{P_1} \tag{4.13}$$

The expansion factor Y (4.14) describes the density change and is affected by the ratio between the port area and the inlet area of the valve body, path of the flow, pressure differential ratio x, Reynolds number, and specific heat ratio γ. When x is lower than the product $F\gamma\, x_T$, the flow coefficient C can be calculated with (4.15), where M is the molecular mass of the flowing fluid, Z the compressibility factor, T_1 and P_1 are, respectively, the inlet absolute temperature and pressure, Q is the volumetric flow rate, and N_9 is a numerical constant from Table 1 of the standard EN 60534-2-1. On the other hand, when x is equal to value $F\gamma\, x_T$, there is a choked flow condition: with fixed inlet conditions, the maximum flow rate Q_{MAX} is at a limit condition, whereas in a choked flow condition, Y is set at 0.667 and the flow coefficient C is computed with (4.16).

$$Y = 1 - \frac{x}{3 \cdot F\gamma \cdot x_T} \tag{4.14}$$

$$C = \frac{Q}{N_9 \cdot P_1 \cdot Y} \sqrt{\frac{M \cdot T_1 \cdot Z}{x}} \tag{4.15}$$

$$C = \frac{Q}{0,667 \cdot N_9 \cdot P_1} \sqrt{\frac{M \cdot T_1 \cdot Z}{F\gamma \cdot x_T}} \tag{4.16}$$

The flow is laminar when the fluid in motion is fluent, otherwise it is turbulent. The motion depends on a combination of different factors such as viscous effects and the inertia of particles, and can be associated with the value of the Reynolds number (Re) calculated with equation (4.17), where N_4 and N_2 are constants from Table 1 of standard EN 60534-2-1, v is the kinematic viscosity, and v-air value is $14.61 \cdot 10^{-6}$ m^2/s. The valve-style modifier F_d and the recovery factor F_L are identified in Table 2 of the same standard and depend on the type of valve.

$$Re = \frac{N_4 \cdot F_d \cdot Q}{v \cdot \sqrt{C_I \cdot F_L}} \left(\frac{F_L^2 \cdot C_I^2}{N_2 \cdot D^4} + 1 \right)^{\frac{1}{4}} \tag{4.17}$$

F_L is the pressure recovery factor of the valve without attached fittings, and it describes the influence of the internal geometry of the valve on its capacity at choked flow. It is defined as the ratio of the actual maximum flow rate under choked flow conditions and the theoretical, nonchoked flow rate when the differential pressure is equal to the difference between the valve inlet pressure and the apparent vena contracta pressure at the choked flow condition. On the other hand, the valve style modifier F_d describes the effect of geometry on the Reynolds number. It is defined by the ratio of the hydraulic diameter of a single flow passage to the diameter of a circular orifice having an area equivalent to the sum of areas of all identical flow passages at a given travel. F_d can be estimated with (4.18) for a single-seated valve with parabolic plug and closing flow, where D_0 is the diameter of the orifice and N_{23} a constant. The factor F_R is associated with the Reynolds number when laminar flow conditions are established through a control valve because of a low differential pressure, high viscosity, very small flow coefficient, or a combination of them.

$$F_d = \frac{1,13 \cdot \sqrt{N_{23} \cdot C \cdot F_L}}{2D_0 \cdot \dfrac{N_{23} \cdot C \cdot F_L}{D_0}} \tag{4.18}$$

$$F_P = \frac{1}{\sqrt{1 + \dfrac{\Sigma \zeta}{N_2}\left(\dfrac{C}{d_2}\right)^2}} \tag{4.19}$$

The piping geometry factor F_P describes the fittings attached upstream and downstream to a control valve body. The F_P factor is the ratio of the flow rate through a control valve installed with attached fittings and the flow rate of the same valve without attached fittings under identical conditions that do not produce choked flow (Figure 4.48). This piping geometry factor F_P can be quantified with (4.19), where N_2 is a numerical constant from Table 1 of

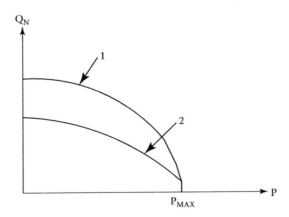

FIGURE 4.48
Qualitative characteristic of a pneumatic valve with (2) and without (1) fittings.

standard EN 60534-2-1, d is the nominal valve size, and the factor $\Sigma\zeta$ is the algebraic sum of the effective velocity head-loss coefficients of the fittings attached to the control valve. The factor $\Sigma\zeta$ may be calculated with (4.20), where ζ_{B1} and ζ_{B2} are the inlet and outlet Bernoulli coefficients, when the piping diameters approaching and leaving the control valve are different (4.21). If the inlet and outlet fittings are short-length concentric reducers, ζ_1 and ζ_2 may be approximated with equation (4.22). Finally, D_1 and D_2 are, respectively, the internal diameters of upstream and downstream piping.

$$\Sigma\zeta = \zeta_2 + \zeta_1 + \zeta_{B1} - \zeta_{B2} \tag{4.20}$$

$$\zeta_{B1} = 1 - \left(\frac{d}{D_1}\right)^4 \ ; \quad \zeta_{B2} = 1 - \left(\frac{d}{D_2}\right)^4 \tag{4.21}$$

$$\zeta_1 = 0.5\left[1 - \left(\frac{d}{D_1}\right)^2\right]^2 \ ; \quad \zeta_2 = \left[1 - \left(\frac{d}{D_2}\right)^2\right]^2 \tag{4.22}$$

$$x_{TP} = \frac{\dfrac{x_T}{F_P^2}}{1 + \dfrac{x_T\zeta_1}{N_5}\left(\dfrac{C}{d^2}\right)^2} \tag{4.23}$$

The flow capacity C can account for the attached fittings with some considerations. The value of x_T is affected by attached fittings in choked flow conditions; thus, x_{TP} can be calculated with (4.23), where x_T is the pressure differential ratio factor of the valve without attached fittings, F_P is the piping geometry factor, N_5 is a constant, and ζ_1 is the upstream velocity head-loss coefficient of fittings, as shown in (4.22). Hence, the flow coefficient, C, should be recalculated with equation (4.15), replacing x_T with the new value x_{TP}, to obtain improved accuracy.

4.5.2   Example of Sizing Calculation

The subject of sizing calculation is a valve with a parabolic plug and with closing flow (NC); hence, the valve-style modifier F_d, the recovery factor F_L, and the pressure differential ratio factor x_T are extracted from the Table 2 of the standard EN 60534-2-1, and the value of the specific heat ratio factor $F\gamma$ for air is 1.

The ratio of the differential pressure and the inlet absolute pressure x (4.24) is less than $F\gamma x_T$, which is equal to 0.70. Therefore, the flow is normal and the flow coefficient, C, is computed with (4.25), where the value of the expansion factor, Y, is 0.762 (4.14). The molecular mass of air, M, is 28.97 kg/kmol, the compressibility factor Z and F_P are approximately equal to one, and the constants N_9, N_2, and N_4 are extracted from Table 1 of the referred standard. Then, the Reynolds number is computed with (4.17), and if the valve Re is greater than 10,000, the flow is turbulent. Hence, the equations 4.21–4.22 are employed to quantify the effective head loss coefficient of the inlet and the outlet reducers $\Sigma\zeta$, and to find a convergent solution for F_P. Finally, x_{TP} can be calculated with equation (4.23) to improve the flow coefficient C with a more accurate estimation.

$$x = \frac{\Delta P}{P_1} = 0.50 \qquad (4.24)$$

$$C = \frac{Q}{N_9 \cdot P_1 \cdot Y \cdot F_P} \sqrt{\frac{M \cdot T_1 \cdot Z}{x}} = 80.8 \qquad (4.25)$$

4.5.3 Time Response

The response time of the pneumatic valve is obtained from control tests with a square-wave signal recording the variation of the output signal (Figure 4.49). The ISO 12238 standard (Pneumatic fluid valves—Directional control valves—Measurement of shifting time) fixes the rules to establish the tests to find the response function of the system, and it defines the main measurable time.

Referred to a positive step of constant current, shown by a digital oscilloscope, the activation time t_1 is the time that passes between the application of the electrical signal and the 10% attainment of the output pressure. P_2 is directed at a horizontal asymptote y_∞; similarly, the deactivation time t_2 due to the response when the signal is removed is defined as the time that passes between the removal of the electrical signal and 90% of the output pressure.

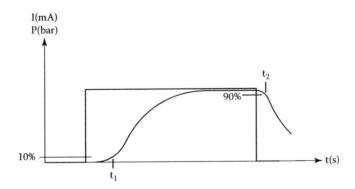

FIGURE 4.49
Qualitative trends of the step control signal and its response.

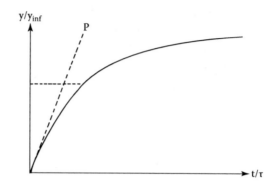

FIGURE 4.50
Step response of a first-order instrument.

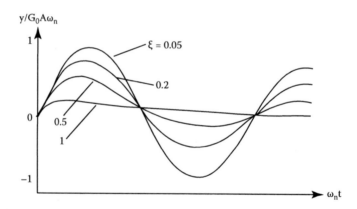

FIGURE 4.51
Step response of a second-order instrument.

The behavior of the valve can be described as a step response of first-(Figure 4.50) or second-order (Figure 4.51) systems. For a first-order system, the input can be given by (4.26), where $u(t)$ is the unary step function, and A is the amplitude of the same step that corresponds to the amplitude value of the constant current I (Figure 4.52) used for the activation. Hence, the transfer function for the output is (4.27), where G_0 is the static gain and τ is the time constant. The time trend of the step response (4.28) is obtained by the inverse transform of the equation (4.27), as shown in (4.29).

$$x(t) = Au(t) \qquad (4.26)$$

$$Y(s) = \frac{G_0 A}{s(1 + s\tau)} \qquad (4.27)$$

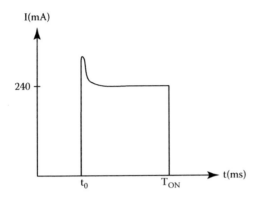

FIGURE 4.52
Step of the real current I.

$$y(t) = G_0 A(1 - e^{-t/\tau}) = y_\infty(1 - e^{-t/\tau}) \tag{4.28}$$

$$y_\infty = G_0 A \tag{4.29}$$

For a second-order system where the input is described by (4.26), the Laplace transfer function, referred to the output, is shown in (4.30), where ξ is the damping ratio that quantifies the damping of the oscillating system and ω_n is the natural frequency of the undamped natural system. Applying the inverse transformation to (4.30), the solution (4.31) describes the trend of the step response with $\xi = 1$. For a value of ξ included between 0 and 1, the step response is defined in (4.32), where α is described in (4.33).

$$Y(s) = \frac{G_0 A \omega_n^2}{s(s^2 + 2\xi\omega_n s + \omega_n^2)} \tag{4.30}$$

$$y(t) = G_0 A \cdot \left[1 - \left(1 + \omega_n t \right) \cdot e^{-\omega_n t} \right] \tag{4.31}$$

$$y(t) = G_0 A \cdot \left[1 - \frac{e^{-\xi\omega_n t}}{\sqrt{1 - \xi^2}} \sin\left(\omega_n \sqrt{1 - \xi^2} t + \alpha \right) \right] \tag{4.32}$$

$$\alpha = \arctan\left(\frac{\sqrt{1 - \xi^2}}{\xi} \right) \tag{4.33}$$

The response time and the activation–deactivation times can be obtained after the acquisition of the signal to characterize the speed of the valve.

4.6   Measures and Observations

According to recommendations in the standards, some static and dynamic tests are performed to examine the real behavior of the pneumatic meso-valve. The static test allows us to determine experimentally the flow rate through the valve under different working conditions and to estimate the loss of load and the flow behavior. On the other hand, the dynamic test permits us to establish the response time and the maximum working frequency, controlling the step response and adjusting the actuation frequency of the device. All the tests are executed at the ambient temperature, conventionally, $T_1 = 293$ K.

4.6.1   Tests in Calm Air

Some tests in calm air are performed in order to verify the seal of the constraints and to find the optimal constant current for the described feeding device. Therefore, no signals are acquired in these tests, but only the displacement of the plug is observed. Changing the potentiometer manually from the feeding circuit, the least current for the martensite phase activation is estimated, that is, the energy to control the NiTi wires is found. These simple tests show that the typical current is not sufficient to realize the total strokes in a period of time of the order of a second. An activation with a current of 50 mA requires about 5 s; besides, this becomes gradual change after 3.5 s.

Increasing the current with a single step to 240 mA for T_{ON} equal to 0.5 s in calm air, the valve opens immediately (50 ms), but the closing of the plug is slow. The cooling time is significantly reduced by the compressed air effect, as described in the next section.

Undesired behaviors can be observed when the activation and deactivation of the valve are repeated with a fixed frequency. A set of 20 square waves with a frequency of 1 Hz, current I_{cost} equal to 240 mA and duty cycle 50% breaks the connections between the NiTi wires and the metallic clamps in the first prototype. The reason is the diameter reduction of the NiTi wires while they change the phase from the martensite to the austenite, and hence, the wires come out of the clamps. If the adhesion is implemented with plastic materials or resinaceous glue, the temperature of the phase change should be properly evaluated.

The new prototype solves the just-described problem, constraining the NiTi wires and the plug with a dap joint using a conical pin and constraining the NiTi wires and the copper wires with a hinge. The tests in calm air are repeated with I_{cost} 240 mA and T_{ON} 0.5 s for 100 cycles at frequency 1 Hz; hence, a behavior similar to that shown in Figure 4.53 can be observed in the last cycles. The NiTi wires in calm air do not completely cool down, therefore, for every cycle after the first, the starting temperature T_0 is equal to the

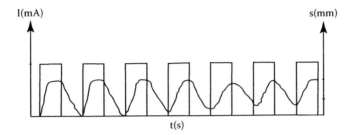

FIGURE 4.53
The current signal and the displacement of the plug with a frequency f of 1 Hz.

value T_1 increased by ε. This situation slows down the recovery and reduces the displacement.

To observe this deleterious effect, after a pause of 5 min, the tests are repeated, doubling the working frequency, and at the 80th cycle the valve begins malfunctioning. As expected, the displacement of the plug becomes shorter and shorter until the NiTi wires receive the electric signal, but the plug does not close the valve. The signal of the displacement has the same shape of a completely damped signal (Figure 4.54). After the limit condition, if the electrical circuit continues to supply energy on the connection between the copper wires and the NiTi wires, some sparks form and at this moment the SMA is definitively damaged. For all these reasons, the valve with the SMA is unable to work in calm air with high frequencies at the ambient temperature; because the cooling is too slow. However, the behavior of the valve is improved by the circulation of compressed air through the body valve. In any case, the SMA valve can work also in calm air, by reducing the duty cycle or changing the frequency. The results of some tests establish that the valve can work in calm air with a frequency of 2 Hz, a current I 240 mA, and for a number of repeated cycles less than 100; a good duty cycle results with T_{ON} equal to 100 ms and T_{OFF} equal to 400 ms.

Different tests are executed, increasing the amplitude of the current up to 500 mA; the NiTi wires increase their speed, but the recovery time becomes

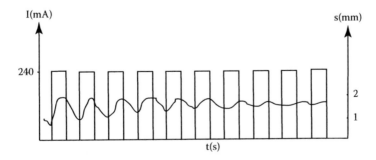

FIGURE 4.54
The current signal and the displacement of the plug with a frequency f of 2 Hz.

worse, whereas good valve behavior can be obtained by reducing T_{ON} and the frequency (i.e., with a frequency of 0.5 Hz and T_{ON} equal to 100 ms).

4.6.2   Static Tests

The static tests for the prototype are executed with the experimental setup shown in Figure 4.55, where the instruments have been described in section 4.4. In relation to Figure 4.55, the valve V is controlled by the electrical signal C, and it has an almost constant pressure during the test. However, when the plug opens the valve, there occurs a slight pressure decrease; therefore, the electronic proportional regulator R should be opportunely calibrated to prevent undesired movements of the plug. The regulator R works also as a transducer of the input pressure P_1, whereas the output pressure P_2 is controlled by the pressure transducer T connected crosswise to the airflow direction, to avoid the velocity load. The flow rate Q is measured by the mass flow F, and the linking pipes are suggested by the standard ISO 6358 (Pneumatic fluid power—Determination of flow-rate characteristics of components using compressible fluids). The standard suggests the use of some pipe diameters (Φ 4, Φ 6, and Φ 8 mm) for an easy connection with the other pipes without pressure loss and for standard roughness conditions. The data acquisition system M records the signal C, the output pressure P_2, the feeding pressure P_1, and the flow rate Q at an ambient temperature T_1 of 293 K, showing the effective flow rate Q through the valve for the fixed upstream pressure P_1 and a measured downstream pressure P_2. Different tests are performed using the NiTi wires with a diameter Φ equal to 50 μm. Controlling the pressure P_1 and the flow rate Q, the relative pressure P_2 is changed from 0.2 bar to the maximum admissible differential pressure.

The recovery time, or rather the cooling time, of the NiTi wires becomes short using compressed air, and the input pressure helps the closing of the plug. Both these effects highlight the inverse proportionality between the pressure P_2 and the recovery time. On the other hand the cooling, owing to compressed air, continues during the activation; therefore, part of the energy supplied during the transient phase between austenite and martensite is lost, i.e., with a pressure P_2 higher than 0.6 bar, the energy supplied by the current

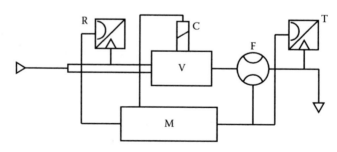

FIGURE 4.55
The experimental setup for the static tests.

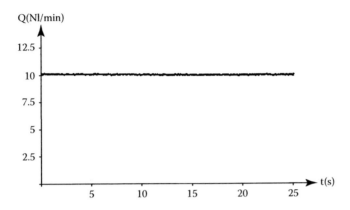

FIGURE 4.56
Flow rate Q due to the overpressure P_1 equal to 0.2 bar.

I_{cost} 240 mA for 200 s is not able to activate the wire. Furthermore, the plug displacement increases when the supplied current is increased, maintaining the same T_{ON}.

The quasi-static tests at regime conditions allow the estimation of the flow rate. When the valve is activated, it is maintained open until the regime condition occurs and recorded for 30 s. The activation current is fixed at 900 mA with an overpressure P_2 of 0.2 bar (corresponding to an absolute pressure of 1.2 bar). The signal from the mass flow is recorded, as shown in the Figure 4.56; hence, the input pressure is changed until it reaches the limit differential pressure.

The procedure described in section 4.5 is used to calculate the flow coefficient C for the resulting values. The ratio of differential pressure x is always lower than the product of $F\gamma$ and x_T, where the differential pressure ratio factor x_T is assumed to be equal to 0.70 according to Table 2 of standard EN 60534-2-1; hence, the flow is nonchoked, while the expansion factor Y and the flow coefficient C can be approximately calculated with (4.14) and (4.15). The Reynolds number, Re, (4.17) establishes the type of the flow and, in our situations, is turbulent; then, the flow coefficient should be corrected with the geometrical factor $\Sigma\zeta$, the algebraic sum of the effective velocity coefficients of head loss due to fittings attached to the valve, which is equal to 0.971. The last value of Table 4.10 is, relative to a differential pressure between upstream and downstream pressure taps, equal to 1.0 bar. The NiTi wires with a diameter of Φ 50 μm are not able to overcome for a long time the effects of the pressure and of the spring. As shown in Figure 4.57, there is a loss of load after 22 s, or rather, the plug closes the valve, although the SMA is in the martensite phase (contracted wires). This is the limit condition in which the wires go beyond the yield point and then the wires are damaged irreparably. The diagrams (Figures 4.58–4.64) represent the average value from the various tests shown in Table 4.10 in quasi-static conditions with a data acquisition time of 30 s.

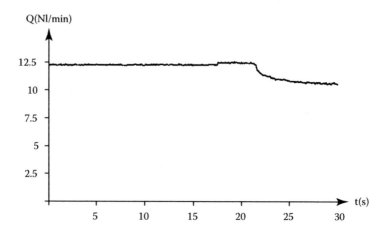

FIGURE 4.57
The flow rate Q due to the overpressure P_1 of 1.0 bar: the boundary condition.

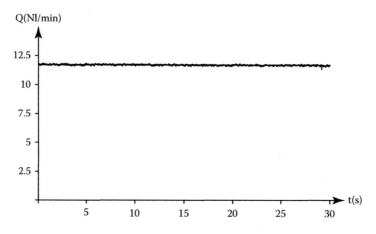

FIGURE 4.58
Test with a pressure P_1 equal to 1.3 bar and an average flow rate Q_{med} equal to 11.5 Nl/min.

The flow rate Q through the mesovalve can be shown in Figure 4.65 as a function of the input pressure P_1 with an output pressure P_2 at 1 bar. The flow rate Q increases progressively until the pressure P_1 becomes equal to 1.6 bar. This is a critical condition associated to the ratio between P_1 and P_2, and over this value, the flow is sonic.

The flow coefficient C is shown in Figure 4.66 and, from a fluid dynamic viewpoint, good performance can be achieved with the relative pressures of 0.2, 0.5, 0.9, and 1.0 bar, whereas a drop appears for pressures equal to 0.4 and 0.7 bar.

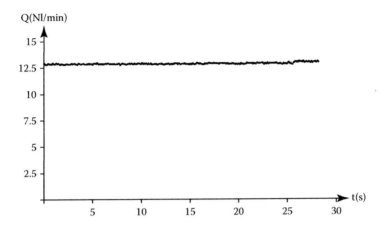

FIGURE 4.59
Test with a pressure P_1 equal to 1.4 bar and an average flow rate Q_{med} equal to 12.8 Nl/min.

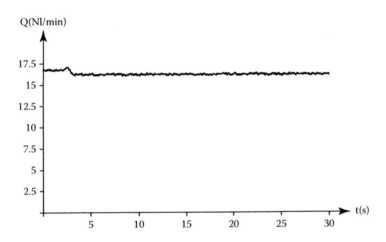

FIGURE 4.60
Test with a pressure P_1 equal to 1.5 bar and an average flow rate Q_{med} equal to 16.2 Nl/min.

To compare the prototype with a commercial valve, some technical data are listed for the mesovalve 2/2 actuated by NiTi wires (Table 4.12) and for a mesovalve 2/2 actuated by solenoid (Table 4.11)

The static characteristics of the SMA valve are generally equivalent to those of the commercial valves, and sometimes exhibit an improved flow coefficient. Furthermore, some technological aspects can be highlighted, that is, the low-cost molding manufacture, low weight, reduced size, and the limited number of components. Using wires with a 50 µm diameter, the maximum input pressure is 2.0 bar and the effective flow rate is under 25 Nl/min.

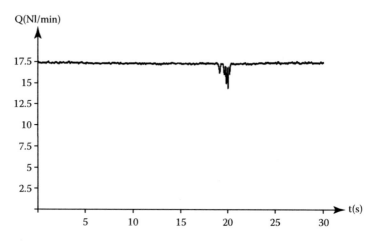

FIGURE 4.61
Test with a pressure P_1 equal to 1.6 bar and an average flow rate Q_{med} equal to 17.3 Nl/min.

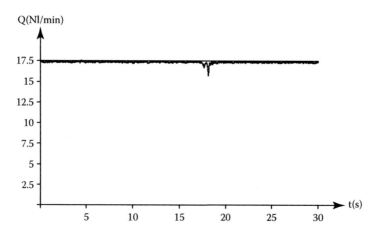

FIGURE 4.62
Test with a pressure P_1 equal to 1.7 bar and an average flow rate Q_{med} equal to 17.5 Nl/min.

Some of these dynamic characteristics are summarized in Table 4.13 and are shown in the next section.

4.6.3 Dynamic Tests

Dynamic parameters can be important for some applications involving a fast system (i.e., the inertial actions cannot be neglected). After performing the static characterization of the mesovalve and its flow, the following tests examine the working velocity and the response, adjusting the input pressure and the electrical inputs, i.e., the current amplitude, the times T_{ON} and T_{OFF} of the feeding square wave, the frequency, and the duty cycle. At the

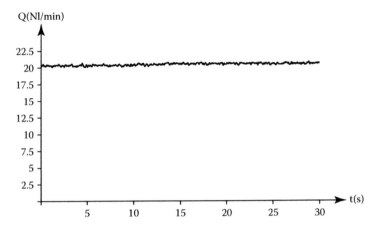

FIGURE 4.63
Test with a pressure P_1 equal to 1.8 bar and an average flow rate Q_{med} equal to 20.5 Nl/min.

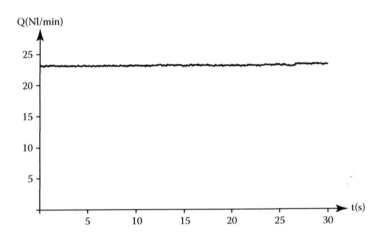

FIGURE 4.64
Test with a pressure P_1 equal to 1.9 bar and an average flow rate Q_{med} equal to 23.1 Nl/min.

beginning, the value of the variables just examined in the calm air tests are set to constant values (I_{cost} 240 mA with a frequency of 1 Hz and a duty cycle equal to 50%). The pressure regulator starts from a relative pressure of 0.2 bar. The cooling of NiTi wires is instantaneous due to the compressed air, although the quantity of the circulating air is low. Furthermore, the displacement of the plug with a step current activation of I_{cost} 240 mA for 100 ms is lower than the displacement observed in calm air tests. Thus, the valve remains almost closed because a lot of Joule energy is dissipated by the air flow. Therefore, in the following tests a higher electric energy is supplied, (i.e., a current step of 500–900 mA, otherwise the duty cycle, the frequency, or the time T_{ON} can be modified).

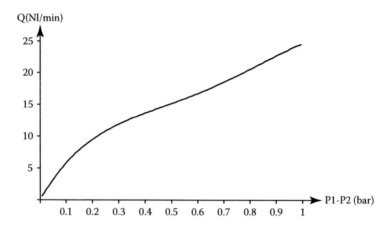

FIGURE 4.65
The P–Q curve of the valve with pressure P_2 of 1 bar.

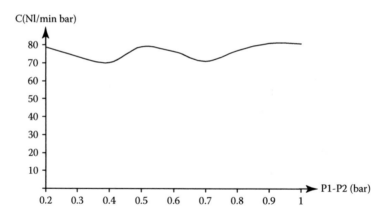

FIGURE 4.66
The flow coefficient C in function of the pressure.

TABLE 4.11

Technical Data of a Solenoid Valve

Valve	Medium	Function	Voltage	Size (mm)	Ø (mm)	C_v	Flow Nl/s ($\Delta p = 1$ bar)
Type 1	Air	2/2 NC	24 V DC	22 × 45	2.5	0.09	1.92
Type 2	Air	2/2 NC	24 V DC	20 × 55	1.6	0.04	0.91

TABLE 4.12

Technical Data of the SMA Valve from the Static Tests

Medium	Function	Operation	Ø (mm)	C_v	K_v	Flow Nl/s ($\Delta p = 1$ bar)
Air	2/2 NC	Direct acting	2.0	0.09	1.35	0.41

TABLE 4.13

Dynamic Properties of the Mesovalve with an Input Pressure P_1 of 1.2 Bar for 100 Cycles

Feeding Parameters		Dynamic Responses of the SMA Valve (ms)	
I_{cost}	500 mA	t_0	100
T_{ON}	200 ms	t_1	15
T_{OFF}	450 ms	t_∞	80
DC	30%	t_R	450
f	1.5 Hz		

The purpose of these tests is to analyze the characteristic response times due to a current step activation. The analytic procedure was described in section 4.5.3, following the activation times and the response times are indicated with the working conditions as the input and output pressure, the step current amplitude I_{cost}, and its duration T_{ON}.

The experimental setup used in these tests is schematically described in Figure 4.67; the input pressure is constant, but when the plug opens the valve, the input pressure tends to undergo a reduction; hence, the pressure regulator R should be properly calibrated to avoid this undesired behavior. If the loss of pressure is excessive, a suitable pneumatic capacitor should be inserted to reduce the pressure variation. The pressure P_2 is measured by

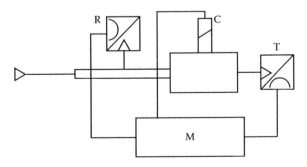

FIGURE 4.67
Electric circuit for dynamic measures.

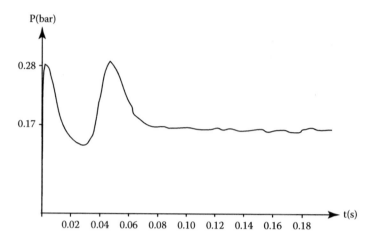

FIGURE 4.68
Pressure response due to a step current with amplitude I_{cost} 900 mA and time T_{ON} 0.9 s.

the transducer T connected directly at the output section of the valve, and the sizes of the pipes are fixed by the standard ISO 6358 (Pneumatic fluid power—Components using compressible fluids—Determination of flow-rate characteristics). The data acquisition system M records the control signal C, the output pressure signal P_2, and the input pressure P_1 at the ambient temperature T_1 293 K. The tracking of the step signal is displayed with these following parameters: I cost 900 mA with a single step for T_{ON} 0.9 s and P_1 equal to 0.2 bar. We can highlight the overshooting of the signal (Figure 4.68), which is acquired with a sampling rate of 1000 data/s. The response of the system is really fast with the proposed configuration; in fact the activation time t_1 (Figure 4.71) is under 5 ms, while the response time t_∞ (after this time the pressure is constant) is equal to 80 ms. Unfortunately the feeding system produces a 900 mA current for an exposition time higher than 0.5 s and this overloading energy leads to wire damages, although the wires are constantly cooled by the compressed air. Due to these damages the NiTi wires lose their functionality remaining in the contracted shape (opened valve); hence the next experiments are performed with a feeding current lower than 900 mA.

Namely, other tests are executed with a feeding current I_{cost} equal to 500 mA for a step time T_{ON} (Figure 4.52) equal to 200 ms. Then the dynamic behavior of the system is shown in the Figure 4.69 after a noise filtration as described in section 4.4.7; the activation time t_1 is equal to 10 ms and the tracking line to the step signal presents an overshoot y_{max} less than 8%. As a matter of fact, the plug is subjected to vibrations with a reduced amplitude. The response time, which can be seen as the approximate contraction time, remains 80 ms; furthermore, adjusting the amplitude current, it should be not higher than 100 ms with a 20% margin of safety for the phase change.

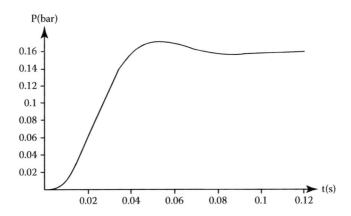

FIGURE 4.69
Step response with P_1 set at 1.2 bar, I_{cost} at 500 mA, and T_{ON} at 200 ms.

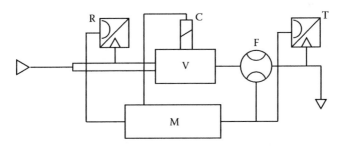

FIGURE 4.70
Experimental setup for the frequency tests.

The response time of a generic commercial valve is almost 15 ms, whereas the SMA mesovalve is slower; hence, a different NiTi alloy is used to obtain a response time of 50 ms.

After the static characterization and the measure of the response time to a step signal of current, the next examined aspect is the frequency behavior of the valve. The previous static-test configuration is used also to observe the behavior of the plug under a fixed command frequency Figure 4.70. The current signal C is characterized by a current amplitude fixed with a potentiometer, and by a duty cycle and by a number of cycles set with a digital electric circuit.

The plug movement is related to the flow through the instrument F at the input pressure P_1 set by the regulator R and represents the most time-consuming phase of the valve activity. Although the connection pipes between the output of the valve and the mass flow are limited as much as possible, there is a dead time t_0 between the electrical activation and the beginning of the wire contraction owing to the gap between the valve and the point of measure of the instrument. Consequently, the activation time

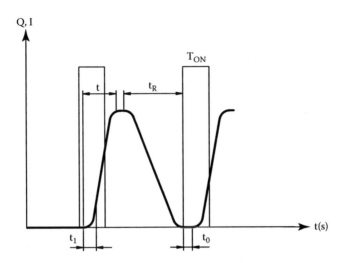

FIGURE 4.71
Main time of the current signal *I* and of the flow rate signal *Q*.

t_1, the response time t_∞, and the plug return time tR also are examined, as shown schematically in Figure 4.71.

An activation with I_{cost} of 500 mA and T_{ON} 200 ms is used to analyze the complete dynamic behavior of the valve, disregarding the suggestions from the last tests. At the beginning, T_{OFF} is fixed at 800 ms with a frequency of 1 Hz, because a slow return of the plug is expected. Then, the time T_{OFF} is optimized up to 450 ms, the working frequency is 1.5 Hz, and the duty cycle DC is almost 30%. The number of cycles is fixed by the resistance R14 at 100 impulses of current. This value is comparable with other different scientific studies (Ikuta, 1990; Johnson, Kraemer, 1998; Leppäniemi, 1998; Velazquez, Hafez, Szewczyk, Pissaloux, 2005) that agree upon the degradation of SMA properties after 100 cycles.

The input pressure P_1 is fixed at 1.2 bar by the digital regulator. In Figure 4.72 some of the main cycles are drawn according to the specified parametric value. Sometimes the opening of the plug is not complete; in fact, its step has a reduced amplitude. This phenomenon specifically happens only during the first 10 cycles; then, the wires reach a regime average temperature that prevents incomplete openings.

Two interesting aspects of the dynamic behavior can be mentioned: the former is positive and is related to the behavior at the activation phase (the opening of the plug), which is repeatable and substantially constant; in fact, the average dead time t_0 of 100 ms, the activation time t_1, and the response time t_∞ are the same as the time extracted with the response tests. The latter is negative and is related to the return of the wires, and the cooling is a complex process. Although the global return time tR is 450 ms, the path corresponding to the plug movement is not repeatable and not linear in the first 100 analyzed cycles.

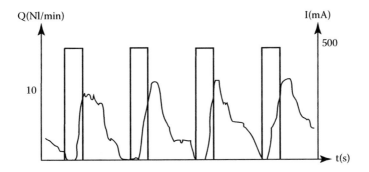

FIGURE 4.72
Some on–off cycles: the light line represents the plug considering the flow rate through the mass flow and the thicker line represents the feeding signal (I_{cost} is 500 mA, T_{ON} is 200 ms, and T_{OFF} is 450 ms).

These tests are performed with two different loads: a mechanical one and a pure thermal one, the former due to the sum of the spring force and the pressure due to the input air flow, and the latter due to the phase changes from martensite to austenite and vice versa. When the SMA is subject to groups of thermomechanical cycles, different properties are influenced. From a metallurgic viewpoint, the fatigue degrades the material by softening it, and a more elastic behavior is noted when the valve is unassembled and reassembled using the wires that have already worked for almost 100 cycles, producing micro-cracks (Velazquez et al., 2005), and this effect is proportional to the quantity of supplied energy. In fact, these phenomena are as intense as the increase of the T_{ON} and I_{cost} values. To avoid or delay this degradation, an in-depth study of the fatigue problem should be performed.

Therefore, the reliability of the realized SMA devices depends on the performance of the global life cycle: time, temperature, stresses, type of application, deformations, and number of cycles are the most important external parameters that bias the cycle life. On the other hand, the biasing internal parameters are the alloy composition and the thermal–mechanical treatments. Hence, to obtain the maximum memory effect, the stresses and the limit deformations should be selected as a function of the number of working cycles (Figure 4.71). D. Stoeckel (Leppäniemi, 1998) shows some data concerning the fatigue of the SMA that can be used as a driveline for the standard NiTi fibers.

References

Airoldi G., Besseghini S., Riva G. (1995) Smart behaviour in a CuZnAl single crystal alloy, Transactions of the International Conference on Martensitic Transformations, *Journal de Physique* IV 5, Part 2, Session 4, Paper 12.

Ampère A. M. (1820) Conclusions d'un mémoire sur l'action mutuelle de deux courans électriques, sur celle qui existe entre un courant électrique et un aimant, et celle de deux aimans l'un sur l'autre, *Journal de Physique, de Chimie, d'Histoire Naturelle et des Arts* 91, Paris, 76–78.

Bansevicius R. Tolocka R. T. (2002) Piezoelectric actuators in Bishop R. H. Ed., *The Mechatronics Handbook*, CRC Press, Boca Raton, FL.

Bellezza F., Lanari L., Ulivi G. (1990) Exact modelling of the flexible slewing link, *Proceedings of the IEEE International Conference on Robotics and Automation*, 734–739.

Beyer J., Koopman B., Besselink P. A., Willemse P. F. (1990) Fatigue properties of a TiNi-6%Cu shape memory alloy, *Proceedings of the 6th International Conference on Martensitic Transformations*, Trans Tech Publications, Clasuthal, Germany, 773–778.

Bhat R. B., Lopez-Gomez A. (2001) *Advanced Dynamics*, CRC Press, Boca Raton, FL.

Bhattacharyya A., Faulkner M. G., Amalraj J. J. (2000) Finite element modeling of cyclic thermal response of shape memory alloy wires with variable material properties, *Computational Materials Science* 17, Elsevier, 93–104.

Biot J. B., Savart F. (1820) Experiéncies électromagnétiques sur la mesure de l'action exercée à distance sur une particule de magnétisme, par un fil conjonctif, *Journal de Physique, de Chimie, d'Histoire Naturelle et des Arts*, Paris, 151.

Bo Z., Lagoudas D. C. (1999a) Thermomechanical modeling of polycrystalline SMAs under cyclic loading, Part I: Theoretical derivations, *International Journal of Engineering Science* 37, Pergamon, 1089–1140.

Bo Z., Lagoudas D. C. (1999b) Thermomechanical modeling of polycrystalline SMAs under cyclic loading, Part II: Material characterization and experimental results for a stable transformation cycle, *International Journal of Engineering Science* 37, Pergamon, 1141–1173.

Bo Z., Lagoudas D. C. (1999c) Thermomechanical modeling of polycrystalline SMAs under cyclic loading, Part III: Evolution of a plastic strain and two-way shape memory effect, *International Journal of Engineering Science* 37, Pergamon, 1175–1203.

Bo Z., Lagoudas D. C. (1999d) Thermomechanical modeling of polycrystalline SMAs under cyclic loading, Part IV: Modeling of minor hysteresis loops, *International Journal of Engineering Science* 37, Pergamon, 1175–1203.

Book W. J. (1984) Recursive Lagrangian dynamics of flexible manipulator arms, *International Journal of Robotics Research* 3 (3), 87–101.

Book W. J. (1990) Modeling, design and control of flexible manipulator arms: A tutorial review, *Proceedings of the IEEE Conference on Decision and Control*, 500–506.

Brailovski V., Trochu F., Daigneault G. (1996) Temporal characteristics of shape memory linear actuators and application to circuit breakers, *Materials and Design* 17, Elsevier, pp. 151–158.

Brinson L. C., Huang M. S. (1996) Simplifications and comparisons of shape memory alloy constitutive models, *Journal of Intelligent Material Systems and Structures*, SAGE Publications, 108–114.

Brinson L. C. (1993) One dimensional constitutive behavior of shape memory alloys: Thermomechanical derivation with non-constant material functions, *Journal of Intelligent Material Systems and Structures* 4 (2), SAGE Publications, pp. 229–242.

Buehler W. J., Wiley R. C. (1965) Nickel-Based Alloys. Technical report, U.S. Patent 3,174,851.

Cannon R. H. Jr., Schmitz E. (1984) Initial experiments on the end-point control of a flexible one-link robot, *International Journal of Robotic Research* 3 (3), pp. 62–75.

Cetinkunt S., Book W. J. (1987) Symbolic modeling of flexible manipulators, *Proceedings of the IEEE International Conference on Robotics and Automation*, pp. 2074–2080.

Chang L. C., Read T. A. (1951) Plastic deformation and diffusionless phase changes in metals: The gold-cadmium beta phase, *Transactions of AIME* 189, pp. 47–52 Cincinnati, OH.

Coulomb C. A. (1785–1789) Memoires sur l'electricite et le magnetisme, Memoires de l'Academie Royale des Science de Paris.

Cowper G. R. (1966) The Shear coefficient in Timoshenko's beam theory, *Journal of Applied Mechanics* 33, pp. 335–340.

Dassa L., Amadori D. (2006) Particular types of flexure hinges—Development and application, *5th International Conference on Advanced Engineering Design*, June 11–14, 2006, Prague, Czech Republic.

Davis, J. R. (1993) Aluminum and aluminum alloys, *ASM Specialty Handbook*, ASM International.

De Araujo C. J., Morin M., Guenin G. (1999) Electro-thermomechanical behaviour of a Ti-45.0Ni-5.0Cu (at.%) alloy during shape memory cycling, *Materials Science Engineering* A 273–275, Elsevier, pp. 305–309.

De Luca A., Siciliano B. (1991) Closed-form dynamic model of planar multilink lightweight robots, *IEEE Transactions on Systems Man and Cybernetics* 21(4), pp. 826–839.

Ding Z., Lagoudas D. C. (1999) Transient heat transfer behavior of one-dimensional symmetric thermoelectric SMA actuators, *Mathematical and Computer Modelling* 29, Pergamon, pp. 33–55.

Einstein A. (1905) Zur Elektrodynamik bewegter Korper, *Annalen der Physik* 17, pp. 891–921.

Eringen, A. C. (1964) Simple microfluids, *International Journal of Engineering Science* 2, pp. 205–217.

Eringen, A. C. (1999) *Microcontinuum Field Theories* 1, Springer-Verlag, New York.

Faraday M. (1821) On some new electro-magnetical motions, and the theory of magnetism, *Quarterly Journal of Science* 12, pp. 75–96.

Faraday M. (1832) Experimental researches in electricity, *Philosophical Transactions of the Royal Society of London* 122, pp. 125–194.

Ferraris G. (1888) Rotazioni elettrodinamiche prodotte per mezzo di Correnti alternate, Torino.

Furuya Y., Shimada H., Matsumoto M., Honma T. (1989) Fatigue and degradation of shape memory effect in NiTi wire, *Proceedings of the MRS International Meeting on Advanced Materials Vol. 9: Shape Memory Materials*, pp. 269–274.

Georgiu H. M. S., Ben Mrad R. (2004) Experimental and theoretical assessment of PZT modelled as RC circuit subject to variable voltage excitations, *Mechatronics* 14, pp. 667–674.

Gilbert W. (1600) Guilelmi Gilberti Colcestrensis De Magnete, magnetiticisque corporibus, et de megnete tellure: physiologia nova, plurimus argumentis, experimentis demonstrata, Londini, excudebat P. Short.

Greninger A. B., Mooradian V. G. (1938) Strain transformation in metastable copperzinc and beta copper-tin alloys, *Transactions of AIME* 128, pp. 337–368.

Grob H., Hamann J., Wiegärtner G. (2000) *Elektrische Vorschubantriebe in der Automatisierungstechnik: Grundlagen, Berechnung, Bemessung*, Publicis MCD Corporate Publishing, chapter 4.

Gyimesi M., Ostergaard D. (1999) Electro-mechanical transducer for MEMS analysis in ANSYS, *Technical Proceedings of the 1999 International Conference on Modeling and Simulation of Microsystems*, pp. 270-273.

Gyimesi M., Wang J.-S., Ostergaard D. (2001) Hybrid p-element and Trefftz method for capacitance computation, *IEEE Transactions on Magnetics*, pp. 3680-3683.

Heartling G. (1994) Rainbow ceramics—A new type of ultra-high-displacement actuator, *American Ceramic Society Bulletin* 73, pp. 93-96.

Henry J. (1831) On account of a large electro-magnet, made for the laboratory of Yale college, *American Journal of Science and Arts* 20, pp. 201–203.

Hertz H. R. (1887) Uber sehr schnelle elektrische Schwingungen, *Annalen der Physik und Chemie* 31, pp. 421–448.

Histand M. B., Alciatore D. G. (1999) *Introduction to Mechatronics and Measurement Systems*, WCB/McGraw-Hill, Boston.

Homna D., Miwa Y., Iguchi M. (1984) Application of shape memory effect to digital control actuator, *Bulletin of the Japan Society of Mechanical Engineers*; 27(230), pp. 1737–1742.

Howell L. L. (2001) *Compliant Mechanisms*, John Wiley and Sons, New York.

Hutchinson J. R. (1981) Transverse vibrations of beams, exact versus approximate solutions, *Journal of Applied Mechanics* 48, pp. 923–928.

Ikuta K. (1988) Application of shape memory alloy actuator for clean gripper, *Proceedings of the 7th RO.MAN.SY Symposium* Udine, Italy.

Ikuta K. (1990) Micro/miniature shape memory alloy actuator, *Proceedings of the IEEE International Conference on Robotics and Automation*.

Ikuta K., Tsukamoto M., Hirose S. (1998) Shape memory alloy servo actuator system for application in an endoscope, *Proc 1998 IEEE International Conference on Robotics and Automation*, pp. 427–430.

Ionescu T. G. (2003) Terminology for the mechanism and machine science: Chapter 1, *Mechanism and Machine Theory* 38, Pergamon Press, pp. 767–776.

Jacobi M. H. (1835) Mémoire sur l'application de l'Electromagnétisme au Mouvement des machines, Mémoires de l'Académie de Saint Pétersbourg.

Jog C. S. (2002) *Foundations and Applications of Mechanics: Continuum Mechanics*, Volume I, CRC Press, Boca Raton, FL.

Johnson A. D., Kraemer J. (1998) State of the art of shape memory actuators, *Proceedings of the 6th UK Mechatronics Forum International Conference (Mechatronics '98)*.

Kasap S. O. (1997) *Principles of Electrical Engineering Materials and Devices*, McGraw Hill, New York.

Kaushal R. S., Parashar D. (2000) *Advanced Methods of Mathematical Physics*, CRC Press, Boca Raton, FL.

Kawashima H. (1996) The shear coefficient for quartz crystal of rectangular cross section in Timoshenko's beam theory, *IEEE Transactions on Ultrasonics, Ferroelectrics, and Frequency Control* 43, pp. 434–440.

Kumar Dwivedy S., Eberhard P. (2006) Dynamic analysis of flexible manipulators, a literature review, *Mechanism and Machine Theory* 41(7), pp. 749–777.

Lagoudas D. C., Kinra V. K. (1993) Design of high frequency SMA actuators, Disclosure of Invention Tamus 803, TAMU, College Station, TX.

Legnani G. (2003) Robotica Industriale, Casa Editrice Ambrosiana.

Legnani G., Tiboni M., Adamini R. (2002) Meccanica degli azionamenti, *Progetto Leonardo*, chapter 2.

Leitmann G., Udwadia F. E., Kryazhimskii A. V. (1999) *Dynamics and Control*, CRC Press, Boca Raton, FL.

Leppäniemi A. (1998) Shape memory alloy: Applications and commercial aspects, *Proceedings of the 6th UK Mechatronics Forum International Conference* (Mechatronics '98).

Liang C., Rogers C. A. (1990) One-dimensional thermomechanical constitutive relations for shape memory materials, *Journal of Intelligent Material Systems and Structures* 1(2), SAGE Publications, pp. 207–234.

Lobontiu N. (2003) *Compliant Mechanisms: Design of Flexure Hinges*, CRC Press, Boca Raton, FL.

Lynch W. A., Truxal J. G. (1962) *Principles of Electronic Instrumentation*, McGraw Hill, New York.

Malek J., Necas J., Rokyta M. (1998) *Advanced Topics in Theoretical Fluid Mechanics*, CRC Press, Boca Raton, FL.

Marion J. B. (1965) *Classical Dynamics*, Academic Press, New York.

Mavroidis C. (2002) Development of advanced actuators using shape memory alloys and electrorheological fluids, *Research in Nondestructive Evaluation* 14, pp. 1-32.

Maxwell J. C. (1865) A dynamical theory of the electromagnetic field, *Philosophical Transactions of the Royal Society of London* 155, pp. 459–512.

Meirovitch L. (1967) *Analytical Methods in Vibrations*, Macmillan, New York.

Meirovitch L. (1986) *Elements of Vibration Analysis*, McGraw Hill, New York.

Mesarovic M. D. and Takahara Y. (1975), *General Systems Theory: Mathematical Foundations*, Academic Press, New York.

Michell J. (1751) *A Treatise of Artificial Magnets*, Cambridge, printed by J. Bentham.

Murray R. M., Li Z., Sastry S. S. (1994) *A Mathematical Introduction to Robotic Manipulation*, CRC Press, Boca Raton, FL.

Niezrecki C., Brei B., Balakrishnan S., Moskalik A. (2001) Piezoelectric actuation: State of the art, *The Shock and Vibration Digest* 33, pp. 269-280.

Oersted H. C. (1820) Experimenta circa effectum conflictus electrici in acum magneticam, *Annales de chimie et de physique Paris*, Tome IV, pp. 417–425.

Olander A. (1932) An electrochemical investigation of solid cadmium-gold alloys, *Journal of American Chemical Society* 56, pp. 3819–3833.

Pacinotti A. (1865) Descrizione di una macchinetta elettromagnetica, *Nuovo Cimento* XIX, p. 378.

Pitteri M., Zanzotto G. (2002) *Continuum Models for Phase Transitions and Twinning in Crystals*, CRC Press, Boca Raton, FL.

Potapov P. L., da Silva E. P. (2000) Time response of shape memory alloy actuators, *Journal of Intelligent Material Systems and Structures* 11, SAGE Publications, pp. 125–134.

Readman M. C. (1994) *Flexible Joint Robots*, CRC Press.

Reynaerts D., Van Brussel H. (1998) Design aspects of shape memory actuators, *Mechatronics* 8(6), Pergamon pp. 635–656.

Rowland H. A. (1876) On a Magnetic Effect of Electric Connect, Physical papers of Henry Augustus Rowland, John Hopkins University.

Shigley J. E., Uicker J. J. (1980) *Theory of Machines and Mechanisms*, McGraw-Hill, New York.

Sittner P., Vokoun D., Dayananda G. N., Stalmans R. (2000) Recovery stress generation in shape memory $Ti_{50}Ni_{45}Cu_5$ thin wires, *Materials Science and Engineering* A286, Elsevier, pp. 298–311.

Smith S. T. (2000) *Flexures, Elements of Elastic Mechanisms: Elements of Elastic Mechanisms*, CRC Press, Boca Raton, FL.

Smits J. G., Dalke S. I., Cooney T. K. (1991) The constituent equations of piezoelectric bimorphs, *Sensors and Actuators A* 28, pp. 41–60.

Stylianou M., Tabarrok B. (1994) Finite element analysis of an axially moving beam: Part I, time integration; Part II, stability analysis, *Journal of Sound and Vibration* 178, pp. 433–481.

Tadikonda S. S. K., Baruh H. (1992) Dynamics and control of a translating flexible beam with prismatic joint, *ASME Journal of Dynamic, Systems, Measurement and Control* 114, pp. 422–427.

Talpaert Y. (2000) *Differential Geometry with Applications to Mechanics and Physics*, CRC Press, Boca Raton, FL.

Tanaka K., Kobayashi S., Sato Y. (1986) Thermomechanics of transformation pseudoelasticity and shape memory effect in alloys, *International Journal of Plasticity* 2, pp. 59–72.

Tesla N. (1888) U.S. Patents 381, 968–381, 969–382, 279–390, 415–390, 820.

Timoshenko S. P. (1921) On the correction for shear of the differential equation for transverse vibrations of bars of uniform cross-section, *Philosophical Magazine*, p. 744.

Tsuchiya K. (1983) Dynamics of a spacecraft during extension of flexible appendages, *Journal of Guidance* 6 (1), pp. 100–103.

Velazquez R., Hafez M., Szewczyk J., Pissaloux E. (2005) Experimental and computational thermomechanical study of a SMA micro-actuator: Aspect of antagonist-type behaviour, *3rd MIT Conference on Computational Fluid and Solid Mechanics*, Boston.

Vinogradov O. G. (2000) *Fundamentals of Kinematics and Dynamics of Machines and Mechanisms*, CRC Press, Boca Raton, FL.

Wang P. K. C., Wie J. D. (1987) Vibrations in a moving flexible robot arm, *Journal of Sound and Vibration* 116, pp. 149–160.

West T. (1993) *Continuum Theory and Dynamical Systems*, CRC Press, Boca Raton, FL.

Wickless W. J. (2004) *A First Graduate Course in Abstract Algebra*, CRC Press, Boca Raton, FL.

Wise S. A. (1998) Displacement properties of RAINBOW and THUNDER piezoelectric actuators, *Sensors and Actuators A* 69, pp. 33-38.

Wu X. D., Fan Y. Z., Wu J. S. (2000) A study on the variations of the electrical resistance for NiTi shape memory alloy wires during the thermo-mechanical loading, *Materials and Design* 21, Elsevier, pp. 511–515.

Yuh J. and Young T. (1991) Dynamic modeling of an axially moving beam in rotation: Simulation and experiment, *ASME Journal of Dynamic Systems, Measurement and Control* 113, pp. 34–40.

Zhou Y.-G., Chen Y.-M., Ding H.-J. (2005) Analytical solutions to piezoelectric bimorphs based on improved FSDT beam model, *International Journal of Smart Structures and Systems* 1, pp. 309–324.

Appendix A: Notation

Symbol	Meaning
$\dfrac{df}{dx_i}$	Total derivative of f with respect to x_i
$\dfrac{\partial f}{\partial x_i}$ $f_{,i}$	Partial derivative of f with respect to x_i
$\dfrac{df}{dt}$ $\dot{f}$	Total derivative of f with respect to time t
$\displaystyle\sum_i f_i$ f_i	Sum of the terms f_i; if the sum is inserted in a member of an equation, the series symbol can be omitted; this omission can be observed when the subscript i appears only in one of the two members of the equation
∇f	Operator transforming the scalar field f in a vector field having, as components, the partial derivatives of f with respect to the spatial variables
$[|*|]$	Jump condition
A	Point of the three-dimensional Euclidean space
a	Column vector $\overrightarrow{OA}$ composed of homogeneous coordinates in the reference frame (0)
a	Scalar acceleration
a	Vector acceleration
$a_{i(j)}$	Column vector $\overrightarrow{O_iA}$ composed of homogeneous coordinates in the reference frame (j)
$\begin{pmatrix} a_1 \\ a_2 \\ a_3 \\ a_4 \end{pmatrix}$	Column vector a in explicit form
$\begin{pmatrix} a_1 & {}_{i(j)} \\ a_2 & {}_{i(j)} \\ a_3 & {}_{i(j)} \\ a_4 & {}_{i(j)} \end{pmatrix}$ or $\begin{pmatrix} a_1 \\ a_2 \\ a_3 \\ a_4 \end{pmatrix}_{i(j)}$	Column vector $a_{i(j)}$ in explicit form
A	Area of the transversal section of a beam
$\underline{a}$	Lagrangian acceleration
a	Eulerian field acceleration
A	Area
A	Area vector
A*	Reduced area
A_s	Austenite start temperature

Symbol	Meaning
A_f	Austenite finish temperature
B	Magnetic field
B	Physical body
B	Left Cauchy–Green tensor
b	Body forces
C	Right Cauchy–Green tensor
c_0	Configuration of body B at time t_0
c_t	Configuration of body B at time t
c	Abbreviation of function cosine
C	Elastic tensor
C	Capacity
C	Compliance matrix
C	Compliance
c	Specific heat
c'	Equivalent specific heat
C_M	Slope of the borderline of the detwinned martensite field in the stress–temperature diagram
C_V	Heat capacity
C_v^i	Heat capacity per volume unit
C	Flow coefficient
C_F	Functional compliance
$\mathbf{C}_S$	Submatrix of **C** that describes the sensitivity of secondary movement
$\mathbf{C}_P$	Submatrix of **C** that describes the correlation between a secondary movement and the functional mechanical action
$\underline{C}_{KL}$	Deformation tensors of microcontinua
$\underline{\in}_{KL}$	Microdeformation tensor
D	Strain rate tensor
D	Elastic tensor
D	Three-dimensional vector of electric displacement
d	Third-order tensor of piezoelectric strain
D	Electric displacement
d**a**	Infinitesimal element of a surface
E	Green–Lagrange strain tensor
E_k	Kinetic energy
E	Young's modulus
E	Electric field
E_c	Potential energy
E	Three-dimensional vector of electric field density
e_r	Rotor's potential linked with rotor's current
e	Vector of electric potential
E	Electromotive tension
E_a	Austenite Young modulus
E_m	Martensite Young modulus
E	Dispersed energy
F	General physical property

Symbol	Meaning
F	Gradient tensor of deformation
F	Resulting forces action
$\mathbf{F}_{(i)}$	Resultant force computed with respect to the origin of the reference system
F_d	Valve style modifier
F_L	Pressure recovery factor
F_R	Reynolds number factor
F_P	Piping geometry factor
f	Direct transmission function of torque
f^{inv}	Inverse transmission function of torque
f	Number of phases
f_r	Natural frequency
f	Reduced friction coefficient of the system
F_0	Initial load
$\mathbf{F}$	Force
${}^3F_{V(i)}$	First three components of $\mathbf{F}_{(i)}$
F_{r1}, F_{r2}	Resistance forces
f_k	Forces per unit of mass
$\mathbf{F}$	Resulting internal force reduced to a surface force
g	Direct transmission function of angular speed
g^{inv}	Inverse transmission function of angular speed
G	Tangential elastic modulus
G	Mass flow rate
$\mathbf{g}$	Voltage matrix
g	Reduced backlash coefficient of the system
G_l	Electromagnetic momentum
h_i	Translation along the axis x_i
h	Helicoidal motion pitch
$\mathbf{H}$	Electromagnetic induction
H	Acceleration matrix
h	Internal heat generated per unit time
h	Thickness
h	Transmission coefficient
h	Convection coefficient
H	Heat convection coefficient
i	Electric current
i_r	Rotor's current
i_s	Stator's current
$I(3)$	Three-dimensional identity matrix
i_L	Load current
i_A	Armature current
i_F	Excitation current
$\mathbf{i}$	Vector of electric current
$\mathbf{I}$	Identity matrix
I_{ij}	Elements of submatrix ${}^3J$

Symbol	Meaning
I	Moment of inertia of the transverse section of a beam
I	Electric current
I_E	Excitation current
$\mathbf{I}$	Direction of wave
$\mathbf{l}$	Matrixes of directional cosines
i_{0KL}	Microinertia symmetric tensors in initial configuration
i_{kl}	Microinertia symmetric tensors during deformation
I_R	Set of static variables
I_D	Set of dynamic variables
$J_{(i)}$	Inertia matrix, or pseudotensor of inertia, referred to frame (i)
$^3\mathbf{J}$	Submatrix of $\mathbf{J}_{(i)}$
J_{ij}	Inertial terms
J_a	Total inertia of the actuating system at the base of the beam
J_L	Inertia moment of the load on the end of the beam
J	Indicator of the inertia of the beam
J_m	Motor's inertia
J_L	Load's inertia
$\mathbf{J}$	Current density
J	Current density
j_{km}	Microinertia tensor of micropolar
K_F	Stiffness
K_{s1}, K_{s2}	Secondary stiffness
K_{pi}	Parasitic stiffness
$\mathbf{K_p}$	Column vector of parasitic stiffness
$\mathbf{K_s}$	Symmetric matrix of secondary stiffness
k_{vert}	Stiffness
K	Kinetic energy per unit mass
K	Factor of conversion
K	Thermal conductivity
K	Constant slope
k	Reduced stiffness coefficient of the system
k_s	Constant of electric-mechanic transformation
L	Inductance
L	Matrix that describes the helicoidal axis of an infinitesimal displacement
L	Speed gradient tensor
L	Lagrangian
L_r	Inductance of rotor
L_s	Inductance of stator
l	Length of beam
$\mathbf{L_{effS}}$	Effective secondary loads
l_m	First-order couples per unit of mass
l_{mk}	Second-order couples per unit of mass
$\mathbf{L}$	Matrix of inductances
$\mathbf{L_{1,1}}$	Statoric inductances

Symbol	Meaning
$L_{2,2}$	Rotoric inductances
$L_{1,2}$	Mutual inductance
$L_{2,1}$	Mutual inductance
L_s	Self-inductance of a stator's winding
L_r	Self-inductance of a rotor's winding
M	Bending torque
m	Mass
M	Mass
m_{eff}	Effective mass
M_s	Martensite start temperature
M_f	Martensite finish temperature
M	Molecular mass
M_{ij}	Position matrix of (j) with respect to (i)
M_L	Mass of load on the end of beam
m_{kl}	Second-order tensor of stress moment
M_s	Mutual inductance of two stator's windings
M_r	Mutual inductance of two rotor's windings
M_{sr}	Mutual inductance between windings of stator and rotor
m_{kln}	Third-order tensor of stress moment
n	Normal unary vector
N	Negatively doped semiconductor
Ni	Magnetomotive force
O	Origin of absolute reference frame
O^i	Origin of i-th reference frame
P	Particle
p	Pitch of helix
P	Centroid of a point particle during deformation
P_0	Centroid of a point particle at initial configuration
p	Coordinates of a geometric point
$p_{j(j)}$	Point of body B_i represented in reference system (j)
p	Number of poles
P	Actuator load
P_t	Transferred power
P	Positively doped semiconductor
P_1	Upstream pressure
$P_{E,tot}$	Electric power
P_E	Net electric power
P_M	Mechanical power
q	Curvilinear abscissa
q_f	Generalized coordinates
q_k	Flow heat per time unit
Q	Matrix of helicoidal motion
Q	Electric charge
Q	Volumetric flow rate
Q_N	Volumetric normal flow rate

Symbol	Meaning
R	Rotation matrix
R	Radius
r	Constrain reaction per unit of area
r	Mechanical friction
R	Electric resistance
R	Matrix of electric resistances
R	Electric resistance
R_0	Resistance in the austenitic phase at the reference temperature T^R
Re	Reynolds numbers
r	Mechanical reduction ratio
R_A	Armature resistance
$\mathbf{R_s}$	Stator winding resistances
$\mathbf{R_r}$	Rotor winding resistances
R_r	Resistance of rotor
R_s	Resistance of stator
s	Curvilinear abscissa
S	Second-order tensor of mechanical strain
S	Surface of continuum body
S	Strain tensor
S	SMA layer
S_f	Part of surface S subjected to action **t**
S_u	Part of surface S subjected to action **r**
s	Vertical displacement
$\mathbf{s^E}$	Fourth-order tensor of elastic compliance
S_k	Poynting vector
s	Abbreviation of sine function
s	Coordinate on the neutral axis of a beam
s	Shift
s_{max}	Maximum realizable stroke
t	Time lapse
T	Second-order tensor of mechanical stress
T_d	Desired actuator torque
T	Generated torque
T_u	User torque
T_{du}	Desired user torque
t	Forces per unit of area
t	Surface forces
$T_{(i)}$	Resultant torque computed with respect to the origin of the reference system
T_m	Motor torque
T_r	Resistance torque reduced to the motor axis
T_{eff}	Torques' effective components
T	Temperature change function
T_0	Temperature of reference configuration
T_0	External temperature

Symbol	Meaning
T_{amb}	Temperature of the environment
T^R	Reference temperature
$\boldsymbol{u}_1, \boldsymbol{u}_2, \boldsymbol{u}_3$	Unary vectors associated with the three Cartesian axes of the absolute reference frame
$\boldsymbol{u}_1^i, \boldsymbol{u}_2^i, \boldsymbol{u}_3^i$	Unary vectors associated with the three Cartesian axes of the i-th reference frame
U	Potential function of conservative forces
$\mathbf{u}$	Speed of the surface ζ
$\mathbf{u}$	Vectorial displacement
$u(s,t)$	Transversal displacement of beam
$\mathbf{u_s}$	Secondary displacements
U_e or U_E	Electric potential energy
v	Scalar speed
$\mathbf{v}$	Vectorial speed
$\boldsymbol{V_R}$	Right stretch tensor
$\boldsymbol{V_L}$	Left stretch tensor
$\underline{\mathbf{v}}$	Lagrangian speed
$\mathbf{v}$	Eulerian field speed
V	Electric potential
V	Volume
V_c	Electromotive force
V_{IN}	Input tension
V	Polarization tension
V_{cc}	Rectified voltage
V_a	Net voltage
v_r	Rotor's electric potential
v_s	Stator's electric potential
W	Speed matrix
W	Spin tensor
W^E	Electromagnetic energy
W	Work
W_E	Electric power
W_M	Mechanical power
w	Translation of beam cross section
X	Set of geometric points
X_0	Reference points
$\boldsymbol{x'}$	Derivative of vector position x(s) in respect to curvilinear abscissa s
x_{gi}	Coordinates of the barycenter of a body
$\mathbf{x}_0$	Position vector of centroid P_0
$\mathbf{x}$	Position vector of centroid P
$\mathbf{x}_0'$	Position vector of a generic point in the neighborhood of a point particle in the initial configuration
$\mathbf{x}'$	Position vector of a generic point in the neighborhood of a point particle during deformation
X	Reactance

Symbol	Meaning
x_{OUT}	Output movement
x	Ratio of pressure differential
Y	Expansion factor
y_∞	Horizontal asymptote
Y	Thermodynamic force
y	Independent constitutive variables
Z	Dependent constitutive variables
Z	Compressibility factor
Z	Impedance
Z_E	Electrical losses
(0)	Absolute reference frame
$\mathbf{0}(3)$	Three-dimensional zero matrix
(i)	i-th reference frame
α_R	Theoretical position after reduction stage
α_L	Position of load
α_e	Theoretic position of rotor
α_0	Initial angle
α_N	Seeback coefficient
α_P	Seeback coefficient
γ	Heat ratio
γ	Creep factor
$\boldsymbol{\gamma}_{(i)}$	Angular momentum
Γ_0	Initial or reference configuration of continuum body
Γ_i	Final configuration of continuum body
$\Gamma_{(i)}$	Matrix of momentum of body B
Γ_{KLM}	Wryness tensor
δ_{Kl}	Shifter
δ_{kL}	Shifter
δ_{kl}	Kronecker symbol
δ	Displacement
Δs_{in}	Input motion
Δs_{out}	Output motion
ΔS	Entropy variation
$\Delta \boldsymbol{T}$	Translation vector
$\Delta \theta$	Angular displacement
$\Delta \alpha$	Angular variation
Δh	Displacement blocking force
ΔF	Blocking stroke force
ΔT	Temperature difference
Δh	Enthalpy
Δl	Extension
ΔP	Differential pressure
ΔR	Absolute change of resistance
$\Delta \rho$	Absolute change of resistivity

Symbol	Meaning
ε	Strain
$\boldsymbol{\varepsilon}_V$	Volumetric deformation tensor
$\boldsymbol{\varepsilon}_D$	Deviatoric tensor
$\boldsymbol{\varepsilon}^{\mathsf{T}}$	Second-order tensor of permeability
$\varepsilon_{\mathsf{KLM}}$	Levi–Civita symbol
ε	Internal energy per mass unit
ε	Dielectric constant
$\varepsilon_{\mathsf{air}}$	Dielectric constant of air
$\varepsilon_{\mathsf{r}}^{*}$	Relative permittivity
ε^{*}	Complex permittivity
ε_0	Vacuum permittivity
ε_{a}	Austenite strain
ε_{m}	Martensite strain
ε_{L}	Maximum residual strain
ζ	Discontinuity surface
η	Entropy density per mass unit
ϑ_{i}	Rotation around axis x_{i}
$\boldsymbol{\theta}$	Tensor of small spin
θ	Revolute degree of freedom
θ	Absolute temperature
θ_{i}	Initial angular position
θ_{f}	Final angular position
θ	Theoretical position of rotor
θ_{R}	Theoretical position after reduction stage
θ_{L}	Position of load
Θ_{a}	Ideal position of actuator
λ	Stretch
μ_0	Magnetic permeability of vacuum
ν	Poisson's coefficient
ν	Kinematic viscosity
ν_{k}	Microgyration tensor of first order
υ_{kl}	Microgyration tensor
$\boldsymbol{\xi}_{0\mathsf{i}}$	Position vectors inside a point particle at initial configuration
$\boldsymbol{\xi}_{\mathsf{i}}$	Position vectors inside a point particle during deformation
$\boldsymbol{\xi}_0$	Position vector inside a point particle of a microcontinuum of grade one at initial configuration
$\boldsymbol{\xi}_{\mathsf{i}}$	Position vector inside a point particle of a microcontinuum of grade one during deformation
ξ	Damping ratio
ξ	Martensite fraction
ξ_{S}	Detwinned martensite fraction
ξ_0	Initial martensite fraction
ρ	Mass density of a material
$\boldsymbol{\rho}_{(i)}$	Momentum

Symbol	Meaning
$^3\boldsymbol{\rho}_{(i)}$	First three components of $\boldsymbol{\rho}_{(i)}$
ρ_E	Electrical resistivity
ρ_0	Resistivity in the austenitic phase at the reference temperature T^R
σ_α	Cauchy's stress
σ	Stress tensor
σ_{MAX}	Maximum stress
σ^E_{kl}	Electromagnetic tensor
σ_a	Austenite stress
σ_m	Martensite stress
σ_S^{cr}	Martensite start critical stress
σ_f^{cr}	Martensite finish critical stress
ζ_{B1}	Inlet Bernoulli coefficient
ζ_{B2}	Outlet Bernoulli coefficient
ς_{lm}	Spin inertia tensor
τ	Mechanical generated couple
τ_e	Theoretical motor torque
τ_L	Applied torque of load
τ_R	Torque transferred
$\boldsymbol{\varphi}$	Microdisplacement vector
$\boldsymbol{\varphi}$	Tensor of small strain
ϕ	Vector of magnetic fluxes
φ	Phase
φ	Electric potential
Φ	Generalized dissipation potential
Φ	Magnetic flux
$\Phi_{(i)}$	Resulting system of forces and torques acting on a body with respect to reference frame (i)
χ	Displacement operator
$\boldsymbol{\chi}$	Microdeformation tensor
$\hat{\boldsymbol{\chi}}$	Inverse microdeformation tensor
χ_K	Director
$\hat{\chi}_k$	Director
$\boldsymbol{\Psi}$	Derivative tensor of the displacement vector
Ψ	Rotation of a bar cross section
ψ	Shape function
ψ	Helmotz free energy function per mass unit
ω	Angular pulsation
ω_i	Natural frequencies
ω_0	Natural angular frequency
ω	Generated angular speed
ω_u	User angular speed
ω_{du}	Desired user speed
ω_d	Desired actuator angular speed
ω_n	Natural frequencies

Symbol	Meaning
Ω	Angular speed
Ω_s	Synchronous speed

Appendix B: System of Units

TABLE B.1

Principal Units of the International System (SI)

Physical Property	Unit Name	Abbreviation
Amount of substance	mole	mol
Electric current	ampere	A
Length	metre	m
Luminous intensity	candela	cd
Mass	kilogram	kg
Temperature	kelvin	K
Time	second	s

TABLE B.2

Unit Multipliers

Prefix Name	Prefix Symbol	Multiplier
yocto	y	10^{-24}
zepto	z	10^{-21}
atto	a	10^{-18}
femto	f	10^{-15}
pico	p	10^{-12}
nano	n	10^{-9}
micro	µ	10^{-6}
milli	m	10^{-3}
centi	c	10^{-2}
deci	d	10^{-1}
—	—	10^{0}
deka	da	10^{+1}
hecto	h	10^{+2}
kilo	k	10^{+3}
mega	M	10^{+6}
giga	G	10^{+9}
tera	T	10^{+12}
peta	P	10^{+15}
exa	E	10^{+18}
zetta	Z	10^{+21}
yotta	Y	10^{+24}

TABLE B.3

Derived Units with Special Names in the International System

Physical Property	Unit Name	Abbreviation	Expression in Principal Units
Temperature	celsius degree	°C	K-273.15
Electric charge	coulomb	C	A s
Electric capacitance	farad	F	$A^2 s^4 kg^{-1} m^{-2}$
Inductance	henry	H	$kg m^2 s^{-2} A^{-2}$
Frequency	hertz	Hz	s^{-1}
Energy	joule	J	$kg m^2 s^{-2}$
Luminous flux	lumen	lm	cd sr
Illuminance	lux	lx	$cd sr m^{-2}$
Force	newton	N	$kg m s^{-2}$
Electric resistance	ohm	Ω	$kg m^2 s^{-3} A^{-2}$
Pressure, stress	pascal	Pa	$kg m^{-1} s^{-2}$
Planar angle	radiant	rad	m/m
Electric conductance	siemens	S	$s^3 A^2 kg^{-1} m^{-2}$
Solid angle	steradian	sr	m^2/m^2
Magnetic flux density	tesla	T	$kg s^{-2} A^{-1}$
Electric tension	volt	V	$kg m^2 s^{-3} A^{-1}$
Power	watt	W	$kg m^2 s^{-3}$
Magnetic flux	weber	Wb	$kg m^2 s^{-2} A^{-1}$

TABLE B.4

Some Important Physical Properties with Their SI Units

Physical Property	Unit Name	Abbreviation	Expression in Principal Units
Acceleration	—	$m s^{-2}$	$m s^{-2}$
Capacitance	farad	F	$A^2 s^4 kg^{-1} m^{-2}$
Thermal conductivity	—	$W m^{-1} K^{-1}$	$kg m K^{-1} s^{-3}$
Electrical conductivity	—	$S m^{-1}$	$s^3 A^2 kg^{-1} m^{-3}$
Convection coefficient	—	$W m^{-2} K^{-1}$	$kg K^{-1} s^{-3}$
Current	ampere	A	A
Current density	—	$A m^{-2}$	$A m^{-2}$
Density	—	$kg m^{-3}$	$kg m^{-3}$
Dynamic viscosity	—	$kg m^{-1} s^{-1}$	$kg m^{-1} s^{-1}$
Electric charge	coulomb	C	A s
Electric field	—	$V m^{-1}$	$kg m s^{-3} A^{-1}$
Electric flux density	—	$C m^{-2}$	$A s m^{-2}$
Energy	joule	J	$kg m^2 s^{-2}$

TABLE B.4 (continued)

Some Important Physical Properties with Their SI Units

Physical Property	Unit Name	Abbreviation	Expression in Principal Units
Force	newton	N	$kg\,m\,s^{-2}$
Heat flow	watt	W	$kg\,m^2\,s^{-3}$
Heat flux	—	$W\,m^{-2}$	$kg\,s^{-3}$
Heat generation per volume	—	$W\,m^{-3}$	$kg\,m^{-1}\,s^{-3}$
Inductance	henry	H	$kg\,m^2\,s^{-2}\,A^{-2}$
Kinematic viscosity	—	$m^2\,s^{-1}$	$m^2\,s^{-1}$
Length	meter	m	m
Magnetic field intensity	—	$A\,m^{-1}$	$A\,m^{-1}$
Magnetic flux	weber	Wb	$kg\,m^2\,s^{-2}\,A^{-1}$
Magnetic flux density	tesla	T	$kg\,s^{-2}\,A^{-1}$
Magnetic permeability	—	$H\,m^{-1}$	$kg\,m\,s^{-2}\,A^{-2}$
Mass	kilogram	kg	kg
Permittivity	—	$F\,m^{-1}$	$A^2\,s^4\,kg^{-1}\,m^{-3}$
Piezoelectric strain matrix **d**	—	$C\,N^{-1}$	$A\,s^3\,kg^{-1}\,m^{-1}$
Piezoelectric stress matrix **e**	—	$C\,m^{-2}$	$A\,s\,m^{-2}$
Piezoresistive stress matrix π	—	Pa^{-1}	$m\,s^2\,kg^{-1}$
Power	watt	W	$kg\,m^2\,s^{-3}$
Pressure	pascal	Pa	$kg\,m^{-1}\,s^{-2}$
Resistivity	—	$\Omega\,m$	$kg\,m^3\,s^{-3}\,A^{-2}$
Specific heat	—	$J\,kg^{-1}\,K^{-1}$	$m^2\,K^{-1}\,s^{-2}$
Stress	pascal	Pa	$kg\,m^{-1}\,s^{-2}$
Time	second	s	s
Velocity	—	$m\,s^{-1}$	$m\,s^{-1}$
Voltage	volt	V	$kg\,m^2\,s^{-3}\,A^{-1}$
Young modulus	pascal	Pa	$kg\,m^{-1}\,s^{-2}$

TABLE B.5

Principal Units of the μMKSV Unit System

Physical Property	Unit Name	Abbreviation
Amount of substance	mole	mol
Electric current	picoampere	pA
Length	micrometre	μm
Luminous intensity	candela	cd
Mass	kilogram	kg
Temperature	kelvin	K
Time	second	s

TABLE B.6

Derived Units with Special Names in the μMKSV Unit System

Physical Property	Unit Name	Abbreviation	Expression in Principal Units
Temperature	celsius degree	°C	K-273.15
Electric charge	picocoulomb	pC	pA s
Electric capacitance	picofarad	pF	$pA^2 s^4 kg^{-1} \mu m^{-2}$
Inductance	terahenry	TH	$kg\, \mu m^2 s^{-2} pA^{-2}$
Frequency	hertz	Hz	s^{-1}
Energy	picojoule	pJ	$kg\, \mu m^2 s^{-2}$
Luminous flux	lumen	lm	cd sr
Illuminance	teralux	Tlx	$cd\, sr\, \mu m^{-2}$
Force	micronewton	μN	$kg\, \mu m\, s^{-2}$
Electric resistance	teraohm	TΩ	$kg\, \mu m^2 s^{-3} pA^{-2}$
Pressure, stress	megapascal	MPa	$kg\, \mu m^{-1} s^{-2}$
Planar angle	radiant	rad	$\mu m / \mu m$
Electric conductance	picosiemens	pS	$s^3 pA^2 kg^{-1} \mu m^{-2}$
Solid angle	steradian	sr	$\mu m^2 / \mu m^2$
Magnetic flux density	teratesla	TT	$kg\, s^{-2} pA^{-1}$
Electric tension	volt	V	$kg\, \mu m^2 s^{-3} pA^{-1}$
Power	picowatt	pW	$kg\, \mu m^2 s^{-3}$
Magnetic flux	weber	Wb	$kg\, \mu m^2 s^{-2} pA^{-1}$

TABLE B.7

Some Important Physical Properties with Their μMKSV Units

Physical Property	Unit Name	Abbreviation	Expression in Principal Units
Acceleration	—	$\mu m\, s^{-2}$	$\mu m\, s^{-2}$
Capacitance	picofarad	pF	$pA^2 s^4 kg^{-1} \mu m^{-2}$
Thermal conductivity	—	$pW\, \mu m^{-1} K^{-1}$	$kg\, \mu m\, K^{-1} s^{-3}$
Electrical conductivity	—	$pS\, \mu m^{-1}$	$s^3 pA^2 kg^{-1} \mu m^{-3}$
Convection coefficient	—	$pW\, \mu m^{-2} K^{-1}$	$kg\, K^{-1} s^{-3}$
Current	picoampere	pA	pA
Current density	—	$pA\, \mu m^{-2}$	$pA\, \mu m^{-2}$
Density	—	$kg\, \mu m^{-3}$	$kg\, \mu m^{-3}$
Dynamic viscosity	—	$kg\, \mu m^{-1} s^{-1}$	$kg\, \mu m^{-1} s^{-1}$
Electric charge	picocoulomb	pC	pA s
Electric field	—	$V\, \mu m^{-1}$	$kg\, \mu m\, s^{-3} pA^{-1}$
Electric flux density	—	$pC\, \mu m^{-2}$	$pA\, s\, \mu m^{-2}$
Energy	picojoule	pJ	$kg\, \mu m^2 s^{-2}$

TABLE B.7 (continued)

Some Important Physical Properties with Their μMKSV Units

Physical Property	Unit Name	Abbreviation	Expression in Principal Units
Force	micronewton	μN	$kg\,\mu m\,s^{-2}$
Heat flow	picowatt	pW	$kg\,\mu m^2\,s^{-3}$
Heat flux	—	$pW\,\mu m^{-2}$	$kg\,s^{-3}$
Heat generation per volume	—	$pW\,\mu m^{-3}$	$kg\,\mu m^{-1}\,s^{-3}$
Inductance	terahenry	TH	$kg\,\mu m^2\,s^{-2}\,pA^{-2}$
Kinematic viscosity	—	$\mu m^2\,s^{-1}$	$\mu m^2\,s^{-1}$
Length	micrometer	μm	μm
Magnetic field intensity	—	$pA\,\mu m^{-1}$	$pA\,\mu m^{-1}$
Magnetic flux	weber	Wb	$kg\,\mu m^2\,s^{-2}\,pA^{-1}$
Magnetic flux density	teratesla	TT	$kg\,s^{-2}\,pA^{-1}$
Magnetic permeability	—	$TH\,\mu m^{-1}$	$kg\,\mu m\,s^{-2}\,pA^{-2}$
Mass	kilogram	kg	kg
Permittivity	—	$pF\,\mu m^{-1}$	$pA^2\,s^4\,kg^{-1}\,\mu m^{-3}$
Piezoelectric strain matrix **d**	—	$pC\,\mu N^{-1}$	$pA\,s^3\,kg^{-1}\,\mu m^{-1}$
Piezoelectric stress matrix **e**	—	$pC\,\mu m^{-2}$	$pA\,s\,\mu m^{-2}$
Piezoresistive stress matrix π	—	MPa^{-1}	$\mu m\,s^2\,kg^{-1}$
Power	picowatt	pW	$kg\,\mu m^2\,s^{-3}$
Pressure	megapascal	MPa	$kg\,\mu m^{-1}\,s^{-2}$
Resistivity	—	$T\Omega\,\mu m$	$kg\,\mu m^3\,s^{-3}\,pA^{-2}$
Specific heat	—	$pJ\,kg^{-1}\,K^{-1}$	$\mu m^2\,K^{-1}\,s^{-2}$
Stress	megapascal	MPa	$kg\,\mu m^{-1}\,s^{-2}$
Time	second	s	s
Velocity	—	$\mu m\,s^{-1}$	$\mu m\,s^{-1}$
Voltage	volt	V	$kg\,\mu m^2\,s^{-3}\,pA^{-1}$
Young modulus	megapascal	MPa	$kg\,\mu m^{-1}\,s^{-2}$

TABLE B.8

Principal Units of the μMKSVfA Unit System

Physical Property	Unit Name	Abbreviation
Amount of substance	mole	mol
Electric current	femtoampere	fA
Length	micrometre	μm
Luminous intensity	candela	cd
Mass	gram	g
Temperature	kelvin	K
Time	second	s

TABLE B.9

Derived Units with Special Names in the μMKSVfA Unit System

Physical Property	Unit Name	Abbreviation	Expression in Principal Units
Temperature	celsius degree	°C	K-273.15
Electric charge	femtocoulomb	fC	fA s
Electric capacitance	femtofarad	fF	$fA^2 s^4 g^{-1} \mu m^{-2}$
Inductance	petahenry	PH	$g \mu m^2 s^{-2} fA^{-2}$
Frequency	hertz	Hz	s^{-1}
Energy	femtojoule	fJ	$g \mu m^2 s^{-2}$
Luminous flux	lumen	lm	cd sr
Illuminance	teralux	Tlx	$cd\ sr\ \mu m^{-2}$
Force	nanonewton	nN	$g \mu m\ s^{-2}$
Electric resistance	petaohm	PΩ	$g \mu m^2 s^{-3} fA^{-2}$
Pressure, stress	pascal	kPa	$g \mu m^{-1} s^{-2}$
Planar angle	radiant	rad	$\mu m / \mu m$
Electric conductance	femtosiemens	fS	$s^3 fA^2 g^{-1} \mu m^{-2}$
Solid angle	steradian	sr	$\mu m^2 / \mu m^2$
Magnetic flux density	teratesla	TT	$g s^{-2} fA^{-1}$
Electric tension	volt	V	$g \mu m^2 s^{-3} fA^{-1}$
Power	femtowatt	fW	$g \mu m^2 s^{-3}$
Magnetic flux	weber	Wb	$g \mu m^2 s^{-2} fA^{-1}$

TABLE B.10

Some Important Physical Properties with Their μMKSVfA Units

Physical Property	Unit Name	Abbreviation	Expression in Principal Units
Acceleration	-	$\mu m\ s^{-2}$	$\mu m\ s^{-2}$
Capacitance	femtofarad	fF	$fA^2 s^4 g^{-1} \mu m^{-2}$
Thermal conductivity	-	$fW \mu m^{-1} K^{-1}$	$g \mu m\ K^{-1} s^{-3}$
Electrical conductivity	-	$fS \mu m^{-1}$	$s^3 fA^2 g^{-1} \mu m^{-3}$
Convection coefficient	-	$fW \mu m^{-2} K^{-1}$	$g K^{-1} s^{-3}$
Current	femtoampere	fA	fA
Current density	-	$fA \mu m^{-2}$	$fA \mu m^{-2}$
Density	-	$g \mu m^{-3}$	$g \mu m^{-3}$
Dynamic viscosity	-	$g \mu m^{-1} s^{-1}$	$g \mu m^{-1} s^{-1}$
Electric charge	femtocoulomb	fC	fA s
Electric field	-	$V \mu m^{-1}$	$g \mu m\ s^{-3} fA^{-1}$
Electric flux density	-	$fC \mu m^{-2}$	$fA\ s\ \mu m^{-2}$
Energy	femtojoule	fJ	$g \mu m^2 s^{-2}$

TABLE B.10 (continued)

Some Important Physical Properties with Their μMKSVfA Units

Physical Property	Unit Name	Abbreviation	Expression in Principal Units
Force	nanonewton	nN	$g\ \mu m\ s^{-2}$
Heat flow	femtowatt	fW	$g\ \mu m^2\ s^{-3}$
Heat flux	-	$fW\ \mu m^{-2}$	$g\ s^{-3}$
Heat generation per volume	-	$fW\ \mu m^{-3}$	$g\ \mu m^{-1}\ s^{-3}$
Inductance	petahenry	PH	$g\ \mu m^2\ s^{-2}\ fA^{-2}$
Kinematic viscosity	-	$\mu m^2\ s^{-1}$	$\mu m^2\ s^{-1}$
Length	micrometer	μm	μm
Magnetic field intensity	-	$fA\ \mu m^{-1}$	$fA\ \mu m^{-1}$
Magnetic flux	weber	Wb	$g\ \mu m^2\ s^{-2}\ fA^{-1}$
Magnetic flux density	teratesla	TT	$g\ s^{-2}\ fA^{-1}$
Magnetic permeability	-	$PH\ \mu m^{-1}$	$kg\ \mu m\ s^{-2}\ fA^{-2}$
Mass	gram	g	g
Permittivity	-	$fF\ \mu m^{-1}$	$fA^2\ s^4\ g^{-1}\ \mu m^{-3}$
Piezoelectric strain matrix **d**	-	$fC\ nN^{-1}$	$fA\ s^3\ g^{-1}\ \mu m^{-1}$
Piezoelectric stress matrix **e**	-	$fC\ \mu m^{-2}$	$fA\ s\ \mu m^{-2}$
Piezoresistive stress matrix π	-	kPa^{-1}	$\mu m\ s^2\ g^{-1}$
Power	picowatt	fW	$g\ \mu m^2\ s^{-3}$
Pressure	kilopascal	kPa	$g\ \mu m^{-1}\ s^{-2}$
Resistivity	-	$P\Omega\ \mu m$	$g\ \mu m^3\ s^{-3}\ fA^{-2}$
Specific heat	-	$fJ\ g^{-1}\ K^{-1}$	$\mu m^2\ K^{-1}\ s^{-2}$
Stress	kilopascal	kPa	$g\ \mu m^{-1}\ s^{-2}$
Time	second	s	s
Velocity	-	$\mu m\ s^{-1}$	$\mu m\ s^{-1}$
Voltage	volt	V	$g\ \mu m^2\ s^{-3}\ fA^{-1}$
Young modulus	kilopascal	kPa	$g\ \mu m^{-1}\ s^{-2}$

TABLE B.11

Conversion of Principal Units between SI and μMKSV

Physical Property		SI Unit		Multiplier	μMKSV Unit
Amount of substance	: 1	mol	=	1	mol
Electric current	: 1	A	=	10^{12}	pA
Length	: 1	m	=	10^6	μm
Luminous intensity	: 1	cd	=	1	cd
Mass	: 1	kg	=	1	kg
Temperature	: 1	K	=	1	K
Time	: 1	s	=	1	s

TABLE B.12

Conversion of Derived Units between SI and μMKSV

Physical Property		SI Unit		Multiplier	μMKSV Unit
Temperature	:1	°C	=	1	°C
Electric charge	:1	C	=	10^{12}	pC
Electric capacitance	:1	F	=	10^{12}	pF
Inductance	:1	H	=	10^{-12}	TH
Frequency	:1	Hz	=	1	Hz
Energy	:1	J	=	10^{12}	pJ
Luminous flux	:1	lm	=	1	lm
Illuminance	:1	lx	=	10^{-12}	Tlx
Force	:1	N	=	10^{6}	μN
Electric resistance	:1	Ω	=	10^{-12}	TΩ
Pressure, stress	:1	Pa	=	10^{-6}	MPa
Planar angle	:1	rad	=	1	rad
Electric conductance	:1	S	=	10^{12}	pS
Solid angle	:1	sr	=	1	sr
Magnetic flux density	:1	T	=	10^{-12}	TT
Electric tension	:1	V	=	1	V
Power	:1	W	=	10^{12}	pW
Magnetic flux	:1	Wb	=	1	Wb

TABLE B.13

Unit Conversion of Some Important Physical Properties between SI and μMKSV

Physical Property		SI Unit		Multiplier	μMKSV Unit
Acceleration	:1	m s^{-2}	=	10^{6}	μm s^{-2}
Capacitance	:1	F	=	10^{12}	pF
Thermal conductivity	:1	W m^{-1} K^{-1}	=	10^{6}	pW μm^{-1} K^{-1}
Electric conductivity	:1	S m^{-1}	=	10^{6}	pS μm^{-1}
Convection coefficient	:1	W m^{-2} K^{-1}	=	1	pW μm^{-2} K^{-1}
Current	:1	A	=	10^{12}	pA
Current density	:1	A m^{-2}	=	1	pA μm^{-2}
Density	:1	kg m^{-3}	=	10^{-18}	kg μm^{-3}
Dynamic viscosity	:1	kg m^{-1} s^{-1}	=	10^{-6}	kg μm^{-1} s^{-1}
Electric charge	:1	C	=	10^{12}	pC
Electric field	:1	V m^{-1}	=	10^{-6}	V μm^{-1}
Electric flux density	:1	C m^{-2}	=	1	pC μm^{-2}
Energy	:1	J	=	10^{12}	pJ

TABLE B.13 (continued)

Unit Conversion of Some Important Physical Properties between SI and µMKSV

Physical Property		SI Unit		Multiplier	µMKSV Unit
Force	: 1	N	=	10^6	µN
Heat flow	: 1	W	=	10^{12}	pW
Heat flux	: 1	W m^{-2}	=	1	pW µm^{-2}
Heat generation per volume	: 1	W m^{-3}	=	10^{-6}	pW µm^{-3}
Inductance	: 1	H	=	10^{-12}	TH
Kinematic viscosity	: 1	m^2 s^{-1}	=	10^{12}	µm^2 s^{-1}
Length	: 1	m	=	10^6	µm
Magnetic field intensity	: 1	A m^{-1}	=	10^6	pA µm^{-1}
Magnetic flux	: 1	Wb	=	1	Wb
Magnetic flux density	: 1	T	=	10^{-12}	TT
Magnetic permeability	: 1	H m^{-1}	=	10^{-18}	TH µm^{-1}
Mass	: 1	kg	=	1	kg
Permittivity	: 1	F m^{-1}	=	10^6	pF µm^{-1}
Piezoelectric strain matrix **d**	: 1	C N^{-1}	=	10^6	pC µN^{-1}
Piezoelectric stress matrix **e**	: 1	C m^{-2}	=	1	pC µm^{-2}
Piezoresistive stress matrix π	: 1	Pa^{-1}	=	10^6	MPa^{-1}
Power	: 1	W	=	10^{12}	pW
Pressure	: 1	Pa	=	10^{-6}	MPa
Resistivity	: 1	Ω m	=	10^{-6}	TΩ µm
Specific heat	: 1	J kg^{-1} K^{-1}	=	10^{12}	pJ kg^{-1} K^{-1}
Stress	: 1	Pa	=	10^{-6}	MPa
Time	: 1	s	=	1	s
Velocity	: 1	m s^{-1}	=	10^6	µm s^{-1}
Voltage	: 1	V	=	1	V
Young modulus	: 1	Pa	=	10^{-6}	MPa

TABLE B.14

Conversion of Principal Units between SI and µMKSVfA

Physical Property		SI Unit		Multiplier	µMKSVfA Unit
Amount of substance	: 1	mol	=	1	mol
Electric current	: 1	A	=	10^{15}	fA
Length	: 1	m	=	10^6	µm
Luminous intensity	: 1	cd	=	1	cd
Mass	: 1	kg	=	10^3	g
Temperature	: 1	K	=	1	K
Time	: 1	s	=	1	s

TABLE B.15

Conversion of Derived Units between SI and μMKSVfA

Physical Property		SI Unit		Multiplier	μMKSVfA Unit
Temperature	: 1	°C	=	1	°C
Electric charge	: 1	C	=	10^{15}	fC
Electric capacitance	: 1	F	=	10^{15}	fF
Inductance	: 1	H	=	10^{-15}	PH
Frequency	: 1	Hz	=	1	Hz
Energy	: 1	J	=	10^{15}	fJ
Luminous flux	: 1	lm	=	1	lm
Illuminance	: 1	lx	=	10^{-12}	Tlx
Force	: 1	N	=	10^{9}	nN
Electric resistance	: 1	Ω	=	10^{-15}	PΩ
Pressure, stress	: 1	Pa	=	10^{-3}	kPa
Planar angle	: 1	rad	=	1	rad
Electric conductance	: 1	S	=	10^{15}	fS
Solid angle	: 1	sr	=	1	sr
Magnetic flux density	: 1	T	=	10^{-12}	TT
Electric tension	: 1	V	=	1	V
Power	: 1	W	=	10^{15}	fW
Magnetic flux	: 1	Wb	=	1	Wb

TABLE B.16

Unit Conversion of Some Important Physical Properties between SI and μMKSVfA

Physical Property		SI Unit		Multiplier	μMKSVfA Unit
Acceleration	: 1	$m\ s^{-2}$	=	10^{6}	$\mu m\ s^{-2}$
Capacitance	: 1	F	=	10^{15}	fF
Thermal conductivity	: 1	$W\ m^{-1}\ K^{-1}$	=	10^{9}	$fW\ \mu m^{-1}\ K^{-1}$
Electric conductivity	: 1	$S\ m^{-1}$	=	10^{9}	$fS\ \mu m^{-1}$
Convection coefficient	: 1	$W\ m^{-2}\ K^{-1}$	=	10^{3}	$fW\ \mu m^{-2}\ K^{-1}$
Current	: 1	A	=	10^{15}	fA
Current density	: 1	$A\ m^{-2}$	=	10^{3}	$fA\ \mu m^{-2}$
Density	: 1	$kg\ m^{-3}$	=	10^{-15}	$g\ \mu m^{-3}$
Dynamic viscosity	: 1	$kg\ m^{-1}\ s^{-1}$	=	10^{-3}	$g\ \mu m^{-1}\ s^{-1}$
Electric charge	: 1	C	=	10^{15}	fC
Electric field	: 1	$V\ m^{-1}$	=	10^{-6}	$V\ \mu m^{-1}$
Electric flux density	: 1	$C\ m^{-2}$	=	10^{3}	$fC\ \mu m^{-2}$
Energy	: 1	J	=	10^{15}	fJ

TABLE B.16 (continued)

Unit Conversion of Some Important Physical Properties between SI and μMKSVfA

Physical Property		SI Unit		Multiplier	μMKSVfA Unit
Force	: 1	N	=	10^9	nN
Heat flow	: 1	W	=	10^{15}	fW
Heat flux	: 1	W m^{-2}	=	10^3	fW μm^{-2}
Heat generation per volume	: 1	W m^{-3}	=	10^{-3}	fW μm^{-3}
Inductance	: 1	H	=	10^{-15}	PH
Kinematic viscosity	: 1	m^2 s^{-1}	=	10^{12}	μm^2 s^{-1}
Length	: 1	m	=	10^6	μm
Magnetic field intensity	: 1	A m^{-1}	=	10^9	fA μm^{-1}
Magnetic flux	: 1	Wb	=	1	Wb
Magnetic flux density	: 1	T	=	10^{-12}	TT
Magnetic permeability	: 1	H m^{-1}	=	10^{-21}	PH μm^{-1}
Mass	: 1	kg	=	10^3	g
Permittivity	: 1	F m^{-1}	=	10^9	fF μm^{-1}
Piezoelectric strain matrix **d**	: 1	C N^{-1}	=	10^6	fC nN^{-1}
Piezoelectric stress matrix **e**	: 1	C m^{-2}	=	10^3	fC μm^{-2}
Piezoresistive stress matrix π	: 1	Pa^{-1}	=	10^3	kPa^{-1}
Power	: 1	W	=	10^{15}	fW
Pressure	: 1	Pa	=	10^{-3}	kPa
Resistivity	: 1	Ω m	=	10^{-9}	PΩ μm
Specific heat	: 1	J kg^{-1} K^{-1}	=	10^{12}	fJ g^{-1} K^{-1}
Stress	: 1	Pa	=	10^{-3}	kPa
Time	: 1	s	=	1	s
Velocity	: 1	m s^{-1}	=	10^6	μm s^{-1}
Voltage	: 1	V	=	1	V
Young modulus	: 1	Pa	=	10^{-3}	kPa

TABLE B.17

Conversion of Principal Units between μMKSV and μMKSVfA

Physical Property		μMKSV Unit		Multiplier	μMKSVfA Unit
Amount of substance	: 1	mol	=	1	mol
Electric current	: 1	pA	=	10^3	fA
Length	: 1	μm	=	1	μm
Luminous intensity	: 1	cd	=	1	cd
Mass	: 1	kg	=	10^3	g
Temperature	: 1	K	=	1	K
Time	: 1	s	=	1	s

TABLE B.18

Conversion of Derived Units between μMKSV and μMKSVfA

Physical Property		μMKSV Unit		Multiplier	μMKSVfA Unit
Temperature	: 1	°C	=	1	°C
Electric charge	: 1	pC	=	10^3	fC
Electric capacitance	: 1	pF	=	10^3	fF
Inductance	: 1	TH	=	10^{-3}	PH
Frequency	: 1	Hz	=	1	Hz
Energy	: 1	pJ	=	10^3	fJ
Luminous flux	: 1	lm	=	1	lm
Illuminance	: 1	Tlx	=	1	Tlx
Force	: 1	μN	=	10^3	nN
Electric resistance	: 1	TΩ	=	10^{-3}	PΩ
Pressure, stress	: 1	MPa	=	10^3	kPa
Planar angle	: 1	rad	=	1	rad
Electric conductance	: 1	pS	=	10^3	fS
Solid angle	: 1	sr	=	1	sr
Magnetic flux density	: 1	TT	=	1	TT
Electric tension	: 1	V	=	1	V
Power	: 1	pW	=	10^3	fW
Magnetic flux	: 1	Wb	=	1	Wb

TABLE B.19

Unit Conversion of Some Important Physical Properties between μMKSV and μMKSVfA

Physical Property		μMKSV Unit		Multiplier	μMKSVfA Unit
Acceleration	: 1	$\mu m\ s^{-2}$	=	1	$\mu m\ s^{-2}$
Capacitance	: 1	pF	=	10^3	fF
Thermal conductivity	: 1	$pW\ \mu m^{-1}\ K^{-1}$	=	10^3	$fW\ \mu m^{-1}\ K^{-1}$
Electric conductivity	: 1	$pS\ \mu m^{-1}$	=	10^3	$fS\ \mu m^{-1}$
Convection coefficient	: 1	$pW\ \mu m^{-2}\ K^{-1}$	=	10^3	$fW\ \mu m^{-2}\ K^{-1}$
Current	: 1	pA	=	10^3	fA
Current density	: 1	$pA\ \mu m^{-2}$	=	10^3	$fA\ \mu m^{-2}$
Density	: 1	$kg\ \mu m^{-3}$	=	10^3	$g\ \mu m^{-3}$
Dynamic viscosity	: 1	$kg\ \mu m^{-1}\ s^{-1}$	=	10^3	$g\ \mu m^{-1}\ s^{-1}$
Electric charge	: 1	pC	=	10^3	fC
Electric field	: 1	$V\ \mu m^{-1}$	=	1	$V\ \mu m^{-1}$
Electric flux density	: 1	$pC\ \mu m^{-2}$	=	10^3	$fC\ \mu m^{-2}$
Energy	: 1	pJ	=	10^3	fJ
Force	: 1	μN	=	103	nN
Heat flow	: 1	pW	=	10^3	fW
Heat flux	: 1	$pW\ \mu m^{-2}$	=	10^3	$fW\ \mu m^{-2}$

TABLE B.19 (continued)

Unit Conversion of Some Important Physical Properties between µMKSV and µMKSVfA

Physical Property		µMKSV Unit		Multiplier	µMKSVfA Unit
Heat generation per volume	: 1	pW µm^{-3}	=	10^3	fW µm^{-3}
Inductance	: 1	TH	=	10^{-3}	PH
Kinematic viscosity	: 1	µm^2 s^{-1}	=	1	µm^2 s^{-1}
Length	: 1	µm	=	1	µm
Magnetic field intensity	: 1	pA µm^{-1}	=	10^3	fA µm^{-1}
Magnetic flux	: 1	Wb	1	1	Wb
Magnetic flux density	: 1	TT	=	1	TT
Magnetic permeability	: 1	TH µm^{-1}	=	10^{-3}	PH µm^{-1}
Mass	: 1	kg	=	10^3	g
Permittivity	: 1	pF µm^{-1}	=	10^3	fF µm^{-1}
Piezoelectric strain matrix **d**	: 1	pC µN^{-1}	=	1	fC nN^{-1}
Piezoelectric stress matrix **e**	: 1	pC µm^{-2}	=	10^3	fC µm^{-2}
Piezoresistive stress matrix π	: 1	MPa^{-1}	=	10^{-3}	kPa^{-1}
Power	: 1	pW	=	10^3	fW
Pressure	: 1	MPa	=	10^3	kPa
Resistivity	: 1	TΩ µm	=	10^{-3}	PΩ µm
Specific heat	: 1	pJ kg^{-1} K^{-1}	=	1	fJ g^{-1} K^{-1}
Stress	: 1	MPa	=	10^3	kPa
Time	: 1	s	=	1	s
Velocity	: 1	µm s^{-1}	=	1	µm s^{-1}
Voltage	: 1	V	=	1	V
Young modulus	: 1	MPa	=	10^3	kPa

Appendix C: Mathematical Derivations

Some mathematical deductions or expressions are particularly long and can interrupt the reading; therefore, we decided to move them into a separate appendix that can be consulted during or after the discussion of the principal subject.

C.1 Spherical Joint

In this paragraph, some expressions on spherical joints are collected as shown in the section 1.2.5, chapter 1. The rototranslation matrix $Q_{(0)}$ is computed with (1.45) using the matrixes shown in (1.66). Every term of matrix $Q_{(0)}$ is expressed as (C.1–C.10), where c_{i0} and s_{i0} are, respectively, the cosine and the sine of the angle α_{i0}, and $c_{i\Delta}$ and $s_{i\Delta}$ are, respectively, the cosine and the sine of the angle $\Delta\alpha_i$.

$$Q_{(0)11} = \left(s_{2\Delta} - s_{2\Delta}c_{3\Delta}\right)c_{20}s_{20} + \left(c_{2\Delta}c_{3\Delta} - c_{2\Delta}\right)c_{20}^2 + c_{2\Delta} \tag{C.1}$$

$$\begin{aligned} Q_{(0)12} = &-c_{20}c_{2\Delta}s_{3\Delta}c_{10} + s_{20}s_{2\Delta}s_{3\Delta}c_{10} + c_{20}c_{2\Delta}c_{3\Delta}s_{10}s_{20} - s_{2\Delta}c_{3\Delta}s_{10} + s_{2\Delta}c_{3\Delta}s_{10}c_{20}^2 + \\ &-s_{10}c_{20}s_{20}c_{2\Delta} - s_{10}c_{20}^2 s_{2\Delta} \end{aligned} \tag{C.2}$$

$$\begin{aligned} Q_{(0)13} = &-c_{20}c_{2\Delta}s_{3\Delta}s_{10} + s_{20}s_{2\Delta}s_{3\Delta}s_{10} - c_{20}c_{2\Delta}c_{3\Delta}c_{10}s_{20} + s_{2\Delta}c_{3\Delta}c_{10} - s_{2\Delta}c_{3\Delta}c_{10}c_{20}^2 + \\ &+c_{10}c_{20}s_{20}c_{2\Delta} + c_{10}c_{20}^2 s_{2\Delta} \end{aligned} \tag{C.3}$$

$$\begin{aligned} Q_{(0)21} = &-s_{20}c_{10}s_{1\Delta}c_{20}c_{2\Delta} + c_{20}c_{10}s_{1\Delta}s_{20}c_{2\Delta}c_{3\Delta} + c_{20}^2 c_{10}s_{1\Delta}s_{2\Delta}c_{3\Delta} + c_{20}c_{10}c_{1\Delta}s_{3\Delta} + \\ &-s_{20}s_{10}c_{1\Delta}c_{20}c_{2\Delta} - c_{20}s_{10}s_{1\Delta}s_{3\Delta} + c_{20}s_{10}c_{1\Delta}s_{20}c_{2\Delta}c_{3\Delta} + c_{20}^2 s_{10}c_{1\Delta}s_{2\Delta}c_{3\Delta} + \\ &+c_{10}s_{1\Delta}s_{2\Delta} - c_{10}s_{1\Delta}s_{2\Delta}c_{20}^2 + s_{10}c_{1\Delta}s_{2\Delta} - s_{10}c_{1\Delta}s_{2\Delta}c_{20}^2 \end{aligned} \tag{C.4}$$

377

$$
\begin{aligned}
Q_{(0)22} =\; & s_{10}c_{20}^2 c_{10}s_{1\Delta}c_{2\Delta} - s_{10}c_{20}c_{10}s_{1\Delta}s_{20}s_{2\Delta} + c_{10}^2 c_{1\Delta}c_{3\Delta} + c_{20}^2 c_{1\Delta}c_{2\Delta} - c_{20}^2 c_{1\Delta}c_{2\Delta}c_{10}^2 + \quad\text{(C.5)}\\
& + s_{1\Delta}s_{3\Delta}s_{20}c_{10}^2 - c_{1\Delta}c_{2\Delta}c_{3\Delta}c_{20}^2 - c_{1\Delta}c_{2\Delta}c_{3\Delta}c_{10}^2 - s_{10}s_{1\Delta}c_{3\Delta}c_{10} + c_{10}s_{1\Delta}c_{20}s_{2\Delta}c_{3\Delta}s_{10}s_{20} + \\
& - s_{10}c_{1\Delta}s_{20}c_{2\Delta}s_{3\Delta}c_{10} - c_{20}c_{1\Delta}s_{20}s_{2\Delta} + c_{20}c_{1\Delta}s_{20}s_{2\Delta}c_{10}^2 - s_{1\Delta}s_{3\Delta}s_{20} - c_{10}^2 s_{1\Delta}c_{20}s_{2\Delta}s_{3\Delta} + \\
& + c_{1\Delta}c_{2\Delta}c_{3\Delta} + c_{1\Delta}c_{2\Delta}c_{3\Delta}c_{10}^2 c_{20}^2 - c_{10}^2 s_{1\Delta}s_{20}c_{2\Delta}s_{3\Delta} - s_{10}c_{1\Delta}c_{20}s_{2\Delta}c_{3\Delta}c_{10} + \\
& + c_{10}c_{1\Delta}s_{3\Delta}s_{10}s_{20} + c_{10}s_{1\Delta}c_{2\Delta}c_{3\Delta}s_{10} - c_{10}s_{1\Delta}c_{2\Delta}c_{3\Delta}s_{10}c_{20}^2 + c_{1\Delta}c_{20}s_{2\Delta}c_{3\Delta}s_{20} + \\
& - c_{1\Delta}c_{20}s_{2\Delta}c_{3\Delta}s_{20}c_{10}^2
\end{aligned}
$$

$$
\begin{aligned}
Q_{(0)23} =\; & -c_{10}c_{20}^2 s_{10}c_{1\Delta}c_{2\Delta} + c_{10}c_{20}s_{10}c_{1\Delta}s_{20}s_{2\Delta} + c_{10}^2 c_{20}s_{1\Delta}s_{20}s_{2\Delta} - c_{10}^2 c_{20}^2 s_{1\Delta}c_{2\Delta} + \quad\text{(C.6)}\\
& + c_{10}^2 s_{1\Delta}c_{3\Delta} + c_{10}c_{1\Delta}c_{3\Delta}s_{10} - c_{10}^2 s_{1\Delta}c_{2\Delta}c_{3\Delta} + s_{10}s_{1\Delta}s_{3\Delta}c_{10}s_{20} + \\
& - s_{10}c_{1\Delta}c_{20}s_{2\Delta}c_{3\Delta}c_{10}s_{20} - c_{1\Delta}c_{20}s_{2\Delta}s_{3\Delta} + c_{1\Delta}c_{20}s_{2\Delta}s_{3\Delta}c_{10}^2 - c_{10}s_{1\Delta}s_{20}c_{2\Delta}s_{3\Delta}s_{10} + \\
& + c_{1\Delta}s_{20}c_{2\Delta}s_{3\Delta}c_{10}^2 - c_{10}s_{1\Delta}c_{20}s_{2\Delta}s_{3\Delta}s_{10} - c_{10}^2 c_{1\Delta}s_{3\Delta}s_{20} - c_{10}^2 s_{1\Delta}c_{20}s_{2\Delta}c_{3\Delta}s_{20} + \\
& + c_{10}^2 s_{1\Delta}c_{2\Delta}c_{3\Delta}c_{20}^2 - s_{10}c_{1\Delta}c_{2\Delta}c_{3\Delta}c_{10} + s_{10}c_{1\Delta}c_{2\Delta}c_{3\Delta}c_{10}c_{20}^2 - s_{1\Delta}s_{3\Delta} - c_{1\Delta}s_{20}c_{2\Delta}s_{3\Delta}
\end{aligned}
$$

$$
\begin{aligned}
Q_{(0)31} =\; & s_{20}c_{10}c_{1\Delta}c_{20}c_{2\Delta} - s_{20}s_{10}s_{1\Delta}c_{20}c_{2\Delta} - c_{20}c_{10}c_{1\Delta}s_{20}c_{2\Delta}c_{3\Delta} - c_{20}^2 c_{10}c_{1\Delta}s_{2\Delta}c_{3\Delta} \\
& + c_{20}s_{10}s_{1\Delta}s_{20}c_{2\Delta}c_{3\Delta} + c_{20}^2 s_{10}s_{1\Delta}s_{2\Delta}c_{3\Delta} + c_{20}s_{10}c_{1\Delta}s_{3\Delta} + c_{20}c_{10}s_{1\Delta}s_{3\Delta} + \quad\text{(C.7)}\\
& - c_{10}c_{1\Delta}s_{2\Delta} + c_{10}c_{1\Delta}s_{2\Delta}c_{20}^2 + s_{10}s_{1\Delta}s_{2\Delta} - s_{10}s_{1\Delta}s_{2\Delta}c_{20}^2
\end{aligned}
$$

$$
\begin{aligned}
Q_{(0)32} =\; & -c_{10}^2 c_{1\Delta}s_{3\Delta}s_{20} - c_{10}^2 s_{1\Delta}c_{2\Delta}c_{3\Delta} - s_{1\Delta}c_{2\Delta}c_{3\Delta}c_{20}^2 - c_{10}^2 c_{20}^2 s_{1\Delta}c_{2\Delta} + \\
& + c_{20}^2 s_{1\Delta}c_{2\Delta} + c_{10}^2 s_{1\Delta}c_{3\Delta} - c_{10}c_{20}^2 s_{10}c_{1\Delta}c_{2\Delta} + c_{10}c_{20}s_{10}c_{1\Delta}s_{20}s_{2\Delta} + \\
& + c_{10}c_{1\Delta}c_{3\Delta}s_{10} + c_{1\Delta}s_{3\Delta}s_{20} + c_{10}^2 s_{1\Delta}c_{2\Delta}c_{3\Delta}c_{20}^2 + s_{1\Delta}c_{2\Delta}c_{3\Delta} + \\
& - c_{10}^2 s_{1\Delta}c_{20}s_{2\Delta}c_{3\Delta}s_{20} - s_{10}c_{1\Delta}c_{20}s_{20}c_{3\Delta}c_{10}s_{20} + c_{1\Delta}s_{20}c_{2\Delta}s_{3\Delta}c_{10}^2 + \quad\text{(C.8)}\\
& + c_{1\Delta}c_{20}s_{2\Delta}s_{3\Delta}c_{10}^2 + s_{10}s_{1\Delta}s_{3\Delta}c_{10}s_{20} + c_{10}^2 c_{20}s_{1\Delta}s_{20}s_{2\Delta} + \\
& + s_{10}c_{1\Delta}c_{2\Delta}c_{3\Delta}c_{10}c_{20}^2 - c_{20}s_{1\Delta}s_{20}s_{2\Delta} + s_{1\Delta}c_{20}s_{2\Delta}c_{3\Delta}s_{20} + \\
& - c_{10}s_{1\Delta}c_{20}s_{2\Delta}s_{3\Delta}s_{10} - s_{10}c_{1\Delta}c_{2\Delta}c_{3\Delta}c_{10} - c_{10}s_{1\Delta}s_{20}c_{2\Delta}s_{3\Delta}s_{10}
\end{aligned}
$$

$$Q_{(0)33} = c_{10}c_{1\Delta}c_{20}s_{2\Delta}s_{3\Delta}s_{10} + c_{10}^2 c_{1\Delta}c_{2\Delta}c_{3\Delta} - c_{10}^2 c_{1\Delta}c_{2\Delta}c_{3\Delta}c_{20}^2 +$$

$$+ c_{10}^2 c_{1\Delta}c_{20}s_{2\Delta}c_{3\Delta}s_{20} + c_{10}c_{1\Delta}s_{20}c_{2\Delta}s_{3\Delta}s_{10} - s_{1\Delta}c_{20}s_{2\Delta}s_{3\Delta} +$$

$$+ s_{1\Delta}c_{20}s_{2\Delta}s_{3\Delta}c_{10}^2 + s_{1\Delta}s_{20}c_{2\Delta}s_{3\Delta}c_{10}^2 - s_{10}s_{1\Delta}c_{20}s_{2\Delta}c_{3\Delta}c_{10}s_{20} +$$

$$- c_{10}^2 s_{1\Delta}s_{3\Delta}s_{20} - s_{10}s_{1\Delta}c_{2\Delta}c_{3\Delta}c_{10} + s_{10}s_{1\Delta}c_{2\Delta}c_{3\Delta}c_{10}c_{20}^2 +$$

$$- s_{1\Delta}s_{20}c_{2\Delta}s_{3\Delta} - s_{10}c_{1\Delta}s_{3\Delta}c_{10}s_{20} + c_{10}^2 c_{20}^2 c_{1\Delta}c_{2\Delta} + c_{1\Delta}c_{3\Delta} +$$

$$+ c_{10}c_{20}s_{10}s_{1\Delta}s_{20}s_{2\Delta} - c_{10}c_{20}^2 s_{10}s_{1\Delta}c_{2\Delta} - c_{10}^2 c_{20}c_{1\Delta}s_{20}s_{2\Delta} +$$

$$- c_{10}^2 c_{1\Delta}c_{3\Delta} + c_{10}s_{1\Delta}c_{3\Delta}s_{10} \qquad (C.9)$$

$$Q_{(0)14} = Q_{(0)24} = Q_{(0)34} = Q_{(0)41} = Q_{(0)42} = Q_{(0)43} = 0; \quad Q_{(0)44} = 1 \qquad (C.10)$$

For small rototranslations, the terms of the matrix $Q_{(0)}$ assume a simplified form as shown in the expressions (C.11–C.19).

$$Q_{(0)12} = -c_{20}\Delta_3 c_{10} + s_{20}\Delta_2\Delta_3 c_{10} - \Delta_2 s_{10} \qquad (C.11)$$

$$Q_{(0)13} = -c_{20}\Delta_3 s_{10} + s_{20}\Delta_2\Delta_3 s_{10} - \Delta_2 c_{10} \qquad (C.12)$$

$$Q_{(0)21} = -c_{20}s_{10}\Delta_1\Delta_3 + c_{10}\Delta_1\Delta_2 + s_{20}\Delta_3 c_{10} + \Delta_2 s_{10} \qquad (C.13)$$

$$Q_{(0)22} = 1 - s_{10}c_{20}\Delta_2\Delta_3 c_{10} - \Delta_1\Delta_3 s_{20} - c_{10}^2\Delta_1 c_{20}\Delta_2\Delta_3 \qquad (C.14)$$

$$Q_{(0)23} = -s_{20}\Delta_3 + c_{20}\Delta_2\Delta_3 c_{10}^2 - c_{20}\Delta_2\Delta_3 - c_{10}\Delta_1 c_{20}\Delta_2\Delta_3 s_{10} - \Delta_1 \qquad (C.15)$$

$$Q_{(0)31} = c_{20}c_{10}\Delta_1\Delta_3 - c_{10}\Delta_2 + c_{20}s_{10}\Delta_3 + s_{10}\Delta_1\Delta_2 \qquad (C.16)$$

$$Q_{(0)32} = s_{20}\Delta_3 + c_{20}\Delta_2\Delta_3 c_{10}^2 - c_{10}\Delta_1 c_{20}\Delta_2\Delta_3 s_{10} + \Delta_1 \qquad (C.17)$$

$$Q_{(0)33} = 1 + s_{10}c_{20}\Delta_2\Delta_3 c_{10} - \Delta_1 c_{20}\Delta_2\Delta_3 + c_{10}^2\Delta_1 c_{20}\Delta_2\Delta_3 - \Delta_1\Delta_3 s_{20} \qquad (C.18)$$

$$Q_{(0)11} = Q_{(0)44} = 1; \quad Q_{(0)14} = Q_{(0)24} = Q_{(0)34} = Q_{(0)41} = Q_{(0)42} = Q_{(0)43} = 0 \qquad (C.19)$$

C.2   Flexure Hinges

Two typologies of flexure hinges are examined in this paragraph, as shown in Figures C.1–C.2.

The behavior of these hinges is described by expressions that relate their bending moments to the associated rotations of the sections. According to the notations shown in Figures C.1–C.2, the bending response M associated with an imposed deformation $\Delta\alpha$ is represented in (C.20) with the inertia moment $J(z)$ and the compliance C expressed in (C.21).

$$\Delta\alpha = \int_{-z^*}^{+z^*} \frac{M}{EJ(z)}\,dz; \quad C \equiv \frac{\Delta\alpha}{M} \tag{C.20}$$

$$J(z) = \frac{bs^3}{12}; \quad C = \frac{12}{Eb}\int_{-z^*}^{+z^*}\frac{1}{s^3(z)}\,dz \tag{C.21}$$

Also, the description of $s(z)$ identified by (C.22) can be derived from the properties listed in (C.23).

$$s(z) = h - 2R\big(\cos(\theta) - \cos(\theta_0)\big) \tag{C.22}$$

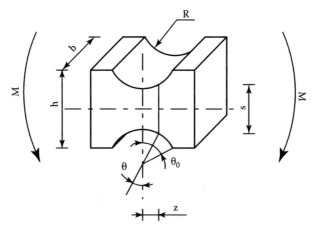

FIGURE C.1
Flexure hinge 1.

$$\sin(\theta) = \frac{z}{R}; \qquad \sin(\theta_0) = \sqrt{1-\left(1+\frac{t}{2R}-\frac{h}{2R}\right)^2}; \qquad z^* = R\sin(\theta_0);$$

$$\cos(\theta) = \sqrt{1-\left(\frac{z}{R}\right)^2}; \qquad \cos(\theta_0) = 1+\frac{t}{2R}-\frac{h}{2R}$$

(C.23)

With the substitutions shown in (C.24), the expression of $s(z)$ can be reorganized (C.25), obtaining the compliance C (C.26).

$$\beta = \frac{t}{2R}; \qquad \gamma = \frac{h}{2R}$$

(C.24)

$$s(z) = 2R\left[(1+\beta) - \sqrt{1-\left(\frac{z}{R}\right)^2}\right]$$

(C.25)

$$C = \frac{12}{Eb} \int_{-R\sin(\theta_0)}^{R\sin(\theta_0)} \frac{dz}{(2R)^3\left(1+\beta-\sqrt{1-\left(\frac{z}{R}\right)^2}\right)}$$

(C.26)

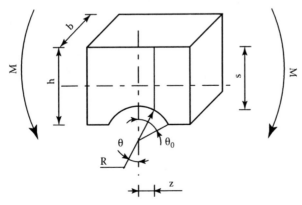

FIGURE C.2
Flexure hinge 2.

The integral expression highlighted in (C.26) can be turned into an equivalent explicit form (C.27).

$$\int \frac{\cos(\theta)}{\left[(1+\beta)-\cos(\theta)\right]^3} d\theta = \frac{1}{(\beta+2)^2} \cdot \frac{\tan(\theta/2)}{\beta+(\beta+2)\cdot\tan^2(\theta/2)} \cdot$$

$$\left[2\cdot\left(1+\frac{1}{\beta}\right)\cdot\frac{1}{\beta+(\beta+2)\cdot\tan^2(\theta/2)}+\left(\frac{3}{\beta}+\frac{3}{\beta^2}+2\right)\right] \tag{C.27}$$

$$+\left[\frac{3}{\beta^{\frac{5}{2}}\cdot(\beta+2)^{\frac{5}{2}}}\cdot(\beta+1)\right]\arctan\left[(\beta+2)\cdot\frac{\tan(\theta/2)}{\sqrt{\beta\cdot(\beta+2)}}\right]$$

The use of the bisection properties represented in (C.28) can be used to transform (C.27) into (C.29).

$$\tan(\theta/2)=\frac{\sin(\theta)}{1+\cos(\theta)}=\frac{\sqrt{1-(1+\beta-\gamma)^2}}{2+\beta-\gamma}; \quad \tan^2(\theta/2)=\frac{1-\cos(\theta)}{1+\cos(\theta)}=\frac{\gamma-\beta}{2+\beta-\gamma}$$

$$\tan(\theta/2)=\frac{1-\cos(\theta)}{\sin(\theta)}=\frac{\gamma-\beta}{\sqrt{1-(1+\beta-\gamma)^2}}; \quad \beta+(\beta+2)\cdot\tan^2(\theta/2)=\frac{2\cdot\gamma}{2+\beta-\gamma}$$

$$\tag{C.28}$$

$$\int \frac{\cos(\theta)}{\left[(1+\beta)-\cos(\theta)\right]^3} d\theta = \frac{1}{\beta^{\frac{5}{2}}\cdot(\beta+2)^{\frac{5}{2}}} \cdot \left\{ \frac{\sqrt{\beta\cdot(\beta+2)}\cdot\sqrt{1-(1+\beta-\gamma)^2}}{2\cdot\gamma} \cdot \left[(\beta+1) \right.\right.$$

$$\left.\frac{\beta\cdot(2+\beta-\gamma)}{\gamma}+3\cdot\beta+3+2\cdot\beta^2\right]+3\cdot(\beta+1)\cdot\arctan\left[\sqrt{\frac{\beta+2}{\beta}}\cdot\frac{\gamma-\beta}{\sqrt{1-(1+\beta-\gamma)^2}}\right]\right\}$$

$$\tag{C.29}$$

Further, (C.29) can be approximated with a series expansion, applying the hypotheses shown in (C.30), to obtain the approximate expression identified in (C.31).

$$\beta\equiv\frac{t}{2\cdot R}\ll 1; \quad \frac{\beta}{\gamma}=\frac{t}{h}\ll\frac{\pi^2}{4}\approx 2.45 \tag{C.30}$$

$$\int_{-\theta_0}^{\theta_0} \frac{\cos(\theta)}{\left[(1+\beta)-\cos(\theta)\right]^3} d\theta = \frac{3\pi}{2^{\frac{5}{2}}\beta^{\frac{5}{2}}} \tag{C.31}$$

Hence, the compliance can be expressed in the first form with (C.32), or in an approximated form with (C.33).

$$C = \frac{3}{2EbR^2} \cdot \frac{1}{\beta^{\frac{5}{2}} \cdot (\beta+2)^{\frac{5}{2}}} \cdot \left\{ \frac{\sqrt{\beta \cdot (\beta+2)} \cdot \sqrt{1-(1+\beta-\gamma)^2}}{2 \cdot \gamma} \cdot \left[\beta(2+\beta-\gamma)\right. \right.$$

$$\left. \cdot \frac{(\beta+1)}{\gamma} + 3 \cdot \beta + 3 + 2 \cdot \beta^2\right] + 3 \cdot (\beta+1) \cdot \arctan\left[\sqrt{\frac{\beta+2}{\beta}} \cdot \frac{\gamma-\beta}{\sqrt{1-(1+\beta-\gamma)^2}}\right] \right\} \tag{C.32}$$

$$C = \frac{9\pi R^{\frac{1}{2}}}{2Ebt^{\frac{5}{2}}} \tag{C.33}$$

A similar procedure can be performed for the second type of flexure hinge (fig. C.2), deriving the response of the structure to an imposed deformation.

$$\Delta\alpha = \int_{-z^*}^{+z^*} \frac{M}{EJ(z)} dz; \quad C \equiv \frac{\Delta\alpha}{M} \tag{C.34}$$

$$J(z) = \frac{bs^3}{12}; \quad C = \frac{12}{Eb} \int_{-z^*}^{+z^*} \frac{1}{s^3(z)} dz \tag{C.35}$$

The expression of $s(z)$ (C.36) can be obtained with the observations listed in (C.36).

$$s(z) = h - 2R(\cos(\theta) - \cos(\theta_0))$$

$$\sin(\theta) = \frac{z}{R}; \quad \sin(\theta_0) = \sqrt{1-\left(1+\frac{t}{R}-\frac{h}{R}\right)^2}; \quad z^* = R\sin(\theta_0); \tag{C.36}$$

$$\cos(\theta) = \sqrt{1-\left(\frac{z}{R}\right)^2}; \quad \cos(\theta_0) = 1+\frac{t}{R}-\frac{h}{R}$$

The substitutions represented in (C.37) can be used to reorganize the expressions of $s(z)$ (C.38) and the compliance C (C.39).

$$\beta = \frac{t}{R}; \quad \gamma = \frac{h}{R} \tag{C.37}$$

$$s(z) = R\left[\left(1+\beta\right) - \sqrt{1-\left(\frac{z}{R}\right)^2}\right] \tag{C.38}$$

$$C = \frac{12}{Eb} \int_{-R\sin(\theta_0)}^{R\sin(\theta_0)} \frac{dz}{R^3\left[1+\beta - \sqrt{1-\left(\frac{z}{R}\right)^2}\right]} \tag{C.39}$$

The expression in (C.29) can be simplified with the variable change highlighted in (C.40), giving the result in (C.41).

$$\frac{z}{R} = \sin(\theta) \tag{C.40}$$

$$C = \frac{12}{EbR^2} \int_{-\theta_0}^{\theta_0} \frac{\cos(\theta)}{\left[\left(1+\beta\right) - \cos(\theta)\right]^3} \, d\theta \tag{C.41}$$

The integral in (C.41) is the same as that obtained for the first type of flexure hinge; therefore, the previous solution can also be used in this situation yielding the result in (C.42) and its approximate expression in (C.43), which is valid under the hypotheses shown in (C.44).

$$k = \frac{12}{EbR^2} \cdot \frac{1}{\beta^{\frac{5}{2}} \cdot (\beta+2)^{\frac{5}{2}}} \cdot \left\{ \frac{\sqrt{\beta \cdot (\beta+2)} \cdot \sqrt{1-(1+\beta-\gamma)^2}}{2 \cdot \gamma} \cdot \left[\beta \cdot (2+\beta-\gamma)\right] \right.$$

$$\left. \cdot \frac{(\beta+1)}{\gamma} + 3 \cdot \beta + 3 + 2 \cdot \beta^2 \right] + 3 \cdot (\beta+1) \cdot \arctan\left[\sqrt{\frac{\beta+2}{\beta}} \cdot \frac{\gamma-\beta}{\sqrt{1-(1+\beta-\gamma)^2}}\right] \right\} \tag{C.42}$$

$$k = \frac{9\pi R^{\frac{1}{2}}}{\sqrt{2}Ebt^{\frac{5}{2}}} \tag{C.43}$$

$$\beta = \frac{t}{R} \ll 1; \quad \frac{\beta}{\gamma} = \frac{t}{h} \ll \frac{\pi^2}{4} \approx 2.45 \tag{C.44}$$

Index

Page references with an f indicate figures; those with a t indicate tables.